Lars Backman
Protein Chemistry

Also of interest

Biomolecules.
From Genes to Proteins
Kaushik, Singh, 2023
ISBN 978-3-11-079375-8, e-ISBN 978-3-11-079376-5

Macromolecular Chemistry.
Natural and Synthetic Polymers
Elzagheid, 2021
ISBN 978-3-11-076275-4, e-ISBN 978-3-11-076276-1

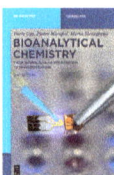

Bioanalytical Chemistry.
From Biomolecular Recognition to Nanobiosensing
Ugo, Marafini, Meneghello, 2025
ISBN 978-3-11-119208-6, e-ISBN 978-3-11-119261-1

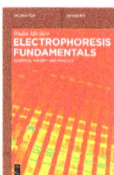

Electrophoresis Fundamentals.
Essential Theory and Practice
Michov, 2022
ISBN 978-3-11-076162-7, e-ISBN 978-3-11-076164-1

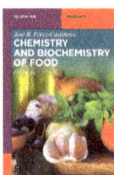

Chemistry and Biochemistry of Food
Pérez-Castiñeira, 2024
ISBN 978-3-11-110834-6, e-ISBN 978-3-11-111187-2

Lars Backman

Protein Chemistry

2nd Edition

DE GRUYTER

Author
Professor Lars Backman
Department of Chemistry, Biochemistry
Umeå University
SE-901 87 Umeå
Sweden
lars.backman@chem.umu.se

ISBN 978-3-11-135066-0
e-ISBN (PDF) 978-3-11-135068-4
e-ISBN (EPUB) 978-3-11-135072-1

Library of Congress Control Number: 2024931622

Bibliographic information published by the Deutsche Nationalbibliothek
The Deutsche Nationalbibliothek lists this publication in the Deutsche Nationalbibliografie;
detailed bibliographic data are available on the Internet at http://dnb.dnb.de.

© 2024 Walter de Gruyter GmbH, Berlin/Boston
Cover image: © Lars Backman, X-ray structure of the actin-binding domain in Rhodamnia argentea
alpha-actinin (7aw8.pdb).
Typesetting: Integra Software Services Pvt. Ltd.
Printing and binding: CPI books GmbH, Leck

www.degruyter.com

Preface to the 2nd edition

Since the release of the first edition of this textbook, many aspects of the protein science field has evolved, perspectives have changed and the knowledgebase has expanded. One of the most important contributions to the field, as well as to life science in general, I believe, is the unveiling of AlphaFold. This AI program, that can predict a protein's 3D-structure from its amino acid sequence with remarkable accuracy, has turned out to be a real game changer.

On the other hand, the basic rules that govern protein behavior and structure are still the same. Therefore, most of the text has only undergone minor revision, to hopefully better the readability whereas the chapters on motor proteins (Chapter 7), enzyme kinetics (Chapter 9), protein folding (Chapter 10) and structure determination (Chapter 12) all have been revised substantially. Finally, a chapter on suitable experimental setups has been added at the end of the book (Chapter 14).

To appreciate the beauty of protein structures, you should use a molecular viewer to display the structure on a large screen. This will allow you to rotate and enlarge certain parts of the structure. I have used ChimeraX to create the protein structures in the book. It can be freely downloaded from (www.cgl.ucsf.edu/chimerax). Protein structures can be downloaded from RCSB Protein Data Bank (www.rcsb.org).

Thanks goes to Karin Sora at De Gruyter, who suggested that it was time to revise the text, and Ria Sengbusch who has helped me with the final touch of the text.

Many thanks to present and former colleagues, and in particular Magnus Andersson, Magnus Wolf-Watz, André Mateus and Karina Persson, who have read and given me valuable feedback on the revised chapters. This has improved the content as well as the readability of the text. I am very grateful for the support and inspiration my wife Anna has given me throughout the work with this text.

Lars Backman
Umeå, January 2024

https://doi.org/10.1515/9783111350684-202

Preface to the 1st edition

This textbook is intended for students and those who need an intelligible text on proteins, their structures and functions. Although the text is intended for students with a basic knowledge of general chemistry, I believe also those with a limited chemical cognizance would benefit by reading it.

The opening chapters give a short account of the cellular organization and the fundamental concepts that govern chemical systems and reactions. Since these concepts are universal, the same set of rules also governs the inner workings of the cell in any organism and consequently all biological life.

The next three chapters present the players (i.e., the building blocks) and their properties and structures as well as how they interconnect to form teams and larger building blocks. This is followed by several chapters discussing proteins from different aspects. One chapter is devoted to the cellular workhorses, the enzymes that make sure that a certain reaction occurs and that it occurs at the right time.

The book ends with a chapter on protein purification, which should prime the reader for practical work with proteins and a chapter giving a short introduction to structure determination of proteins.

There are many images in the book that display proteins but the format makes it difficult to appreciate the images fully. Therefore, I suggest the reader to download the protein structure file from RCSB Protein Data Bank (www.rcsb.org) and use a molecular viewer to display the structure on a computer screen. A good and easy-to-use molecular viewer is the freely available UCSF Chimera. All protein structures in the book have been produced with Chimera.

The content of this book is based on a 9-week course I have given for several years to students enrolled in a master's program in biotechnology at Umeå University during their second year of studies.

I am also thankful to Oleg Lebedev at de Gruyter, who talked me into writing this book, and to Lena Stoll and Ria Fritz who succeeded him and has turned my text into something that reads smoothly.

Many thanks to my colleagues Per-Olof Westlund, Magnus Andersson, Magnus Wolf-Watz, Tobias Sparrman, Karina Persson, Michael Hall and Mikael Oliveberg, who all have read and given me feedback on various parts of the text, which undoubtedly have improved the text. Special thanks go to my wife Anna, who has pushed and inspired me all along the road to reach the final full stop.

Lars Backman
Umeå, August 2019

https://doi.org/10.1515/9783111350684-203

Contents

1 The ballpark where it all occurs

The minimal functional unit of any living organism is the cell. Inside this confined space the machinery required to produce everything necessary to survive (sustain life) is present. The presence of genetic material allows each cell to divide and thereby produce a progeny. Cells come in all shapes as well as sizes; some are extreme like some human neurons that can be close to a meter long, but most are rather small, with diameters around 10 µm or less. The number of cells in an organism also differs, from a single cell in bacteria to some 10^{13}–10^{14} cells in adult human (Figure 1.1).

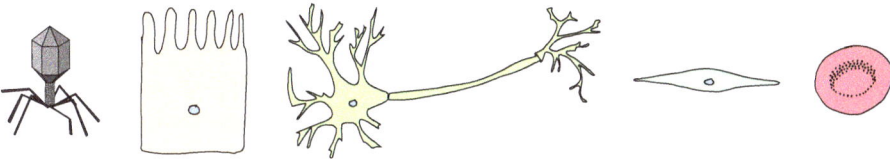

Figure 1.1: Some typical cell shapes showing a bacteriophage, an endothelial cell, a nerve cell, a fibroblast and a red blood cell.

Irrespective of shape or size, all cells have certain common characteristics. They are all surrounded by a barrier, a membrane (sometimes called plasma membrane), separating the interior of the cell from the exterior. In the membrane, there are usually "doors" or "ports" that allow import as well as export of molecules and ions. Some molecules can diffuse through the membrane, whereas others require intricate systems to pass the barrier. There are receptors on the exterior surface of the membrane that accept external signals and transmit them across the membrane, to an internal acceptor, which in turn activates an intracellular process. Similar mechanisms are used to transmit signal the opposite way, from the inside of the cell to its outside. The transport across the membrane as well as all other processes occurring in the cell relays on proteins.

The number of proteins differs greatly from organism to organism. One of the smallest organisms, the parasite *Mycoplasma genitalium* (causes urethritis) contains 482 protein-coding genes, but only 382 of these genes are essential for survival. The genome of the endosymbiont *Candidatus Carsonella ruddii* is even smaller and codes for 182 proteins. This is in contrast to the human genome that consists of ca. 20,000–21,000 protein-coding genes that due to alternative splicing may give rise to many more proteins (more than one protein with different functions). The human genome also contains some 16,000 to 21,000 noncoding genes.

https://doi.org/10.1515/9783111350684-001

1.1 The tree of life

It is believed that life on Earth occurred about 3.7–4 billion years ago. In spite of tremendous efforts, the nature of the initial life is unknown, but it can be assumed that it was very simple. With time, functions were gained and evolved and life developed into something that constitutes the last universal common ancestor (LUCA) to life of today.

This common ancestor gave rise to two branches in the tree of life. One branch that includes all bacteria and one that branched off with time into two separate branches: archaea and eukaryotes. The placement of the last eukaryotic common ancestor (LECA) is still an open question. It is suggested that a proto-eukaryote arose from an archaea that acquired a bacterial endosymbiont.

During evolution, each of these three major groups of has evolved and given rise to numerous organisms. It is estimated that there may be up to 10^{12} difference bacterial and archaeal species on Earth.

At the top of the tree of life, animals are located, thus being very recent inventions. For instance, the first human population diverged only about 300,000 years ago, probably in Africa (Figure 1.2).

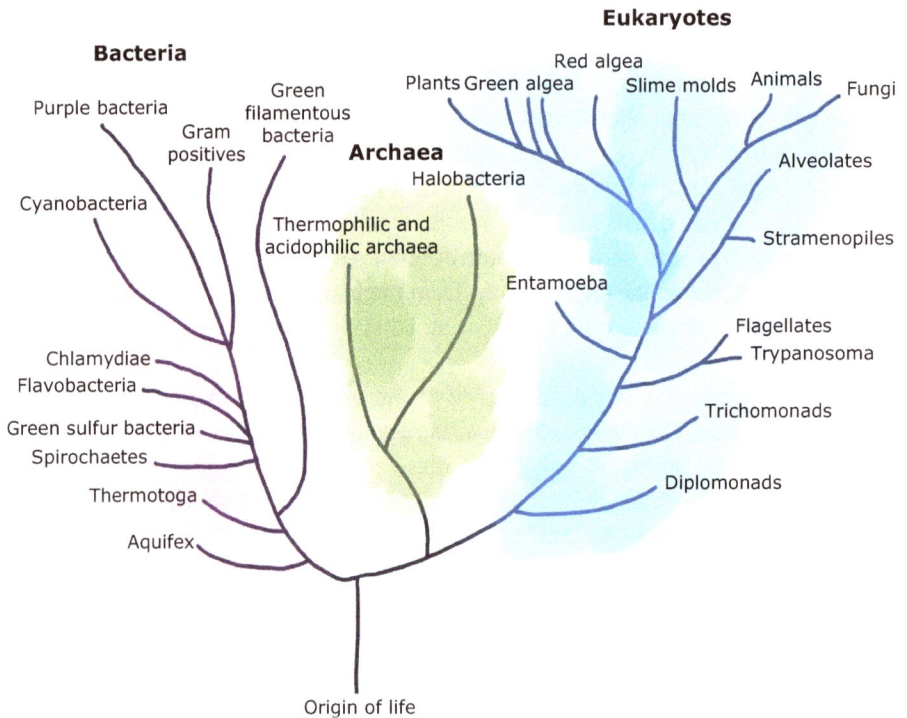

Figure 1.2: The tree of life.

1.2 Bacteria

Bacteria can be divided into three groups depending on their shape: spherical (coccus), rod-like (bacillus) and curved (vibrio, spirillum or spirochete). Although bacteria are unicellular, they may communicate with each other when forming biofilms through quorum sensing by using pheromones.

Bacteria are generally very small. The rod-shaped *Escherichia coli* is about 2 μm long and 0.5 μm in diameter. However, there are bacteria that are smaller and some that are larger.

Bacteria, like all other organisms, are surrounded by a membrane that controls the flow of molecules in and out of the cell, and thereby prevents the loss of cell constituents and maintain proper intracellular milieu. The membrane is covered by a rigid cell wall or envelope, made of a peptidoglycan. In Gram-positive bacteria, the peptidoglycan is very thick and retains the Gram stain. The peptidoglycan layer in Gram-negative bacteria is very thin and therefore does not retain the Gram stain (Figure 1.3).

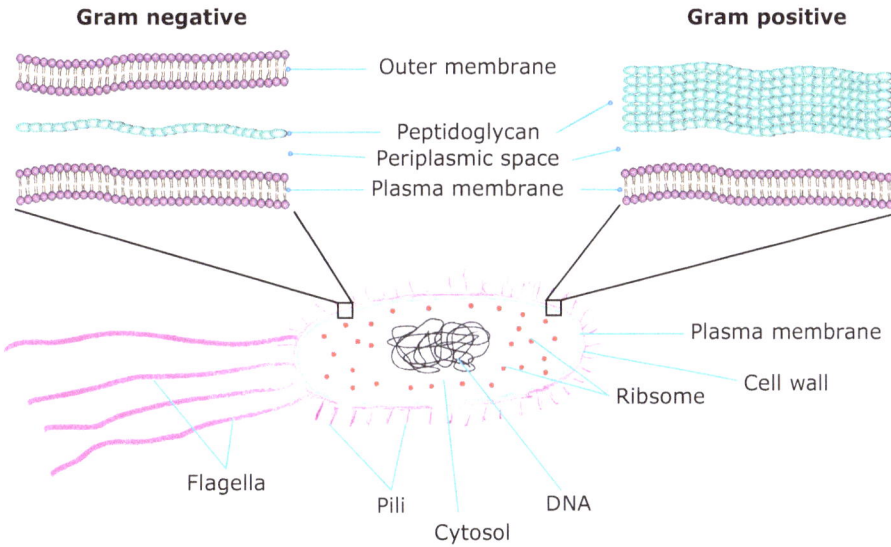

Figure 1.3: A typical bacterial cell.

The bacterial genome is localized to the nucleoid that often contains a single circular DNA molecule. The genome of *E. coli* consists of ca. 4,600,000 bp that under proper growth conditions are translated to proteins by more than 50,000 ribosomes. On the surface of the envelope, there are flagella and pili. Flagella allow the cell to swim in solution, and pili provide adhesion points to the surface of animal cells.

1.3 Archaea

Archaea are the only organisms that inhabit extreme environments, such as hot springs, extremely acidic or alkaline waters. A well-studied archaeon is *Halobacterium* that thrives in extremely saline solution.

The shapes of archaea are similar to bacteria. Both archaea and bacteria move by means of flagella and divide by binary fission. The cell wall of archaea lacks peptido-glycan, and the lipids of the membrane are different as they contain hydrocarbons rather than fatty acids that are linked to the glycerol moiety by ether bonds and not ester bonds as in bacterial and eukaryotic membrane lipids. It is also evident that metabolic pathways in archaea are distinct.

Based on genomic and biochemical studies, it is suggested that archaea are closer related to eukaryotes than to bacteria, particularly as one group of archaea, Asgardarchaeota, harbor several eukaryotic signature proteins. This suggests that the first eukaryote(s) evolved from an archaeon.

1.4 Eukaryotes

Eukaryotes are distinguished from bacteria and archaea by the presence of membrane-bounded organelles. In particular, the genetic material is contained within a distinct nucleus enclosed by a nuclear membrane or envelope. Eukaryotes can be classified as animals, plants, fungi or protists. Any organisms with a nucleus that are not animal, plant or fungus are classified as protists. Another classification separates eukaryotes into Excavate, Chromalveolata, Rhizaria, Archaeplastida and Unikonts. In this classification, animals and fungi, along with some protists, are placed into Unikonta and plants into Archaeplastida whereas Excavata, Chromalveolata and Rhizaria include all other protists (Figure 1.4).

1.4.1 Nucleus

Most of the genetic material in a cell is contained in the cell nucleus. The nucleus is surrounded by the nuclear envelope that consists of an outer and inner nuclear membrane. In the envelope, there are nuclear pores that allow transport of molecules in both directions across the envelope. The nucleolus is a region of the genetic material that is particularly active and coding for ribosomal proteins and RNAs (Figure 1.5).

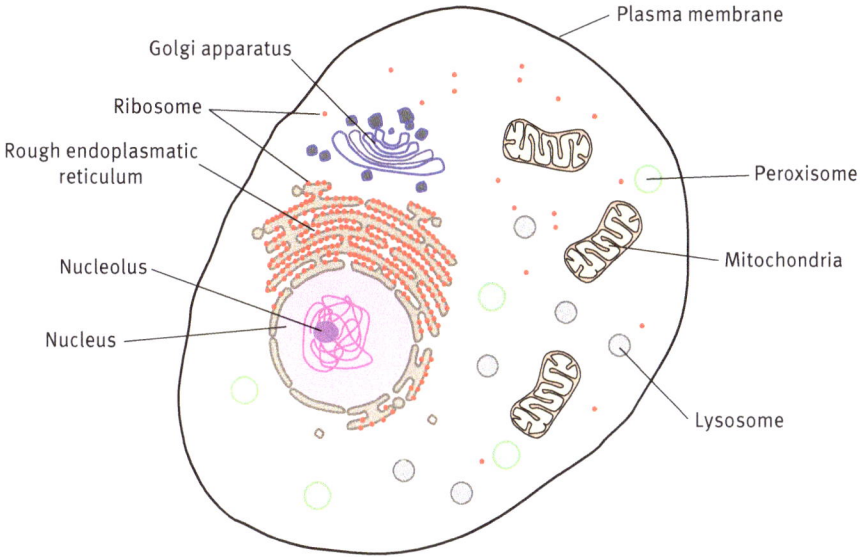

Figure 1.4: A general animal cell with cell organelles.

1.4.2 Endoplasmatic reticulum

The endoplasmatic reticulum, or ER, is an interconnected network of tubular membranes and flattened sacs, also called cisternae. The interior of the ER constitutes the lumen. As the ER is continuous with the outer nuclear membrane, the lumen and the volume between the two nuclear membranes are connected. The outer surface of the rough ER is studded with ribosomes, the site of protein synthesis. Proteins synthesized by ribosomes attached to the rough ER are generally destined for membranes or for export via the Golgi apparatus. Soluble proteins, present in the cytoplasm or in cell organelles, are usually synthesized by free ribosomes.

The smooth ER has no ribosomes attached and is therefore not involved in protein synthesis. However, the smooth ER is important for the production of lipids and steroids (Figure 1.5).

1.4.3 Golgi apparatus

Similar to the ER, the Golgi apparatus consists of a series of membrane sacs. Named after the Italian discoverer Camillio Golgi, the Golgi apparatus is the site for packaging and modification (primarily adding carbohydrates) of proteins that are to be exported out of the cell. Proteins destined for export initially accumulate in the ER lumen be-

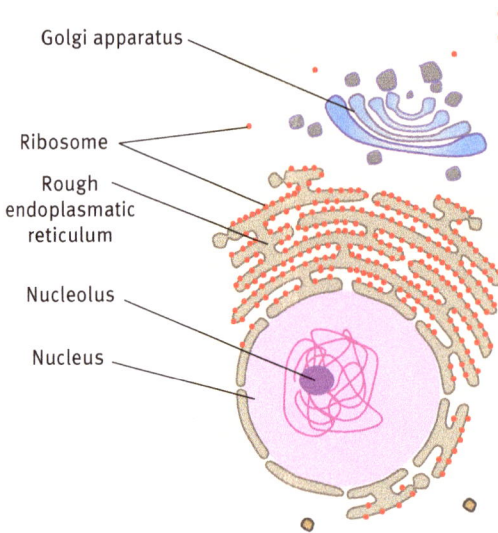

Figure 1.5: Nucleus, endoplasmatic reticulum and Golgi apparatus.

fore budding of in a vesicle that will fuse with the Golgi apparatus, where it will be processed further and packed in a secretory vesicle (Figure 1.5).

1.4.4 Mitochondria

The mitochondria are important for energy conversion. Oxidation processes, such as glycolysis, citric acid cycle and fat oxidation, cause reduction of NAD^+ and FAD to NADH and $FADH_2$, respectively. Reduced NADH and $FADH_2$ are oxidized (and can be reused) when electrons are transferred to acceptor complexes in the electron transport chain. As electrons are transported along the chain and finally reduce molecular oxygen, protons are transferred from the interior of the mitochondria to the space between the outer and inner mitochondrial membrane. The proton gradient that arises is then used to fuel the ATP synthase to generate ATP from ADP and inorganic phosphate. Together these two processes, electron transport and oxidative phosphorylation, constitute cellular respiration.

The inner mitochondria membrane is folded, creating cristae. Most of the components of the electron transport chain are located in or on the cristae. The semifluid matrix constitutes the interior of the organelle. It is in the matrix that the reactions of the citric acid cycle and fat oxidation occur (Figure 1.6).

A small amount of genetic material as well as everything required to transcribe and translate the genetic information is also present in the mitochondria.

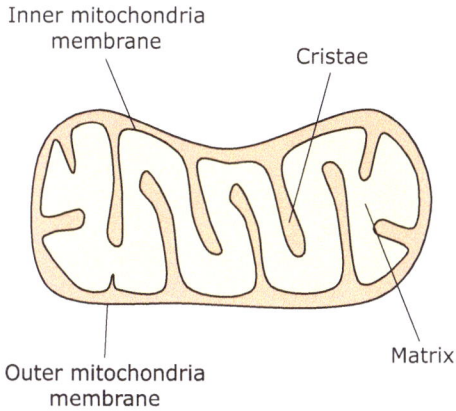

Inner mitochondria membrane

Cristae

Outer mitochondria membrane

Matrix

Figure 1.6: A mitochondrion.

1.4.5 Lysosome

All eukaryotic cells are equipped with a waste disposal site. The lysozyme contains enzymes that can degrade any biological molecules. Therefore, it is important that the activities of these enzymes are confined inside the lysosome from the rest of the cell. This is accomplished in two ways. First, lysosomal enzymes are processed through the ER and Golgi apparatus and transported to the lysosome sequestered in vesicles. Secondly, the acidic milieu of the lysosome (pH < 5) is required for full activity, whereas in the cytoplasm, with a pH about 7.2, their activity would be much lower.

1.4.6 Peroxisome

Like the lysosome, the peroxisome is enclosed by a single membrane and also about the same size. The peroxisome has an important role in breaking down long chained fatty acids to medium chain fatty acids that can be processed further by the mitochondrion. The peroxisome is also important for the detoxification of hydrogen peroxide as well as other toxic substances, such as ethanol. In plant cells, specialized peroxisomes called glyoxysomes have a prominent role in converting fatty acid to carbohydrates during seed germination.

1.4.7 Chloroplast

In contrast to animal cells, plant cells contain chloroplasts and vacuoles. Like the mitochondrion, the chloroplast is surrounded by a double membrane and inside the chloroplast there is a membrane system called the thylakoid membrane (Figure 1.7). The thylakoid membrane is the site of photosynthesis, where sunlight is captured and

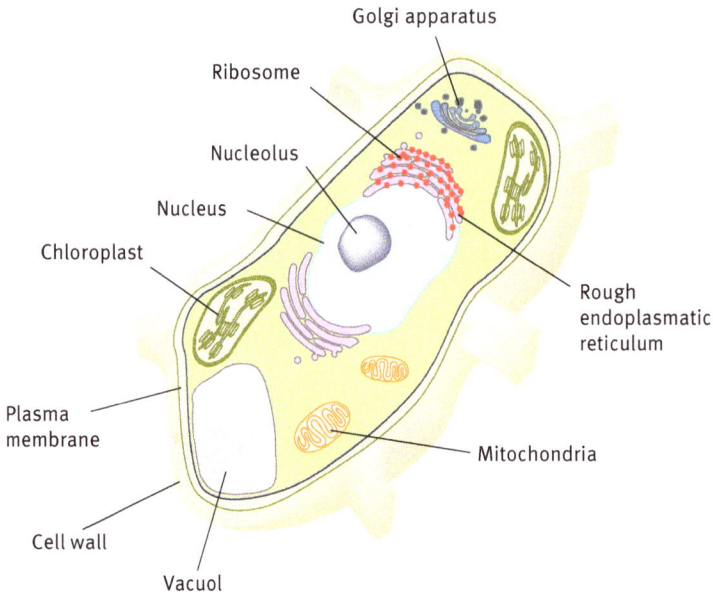

Figure 1.7: Schematic drawing of a plant cell.

used to produce oxygen from water and at the same time converting the trapped energy into chemical useful energy in the form of ATP and NADPH that is used for carbon fixation.

The compartment between the inner chloroplast membrane and the thylakoid membrane is called stroma, and the volume inside the thylakoid membrane is called the lumen. The thylakoid membrane form grana stacks that are connected by stroma thylakoids (Figure 1.8).

Similar to the mitochondrion, the chloroplast harbors a full machinery to process genetic information. Several of the proteins required for photosynthesis and carbon fixation are synthesized by the chloroplast itself.

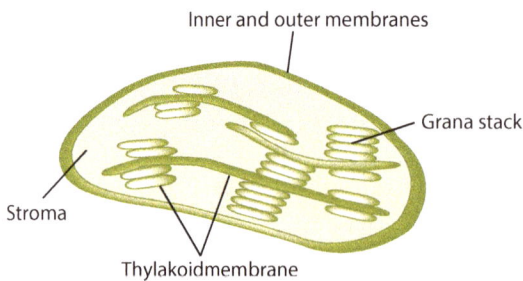

Figure 1.8: Schematic drawing of a chloroplast.

1.4.8 Vacuol

Plants usually contain a single vacuole that grows with time. The plant vacuole is a storage site for waste products, such as phenols and waxes. The vacuole maintains also the turgor pressure that keeps the plant from wilting (Figure 1.7).

1.4.9 Cytoplasm, cytosol and cytoskeleton

The volume inside a cell, with organelles and everything else present in the cell, is called the cytoplasm. The soluble part of the cytoplasm that is the compartment between the plasma membrane and all cell organelles constitutes the cytosol.

The cytosol contains an intricate network of proteins called the cellular cytoskeleton. This network, composed to varying degrees of actin filaments, intermediate filaments and microtubules, is essential for many cellular processes (Figure 1.9).

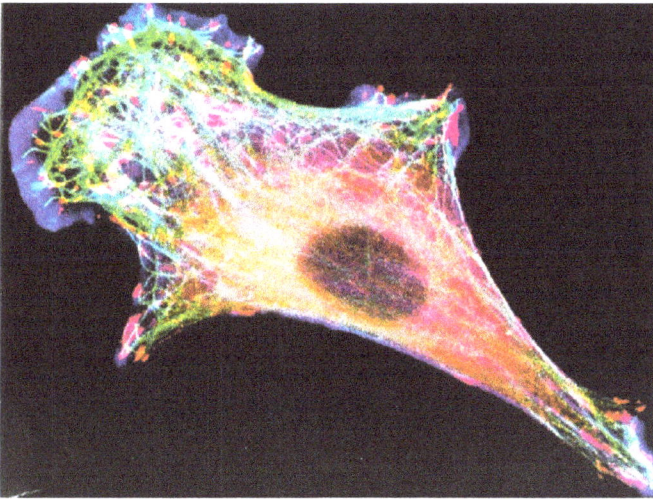

Figure 1.9: The cytoskeleton of a fibroblast. This is one of the first images showing a protein network or cytoskeleton in a cell. Blue indicates the distribution of actin, the main protein of microfilaments. Red marks the distribution of the actin-binding protein vinculin, a protein that anchors filamentous actin to the membrane. Green shows the distribution of tubulin, the main protein of microtubule. Courtesy of Victor Small, Austrian Academy of Science.
From "The molecules of the cell matrix" by Klaus Weber and May Osborne (1985) *Scientific American* 253: 92–102.

1.5 Water

Walter is the most abundant substance in all living organisms, making up more than 70% of the body weight. The presence of water creates a fluid environment that allows a molecule to move and therefore to interact with other molecules. The chemical properties of water also make it an excellent solvent.

The electronegative oxygen atom in a water molecule attracts the hydrogen electrons, giving the oxygen partial negative charge and the two hydrogen atoms a partial positive charge. This makes the water molecule a dipole, with one side negatively charged and the other positively charged. Therefore, a water molecule can interact electrostatic with other water molecules as well as other polar molecules and thereby form hydrogen bonds (Figure 1.10).

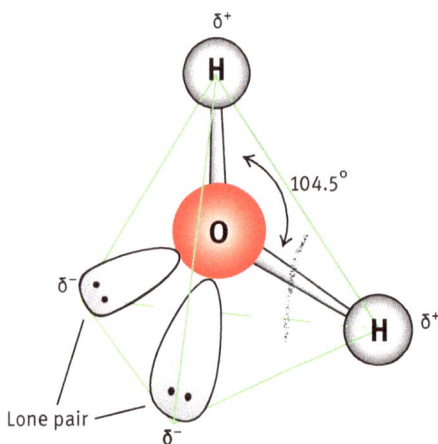

Figure 1.10: The arrangement of bonds and lone pairs can be described as a distorted tetrahedral. The lone pairs affect the hydrogen atoms and push them closer to each other. In a perfect tetrahedral arrangement, all angles should be 109.5°.

A water molecule can form four hydrogen bonds, the two hydrogen atoms as donators and the two lone pair electrons as acceptors. However, the actual number of hydrogen bonds depends on the state of water. Water molecules in ice are perfectly coordinated, as in other solids, and each water molecule has hydrogen bonds to four neighboring molecules (Figure 1.11). In water, thermal molecular motion causes hydrogen bonds to break, thereby decreasing the average number of hydrogen bonds of each molecule. The perfect coordination is lost. Up to a certain temperature, which happens to be 4 °C, the molecules are more closely packed. At higher temperatures, thermal motion leads to an expansion; the water "swells." A very important consequence of this behavior is that the density of water is greater than that of ice. This is the reason why ice floats on water. It is easy to imaging what would happen if ice would not float.

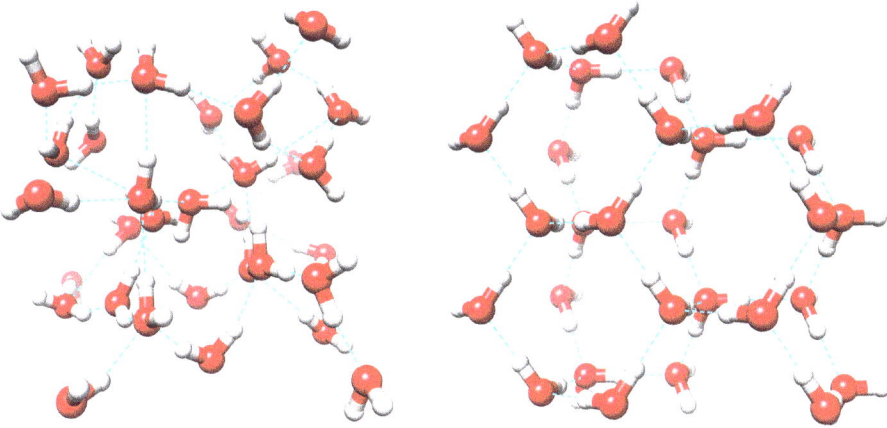

Figure 1.11: In hexagonal ice the water molecules are perfect coordinated, each water molecule is hydrogen bonded to four other water molecules. When ice melts, the perfect coordination is lost and each water molecule is no longer bonded to four other molecules.

The anomaly properties of water are also due to hydrogen bonding. Compared to other hydrides, the melting and boiling points of water are around 100 °C higher than expected (Figure 1.12).

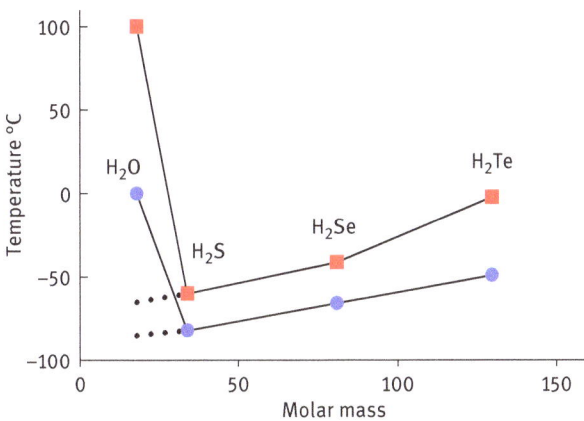

Figure 1.12: Melting and boiling points of hydrides. The normal behavior for atoms in the same group in the periodic system is that the melting and boiling points would increase with molar mass. The dotted line indicates the expected melting and boiling points of water.

The high surface tension and viscosity of water is also due to hydrogen bonding. The capillary force that allows plants to bring water up through the root system is dependent on the surface tension of water. Also the large heat capacity of water is related to the ability to form hydrogen bonds.

The polar property makes water an excellent solvent for other polar substances. Thus, any substance that form ions in solution, such as a salt like sodium chloride, or can form hydrogen bonds can be dissolved in water (Figure 1.13).

Water is indeed a remarkable solvent!

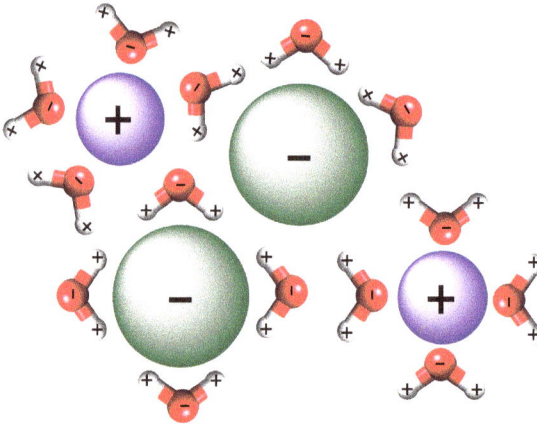

Figure 1.13: Solvation of sodium chloride in water. When sodium chloride dissolves in water, the water molecules form a shell around each ion, turning the partial negative oxygen toward the sodium ion and the positive hydrogens toward the negative chloride ion.

Further reading

Bakshi, S., Siryaporn, A., Goulian, M. and Weisshaar, J.C. (2012). Superresolution imaging of ribosomes and RNA polymerase in live Escherichia coli cells. *Mol Microbiol* 85:21–38.

Baum, D.A. and Baum, B. (2014). An inside-out origin for the eukaryotic cell. *BMC Biol* 12:76.

Eme, L., Spang, A., Lombard, J., Stairs, C.W. and Ettema, T.J.G. (2017). Archaea and the origin of eukaryotes. *Nat Rev Microbiol* 15:711–723.

Ezkurdia, I., Juan, D., Rodriguez, J.M., Frankish, A., Diekhans, M., Harrow, J., Vazquez, J., Valencia, A. and Tress, M.L. (2014). Multiple evidence strands suggest that there may be as few as 19,000 human protein-coding genes. *Hum Mol Genet* 23:5866–5878.

Forterre, P. (2013). The common ancestor of archaea and eukarya was not an archaeon. *Archaea* 2013:372396.

Glass, J.I., Assad-Garcia, N., Alperovich, N., Yooseph, S., Lewis, M.R., Maruf, M., Hutchison, C.A. 3rd, Smith, H.O. and Venter, J.C. (2006). Essential genes of a minimal bacterium. *Proc Natl Acad Sci USA* 103:425–430.

Locey, K.J. and Lennon, J.T. (2016). Scaling laws predict global microbial diversity. *Proc Natl Acad Sci USA* 113:5970–5975.

Nilsson, A. and Pettersson, L.G. (2015). The structural origin of anomalous properties of liquid water. *Nat Commun* 6:8998.

Plopper, G., Sharp, D., Sikorsk, E. and Lewin, B. (2015). Lewin's Cells. Jones and Bartlett Publishers. Sudbury, Mass. Sudbury, Mass.

Sharp, K.A. (2001) "Water: Structure and properties". Encyclopedia of Life Sciences. doi: 10.1038/npg. els.0003116.

Spang, A., Mahendrarajah, T.A., Offre, P. and Stairs, C.W. (2022). Evolving perspective on the origin and diversification of cellular life and the virosphere. *Genome Biol Evol* 14:evac034.

Tamames, J., Gil, R., Latorre, A., Pereto, J., Silva, F.J. and Moya, A. (2007). The frontier between cell and organelle: Genome analysis of Candidatus carsonella ruddii. *BMC Evol Biol* 7:181.

Williams, T.A., Cox, C.J., Foster, P.G., Szollosi, G.J. and Embley, T.M. (2020). Phylogenomics provides robust support for a two-domains tree of life. *Nat Ecol Evol* 4:138–147.

2 The rules

For a spontaneous process, the change in Gibbs free energy (ΔG) is always negative (is less than zero). Changes in both enthalpy (ΔH) and entropy (ΔS) contribute to Gibbs free energy as $\Delta G = \Delta H - T\Delta S$, where T is the absolute temperature in Kelvin. The change in enthalpy is related to breaking and forming bonds, whereas entropy can be seen as a measure of disorder.

The second law of thermodynamics states that the entropy of an isolated system will increase over time for a spontaneous process. If we consider protein folding, obviously a spontaneous process, the unfolded protein is highly disordered in contrast to the fully folded and functional protein; the entropy decreases. This appears to violate the second law of thermodynamics. However, we need to consider that the folding occurs in water.

2.1 The hydrophobic effect

Proteins contain both polar and nonpolar groups (i.e., amino acid residues). A polar group, such as a carboxyl or amino group, can form hydrogen bonds with water molecules as well as with other polar groups. On the other hand, nonpolar groups cannot form hydrogen bonds and causes a disturbance in the water structure. Therefore, water molecules form a cage-like structure around nonpolar groups; these water molecules are ordered and the entropy decreases. During the folding process, these nonpolar groups are removed from contact with water by being placed in the interior of the fully folded protein, "releasing" the caged water and avoiding further contact with the surrounding water, thus increasing the disorder of the water molecules and therefore also the entropy.

As protein folding occurs spontaneously the entropic contribution from the behavior of water (ΔS_{water}) must be larger than that from to the actual folding process ($\Delta S_{protein}$); therefore, we can express the change in entropy as $\Delta S = \Delta S_{water} - \Delta S_{protein}$ and ΔS must be positive (be larger than zero; Figure 2.1).

This effect, called the hydrophobic effect, is the driving force for protein folding as well as for membrane formation. The hydrophobic effect implies that without nonpolar groups or, as it is usually called a hydrophobic core, a protein will not be able to fold.

https://doi.org/10.1515/9783111350684-002

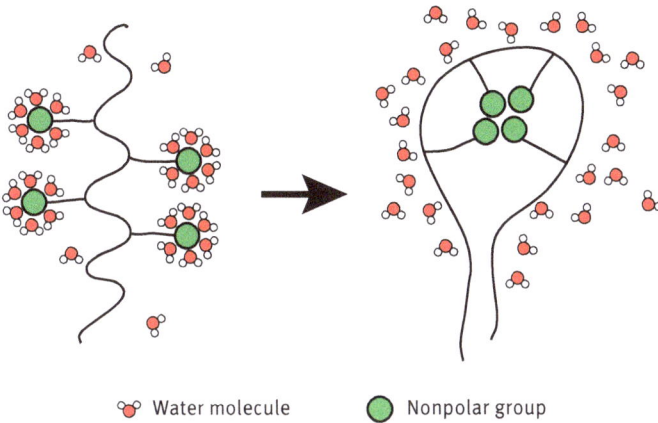

Figure 2.1: Protein folding places nonpolar inside the protein, without contact with the surrounding polar water. Thereby increasing the entropy, and driving the folding process.

2.2 Covalent bonds

Covalent bonds are formed when two atoms share electron pairs. The peptide bond, the covalent bond formed between two amino acids during protein synthesis, is strong and static; once it is formed nothing more can happen.

Many cellular processes require transient contacts between proteins or between a protein and a small molecule (ligand). Initially, the interacting molecules are juxtaposed, bonds are formed to keep them attached to each other for a certain time and then the bond(s) need to be broken to allow the complex to dissociate. From an energy point of view, covalent bonds are costly to break or make; the energy required to break a single carbon–carbon (346 kJ/mol) or oxygen–carbon (358 kJ/mol) bond is about 350 kJ/mol. Therefore, covalent bonds are not well suited to stabilize molecules that undergo continuous conformational changes or to keep components attached and stabilize short-lived complexes due to the high energy cost.

Noncovalent bonds are much better suited for dynamic purposes.

2.3 Noncovalent bonds

It may be surprising but proteins are not very stable structures. Many proteins are stable by only 40–50 kJ/mol. Since the energy required to break a single hydrogen bond is about 20 kJ/mol, it may be enough to break a few weak bonds to unfold or denature a protein and make it nonfunctional. It is important to realize that this does not mean that there are only a few noncovalent bonds in a protein but rather that the energy gain by forming intramolecular hydrogen bonds compared to hydrogen bonds with water is small.

2.3.1 Ion–ion interactions

A permanent ion may attract an ion of opposite charge and form a salt bridge. Although ions of the same charge will not form a bond, the repulsive force can still be important.

According to Coulomb's law the force F between two charged ions, with charges q_1 and q_2 Coulombs (C), separated by the distance r is defined as

$$F = k \cdot \frac{q_1 \cdot q_2}{r^2}$$

where k is the Coulomb's constant, equal to $9 \cdot 10^9$ J·m·C^{-2} in vacuum. Since the force depends on the medium, the Coulomb's constant can be rewritten to give

$$F = \frac{1}{4\pi\varepsilon_0} \cdot \frac{q_1 \cdot q_2}{\varepsilon_r(T) \cdot r^2}$$

$$F = \frac{1}{4\pi\varepsilon_0} \cdot \frac{q_1 \cdot q_2}{\kappa \cdot r^2}$$

where ε_0 is the permittivity in vacuum. $\varepsilon_r(T)$ is the relative permittivity and κ is the dielectric constant that reflects what effect the medium has on the force. The dielectric constant is strongly influenced by the nature of the medium. In a nonpolar solvent, such as benzene or hexane, the dielectric constant is 2–3, whereas in a polar solvent like water the dielectric constant is around 80. For an ion pair separated by 4 Å, the energy of interaction will be −39 kJ/mol in water, whereas in a nonpolar solvent it would be nearly 40 times larger. In other words, electrostatic interactions are weakened by polar media.

In water, a solvated sodium ion will attract not only a negatively charged chloride ion but also water molecules. The water dipole will reorient in such a way that the partially negatively charged oxygen will face the positive sodium ion and the partially positively charged hydrogen will face negative chloride ion. This polarization creates a layer of water molecules surrounding the ions, causing a screening effect that reduces the Coulomb interaction between the sodium and chloride ions (Figure 2.2). In a nonpolar medium, there are no solvent molecules that can interact with the ions and the charges are not screened, and there are no favorable interactions with the medium.

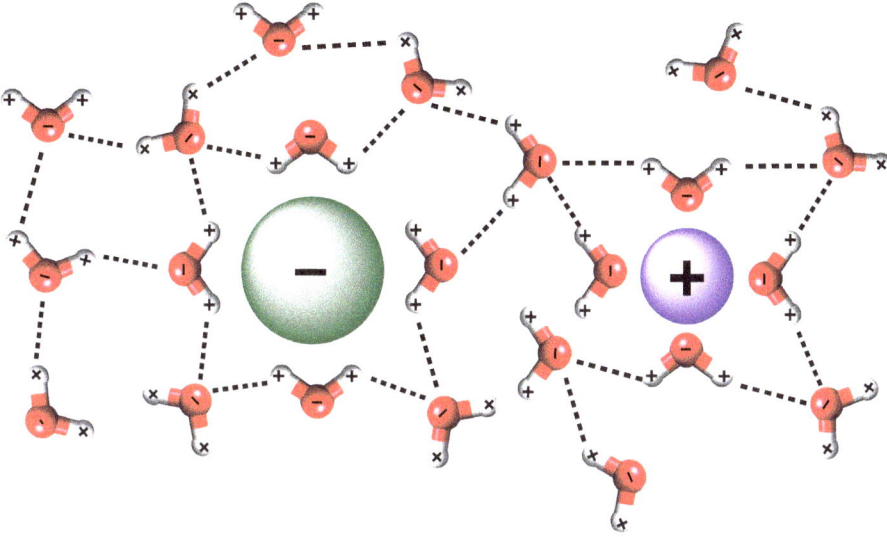

Figure 2.2: Screening of ion charges by layers of water dipoles. Dotted lines represent hydrogen bonds.

2.3.2 Hydrogen bonds

A hydrogen bond forms between a hydrogen donor and an acceptor. In proteins and other biomolecules, the hydrogen donor is usually an electronegative nitrogen or oxygen with a covalent bound hydrogen. Fluorine may also participate in hydrogen bonds, but is scarcely present in biomolecules. However, any atom that has a sufficient electronegativity that induces a partial positive charge on the hydrogen atom can form hydrogen bonds. The acceptor is usually an atom with lone pair of electrons (such as nitrogen or oxygen), even though other fully or partially negatively charged groups can function as an acceptor (Figure 2.3).

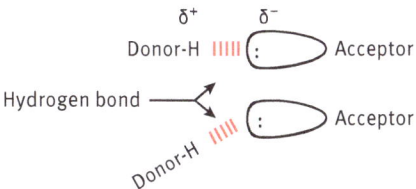

Figure 2.3: An electronegative atom with a covalent bound hydrogen is donor in a hydrogen bond and an atom with lone pair of electrons, such as nitrogen or oxygen, is the acceptor in a hydrogen bond.

The strength of a hydrogen bond depends on both geometry and distance of the bond as well as donor atoms. The energy of interaction for a hydrogen bond with nitrogens and/or oxygens is in the range of −10 to −40 kJ/mol and is highly dependent on the solvent. The energy of a hydrogen bond in an α-helix in gas phase is around −23 kJ/mol, but only −8 kJ/mol in water. The reduction can be attributed to the hydrophobic effect of water, involving contributions from the entropy. The strength of a bond with C–H as donor is weaker (less than −10 kJ/mol and only present in nonpolar environments). The distance between the donor atom and acceptor atom varies, but is usually ~ 3 Å; a shorter distance gives stronger bonds. If the donor and acceptor are separated by less than ~ 2.5 Å and have similar pK_as, the energy required to transfer the hydrogen from donor to acceptor is very low. Such a low barrier hydrogen bond is very strong and the strength is estimated to be close to half of the strength of a covalent bond.

The geometry is also important. A linear hydrogen bond, where donor-H is in line with the acceptor's lone pair orbital is stronger than a bond where there is an angle between the donor-H and acceptor (Figure 2.4).

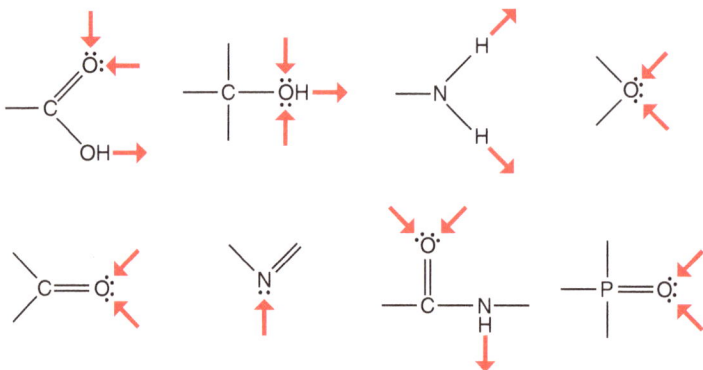

Figure 2.4: Common functional groups in proteins that are important hydrogen bond donors or acceptors.

2.3.3 Van der Waals interactions

A permanent dipole with its asymmetric distribution of electrons may attract another dipole if they are close enough. Such a permanent dipole–permanent dipole (or Keesom) interaction is weak (less than 3–4 kJ/mol) and highly dependent on distance and temperature. The energy of interaction, assuming that the dipoles are constantly rotating, is averaged over all possible orientations of the dipoles and decreases with the inverse sixth power of distance:

$$E_{vdW} = -\frac{B}{r^6}$$

where r is the distance between the dipoles and B is a positive constant of the attractive force. The strongest van der Waals interactions occur when the two dipoles are slightly more separated than the sum of their van der Waals radii.

A permanent dipole can induce a dipole moment in a closely placed molecule or group that leads to a mutual electrostatic interaction. A permanent dipole–induced dipole interaction or Debye force is weaker than that between two permanent dipoles.

Motions of the electrons in one molecule may influence the electrons in another molecule and induce a correlated motion. Such London dispersion forces depend on random fluctuations of electrons. The strength of London dispersion forces is weaker than that of Keesom and Debye forces.

2.4 pH

When an acid (HAc) is added to water, a fraction of the acid will dissociate and increase the concentration of H^+ (rather H_3O^+, a hydronium ion) and Ac^-

$$HAc + H_2O \rightleftharpoons Ac^- + H^+$$

The equilibrium constant, K_a, is defined as

$$K_a = \frac{[Ac^-] \cdot [H^+]}{HAc}$$

As pH is defined as

$$pH = -\log[H^+]$$

the increased concentration of H^+ will lower the pH. The stronger the acid, the larger the fraction that will dissociate and the lower the pH will be.

The strength of a weak acid is expressed by its pK_a-value that is defined as the negative logarithm of the acid dissociation constant

$$pK_a = -\log10(K_a)$$

The relation between pH and pK_a is given by the Henderson–Hasselbalch equation

$$pH = pK_a + \log\frac{[Ac^-]}{[HAc]}$$

The expression signifies that the pH equals the pK_a-value when exactly half of the weak acid is dissociated (or ionized). In pure water that is considered as a neutral solution, the concentration of H^+ is $1 \cdot 10^{-7}$ M due to autoprotolysis and therefore the pH is 7. A solution with a lower pH is acidic, whereas at a higher pH it is basic.

2.5 Numbers are important

Proteins are mobile; they move around all the time in the cell. Some are recruited to the plasma membrane whereas others have to find a certain protein to interact with and form a functional complex. Yet other proteins are activated or deactivated by binding of small molecules. Thus it is necessary that the two (or more) molecules meet in order to form the functional complex. In the cell soup, proteins move back and forth in three dimensions randomly. To form the functional complex, the two proteins have to "collide" in the proper way to find the correct attachment points and bind to each other. There are two main factors that will affect the success rate of these collisions: number of molecules and diffusion rate.

Many proteins are present in the cell at concentrations around or less than 1 μM. Such a concentration corresponds to around $\sim 6 \cdot 10^{17}$ protein molecules per liter. With this estimate, we can calculate the copy number of a protein within a single cell. If we take a cell like the rod-shaped *Escherichia coli* that is about 2 μm long and 0.5 μm in diameter, this cell has a volume close to 0.4 μm^3 or $0.4 \cdot 10^{-15}$ l. At a protein concentration of 1 μM or less, we would expect to find less than 240 copies of this particular protein. Obviously, the copy number of a certain protein at a micromolar concentration is rather limited.

The diffusion rate is more difficult to estimate due to the crowding effect. In an *Escherichia coli* cell there are about 40,000 ribosomes and some 4,200 different proteins at varying concentrations as well as numerous small molecules. Thus, it is rather obvious the cell is packed with molecules that occupy space; the cell is undoubtedly crowded. This crowding reduces the diffusion rate of proteins and other large molecules whereas small molecules are probably not influence to any major degree.

Although crowding has a negative effect on the diffusion rate, it has at the same time a positive effect; as the active concentration of any molecule will be larger, any complex formation will be enhanced.

Further reading

Atkins, P. (2010). The laws of thermodynamics: A very short introduction. Oxford University Press. Oxford.

Atkins, P., de Paula, J. and Keeler, J. (2022). Atkins' physical chemistry. Oxford University Press. Oxford.

Kemp, M.T., Lewandowski, E.M. and Chen, Y. (2021). Low barrier hydrogen bonds in protein structure and function. *Biochim Biophys Acta Proteins Proteom* 1869:140557.

Cleland, W.W. (2000). Low-barrier hydrogen bonds and enzymatic catalysis. *Arch Biochem Biophys* 382:1–5.

Milo, R. and Phillips, R. (2015). Cell biology by the numbers.Garland Science. Taylor & Francis Group. LLC. Abingdon.

Po, H.N. and Senozan, N.M. (2001). The Henderson-Hasselbalch equation: Its history and limitations. *J Chem Edu* 78:1499–1503.

Sarkar, A. and Kellogg, G.E. (2010). Hydrophobicity--shake flasks, protein folding and drug discovery. *Curr Top Med Chem* 10:67–83.

Sheu, S.Y., Yang, D.Y., Selzle, H.L. and Schlag, E.W. (2003). Energetics of hydrogen bonds in peptides. *Proc Natl Acad Sci USA* 100:12683–12687.

Steiner, T. (2002). The hydrogen bond in the solid state. *Angew Chem Int Ed Engl* 41:49–76.

3 The players

It is estimated that there are tens of thousands of different proteins in a human cell, all performing a different tasks. Still, they all are built in the same way from 20 simple molecules: 20 different α-amino acids.

All these 20 α-amino acids contain a carboxyl group, an amino group and a hydrogen atom, covalently bound to the so-called α-carbon. What distinguishes one amino acid from another is the side chain, R. If the side chain would be a methyl group (-CH₃), then this carbon would be the β-carbon (Figure 3.1).

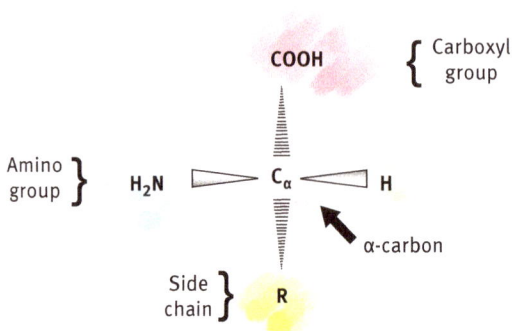

Figure 3.1: A typical α-amino acid.

However, when an α-amino acid is dissolved in water (or a neutral buffer) both the amino and carboxyl groups ionize. As the carboxyl group is a weak acid, it will lose a proton (deprotonated) and become negatively charged. In the same manner, the amino group, being a weak base, will take up a proton (protonated) and become positively charged. A molecule with both a negative and positive charge is called a zwitterion. Although such an amino acid is charged, its net charge is zero.

As long as the pH of the buffer is above the pK_a of the carboxyl group and below that of the amino group, both ionize and are charged. The carboxyl group will be protonated when pH drops below the pK_a of the carboxyl group and the amino group will be deprotonated when pH increases above the pK_a of the amino group. The charged of the amino acid will vary depending on the pH of the solvent (Figure 3.2).

All α-amino acids, except for glycine, are optically active, meaning that they rotate plane-polarized light clockwise or counter-clockwise. An optically active molecule contains an asymmetric or chiral center which is signified by a tetrahedral carbon with four different substituents. Such a chiral molecule cannot be superimposed on its mirror image. As long as the side chain is anything else than a proton (as in glycine), the α-carbon constitutes a chiral center (Figure 3.3).

https://doi.org/10.1515/9783111350684-003

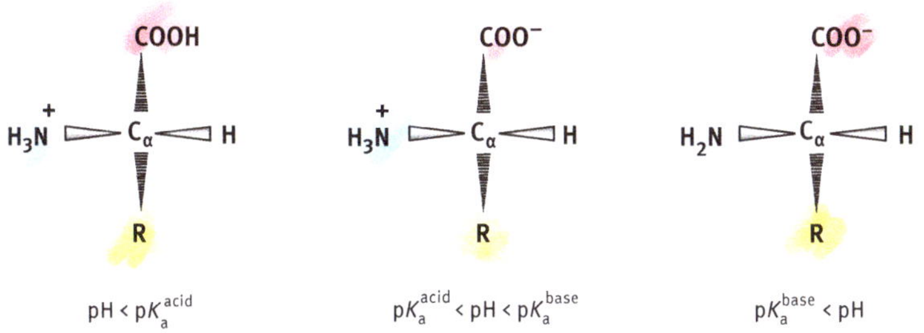

Figure 3.2: The pH of the solvent will affect the ionization of the carboxyl and amino groups of the amino acid. At pH < pK_a of the carboxyl group, the carboxyl is protonated and uncharged and at pH > pK_a of the amino group, it is deprotonated and uncharged.

L-α-amino acid   D-α-amino acid

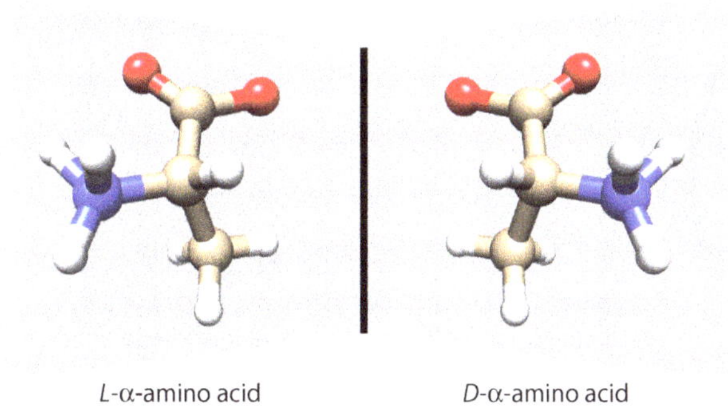

Figure 3.3: L- and D-enantiomers of a typical α-amino acid.

The optical activity is related to the reference molecules L- and D-glyceraldehyde (Figure 3.4). L-glyceraldehyde rotates plane-polarized light counter-clockwise and is said to be levorotatory. D-glyceraldehyde is dextroratatory as it rotates plane-polarized light clockwise. Laevus and dexter are the Latin words for left and right, respectively.

Whether an amino acid is an L- or D-enantiomer depends on the similarity to glyceraldehyde. In a Fisher projection of glyceraldehyde, the hydroxyl group is placed to the left in the L-enantiomer and to the right in the D-form. Therefore, when the amino group is placed to the left, it is an L-α-amino acid. This does not necessarily imply that an L-α-amino acid rotates plane-polarized light counter-clockwise. When there are several chiral centers, the position of the hydroxyl (or analogous) group furthest away from the aldehyde or keto group determine whether it is an L- or a D-configuration. The D/L system is not very useful for describing the absolute configuration of chiral centers. It should be remembered that α-amino acids present in proteins are all in the L-configuration (Figure 3.5).

A convention to indicate the absolute configuration of chiral centers is based on the "weight" or priority of the substituents at the chiral center. In this system, each substitu-

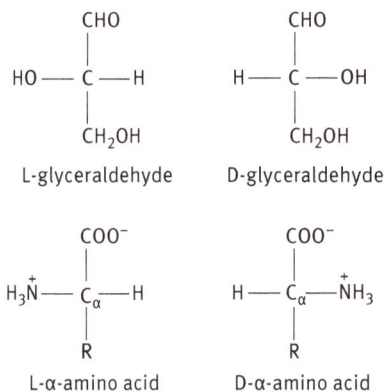

L-glyceraldehyde D-glyceraldehyde

L-α-amino acid D-α-amino acid

Figure 3.4: Fisher projections of glyceraldehyde and an α-amino acid. In the Fisher convention, horizontal bonds are pointing out of the plane and vertical bonds are pointing in to the plane. In the L-enantiomer, the hydroxyl and amino group are to the left.

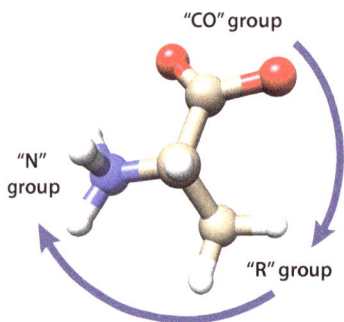

"CO" group

"N" group

"R" group

Figure 3.5: A simple mnemonic is the CORN rule. By rotating the molecule, such that the hydrogen is in front of the α-carbon, reading CO-R-N in the counter-clockwise direction indicates an L-α-amino acid.

ent is given a rank depending on the atomic number of the atom or group bound to the chiral center. If two substituents have the same priority, the atomic numbers of the next bound atoms are compared. The priority of some common groups found in amino acids is: $SH > OH > NH_2 > COOH > CH_2OH > CH_3 > H$. To determine the absolute configuration, the priority of the four substituents is ranked from the highest to the lowest and the molecule is oriented such that the group or atom with the lowest priority is pointing away from the chiral center. If going from the group with highest priority to the next in rank is clockwise, the chiral center has an R-configuration (R for rectus, Latin for right). If the order is counter-clockwise the chiral center has an S-configuration (S for sinister, Latin for left). Thus, the α-carbon in the L-α-amino acid in Figure 3.5 has an S-configuration. In fact, all but one (cysteine) of the α-amino acid has an α-carbon in the S-configuration.

3.1 Twenty different α-amino acids

Since three of the substituents bound the α-carbon are the same, it is obvious that the different chemical and physical properties of each of the 20 α-amino acids depends on the side chain. The size of the side chain differs; the smallest side chain is only a proton (-H).

Some side chains contain a carboxyl or an amino group, and can be ionized depending on the pH of the solvent. Some side chains are very nonpolar, whereas other are polar, which will influence their solubility in water. Some contain a conjugated ring system, giving them useful optical properties, whereas some side chains are branched. The 20 amino acids are usually classified as nonpolar, polar-uncharged, polar-charged or aromatic. It should be noted that some amino acids fit in more than one group; a nonpolar-uncharged amino acid may have some polar properties (Figure 3.6).

3.1.1 Nonpolar amino acids

The nonpolar amino acids are glycine (Gly; G), alanine (Ala; A), valine (Val; V), leucine (Leu; L), isoleucine (Ile; I), methionine (Met; M) and proline (Pro; P). These amino acids are characterized by an aliphatic side chain of varying length whose hydrophobicity increases with increased chain length. The side chains of these amino acids cannot participate in hydrogen bonding and have a propensity to avoid contact with water. Therefore, the amino acids of this group are usually found in the interior of proteins, shielded from water contact. Since the interior of a protein is rather crowded, these side chains will be in close proximity to other side chains and possibly participate in van der Waals interactions.

Glycine with only a hydrogen atom in the side chain can probably not participate in any type of interactions. However, the small size allows glycine to be placed in regions where larger side chains do not fit. A branched side chain, as in valine, leucine and isoleucine, is less flexible and has a larger packing sphere than a straight hydrocarbon chain. It should also be noted that glycine is the only α-amino acid with an achiral α-carbon.

The sulfur-containing side chain of methionine is considered as one of the most hydrophobic side chains. Although the electron configuration of sulfur is similar to that of oxygen, the electronegativity is much lower (2.58 compared to 3.44 for oxygen) and close to that of carbon (2.55). The small difference in electronegativity between sulfur and carbon prevents the formation of a dipole. Despite methionine's nonpolar property, it can interact weakly with polar molecules and cationic metals, such as iron, zinc and copper. Methionine is also important in the biosynthesis of metabolic molecules such as taurine (present in bile) and S-adenosyl methionine (donor of methyl groups in several metabolic processes).

Proline differs from the other amino acids, in that the side chain is covalently bound to the α-amino group, forming a pyrrolidone ring. The cyclization reduces the conformation space of proline; both the α-amino group and the side chain are locked in space. This has important consequences for protein structure as will be discussed below.

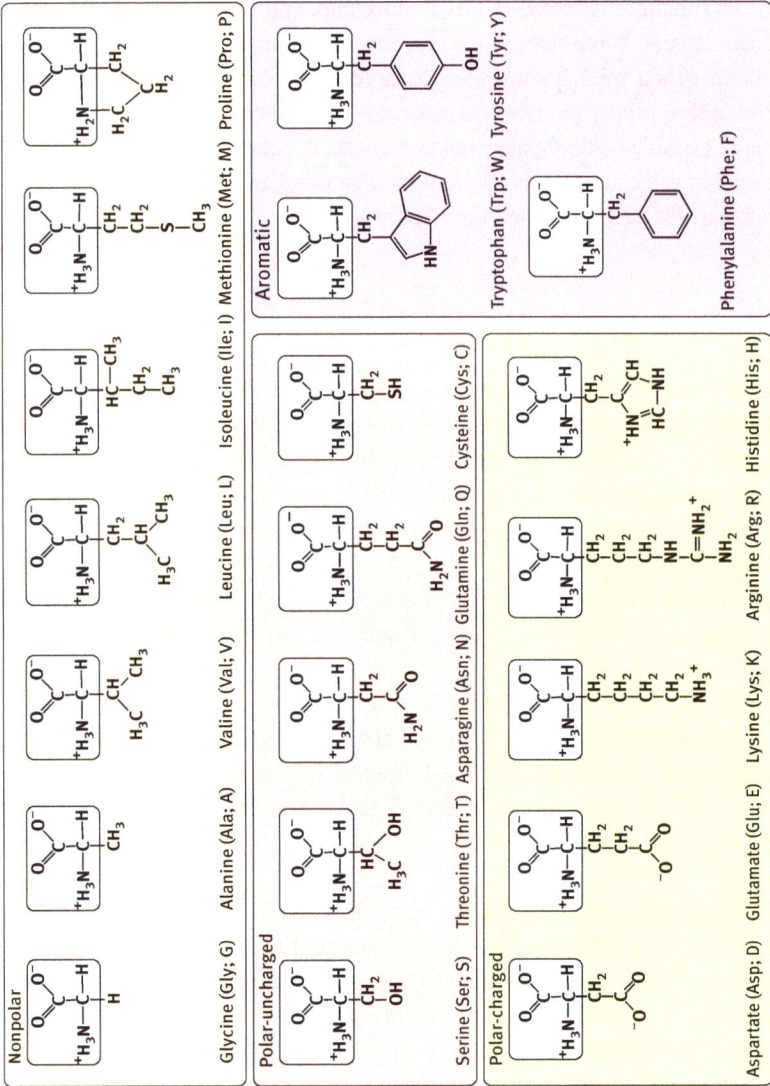

Figure 3.6: The 20 α-amino acids found in proteins in their ionized form at neutral pH. The structure common for all amino acids are boxed in blue. The ionizations of aspartate, glutamate, arginine and histidine are assigned to a single atom but in reality the charges are delocalized.

3.1.2 Polar-uncharged amino acids

Polar amino acids are soluble in water as they can interact favorably with the water molecules. The hydroxyl group in the side chains of serine (Ser; S) and threonine (Thr; T) and the amide group in the side chains of asparagine (Asn; N) and glutamine (Gln; Q) form hydrogen bonds with water. Neither the hydroxyl group nor the amide group is charged under most cellular conditions.

Both serine and threonine can undergo post-translational modifications. Phosphorylation of the hydroxyl by ATP (adenosine triphosphate) is a major mechanism in signal transduction as well as a control mechanism of enzyme activity (Figure 3.7).

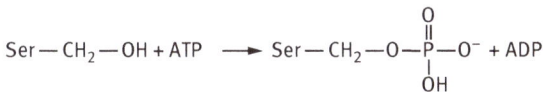

$$Ser-CH_2-OH + ATP \longrightarrow Ser-CH_2-O-\underset{\underset{OH}{|}}{\overset{\overset{O}{\|}}{P}}-O^- + ADP$$

Figure 3.7: Phosphorylation of serine. In the enzymatic reaction a phosphate group from ATP is transferred to the hydroxyl group of serine. When dephosphorylated, the hydroxyl-bound phosphate is released as a free phosphate group, thus ATP is not recovered.

Serine and threonine can also be glycosylated by attachment of a carbohydrate unit to the hydroxyl oxygen forming an *O*-glycoside bond.

Under certain conditions, the hydroxyl group of serine participates in catalysis. In a group of proteases (enzymes that digest proteins), called serine proteases, the serine hydroxyl is activated by deprotonation, making a good nucleophile (-O⁻). This only occurs when the serine hydroxyl is close to certain residues that facilitate the deprotonation by lowering the pK_a of the hydroxyl.

The amide groups of asparagine and glutamine can form hydrogen bonds, both as hydrogen donator and acceptor. Similar to serine and threonine, asparagine and glutamine can be glycosylated, but in this case the carbohydrate unit is linked by an *N*-glycoside bond to the amide nitrogen.

The side chain of cysteine (Cys; C) contains a thiol group (or sulfhydryl). Due to the electron configuration, the sulfur has a partial negative charge, which makes the side chain of cysteine slightly polar. Since it is the electron configuration and not the electronegativity of sulfur that causes the partial charge, the hydrogen is not affected as in a hydroxyl or carboxyl group. Thus, the thiol group is not involved in hydrogen bonding.

The thiol group of cysteine undergoes redox reactions (Figure 3.8). Under oxidizing conditions, cysteine can react with another cysteine in the proximity and form a covalent bond or a disulfide bond, also called a disulfide bridge or S−S bond. Since the intracellular milieu is reducing (due to the high concentration of NADPH and glutathione), disulfide bonds are usually found in the interior of proteins. Cysteines located to the surface of proteins tend to be in the reduced state. During purification of

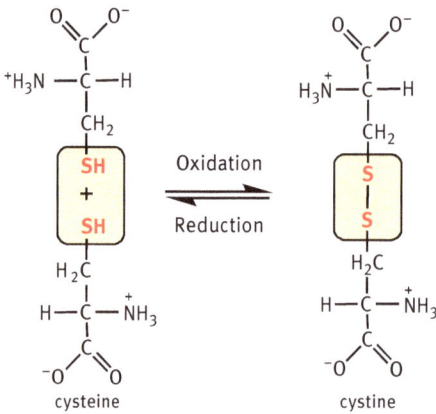

Figure 3.8: The redox reaction of cysteine.

proteins, it is common to include a reducing agent in the media to keep cysteines in the reduced state, avoiding the formation of covalent disulfide bonds.

3.1.3 Polar-charged amino acids

The side chains of aspartate (Asp; D), glutamate (Glu; E), lysine (Lys; K), arginine (Arg; R) and histidine (His; H) are ionized at neutral pH. This implies that they can all participate in ion–ion interactions and in hydrogen bonds, as acceptors as well as donators. Therefore, these residues are often found on the surface of proteins where they can interact with other residues or with the surrounding water. Whenever a polar (charged or uncharged) residue is located in the interior of a protein, it must have another polar residue (or water molecule) close enough to form a hydrogen bond; otherwise, the placement would be energetic unfavorable.

At neutral pH, the side chains of aspartate and glutamate function are negatively charged; they act as acids in that they release a proton. The side chains of lysine, arginine and histidine are positively charged under the same conditions; they act as bases in that they take up a proton.

The pK_a values of the side chain carboxyls in aspartate (~3.5) and glutamate (~4.2) are higher than those of the α-carboxyls (~2). In addition, the pK_a values of the side chain amino groups of lysine (~10.5) and arginine (~12.5) are higher than those of the α-amino groups (~9.5). The reason being that the different chemical surrounding in the side chains influences the pK_a values and thus the ability to be deprotonated or protonated (Table 3.1).

The side chain of arginine contains a guanidine group (or guanidinium group in the protonated form). The three nitrogen atoms in the guanidinium group can all participate in ion–ion interactions as the charge is delocalized over the whole group. The guanidinium group has also several possibilities to act as acceptor and donor in hydrogen bond formation.

The side chain amino group (ε-amino group) in lysine is rather reactive; it can form a Schiff base with aldehydes, and it can be oxidized as well as hydroxylated.

Table 3.1: Properties of α-amino acids.

	Free amino acid			Protein
	Residue mass[a]	pK_a–COOH[b]	pK_a–NH$_2$[b]	pK_a side chain[b]	pK_a side chain[c]
Nonpolar					
Glycine	57.05	2.34	9.60		
Alanine	71.08	2.34	9.69		
Valine	99.13	2.32	9.62		
Leucine	113.16	2.36	9.60		
Isoleucine	113.16	2.36	9.60		
Methionine	131.20	2.28	9.21		
Proline	97.12	1.99	10.60		
Polar-uncharged					
Serine	87.08	2.21	9.15		
Threonine	101.11	2.09	9.10		
Asparagine	114.11	2.02	8.80		
Glutamine	128.13	2.17	9.13		
Cysteine	103.15	1.96	10.28	8.18	6.8 ± − 2.7
Polar-charged					
Aspartate	115.09	1.88	9.60	3.65	3.5 ± 1.2
Glutamate	129.12	2.19	9.67	4.25	4.2 ± 0.9
Lysine	128.18	2.18	8.95	10.53	10.5 ± 1.1
Arginine	156.19	2.17	9.04	12.48	
Histidine	137.14	1.82	9.17	6.00	6.6 ± 1.0
Aromatic					
Tryptophan	186.22	2.83	9.39		
Tyrosine	163.18	2.20	9.11	10.07	10.5 ± 1.2
Phenylalanine	147.18	1.83	9.13		

[a]The residue mass is the mass of the amino acid with an uncharged side chain in a protein. The molecular mass of an amino acid is obtained by adding 18 (the mass of water) to the residue mass.
[b]pK_a values of free amino acids are from CRC Handbook of Chemistry and Physics, 73rd edition.
[c]pK_a values of some side chains when present in proteins are from G.R Grimsley, J.M. Scholtz and N. Pace (2009) A summary of the measured pK values of the ionizable groups in folded proteins. *Prot. Sci.* 18:247–251.

Histidine is the only amino acid with a pK_a close to neutral pH; the pK_a of histidine's side chain is 6.0. Therefore, even relatively small changes in pH will change the average charge (Figure 3.9). When the side chain is protonated, the positive charge is distributed between the two nitrogen atoms. When deprotonated, the hydrogen can be bound to either of the nitrogen atoms in the planar ring. The imidazole group has aromatic properties independent of the protonation state and can form π–stacking interactions. In contrast to the other aromatic amino acids (tryptophan, tyrosine and phenylalanine) histidine does not absorb light in the near UV-region (280 nm) in any protonation state.

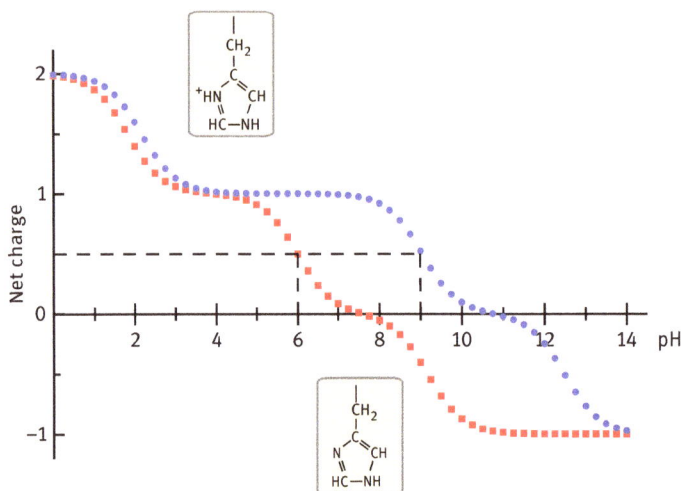

Figure 3.9: pH-titration of histidine (red square) and arginine (blur circle). The dashed line indicates the charges at a pH equal to the pK_as of the side chains. Insert: the side chain of histidine in the charged form below pH 6 and in the uncharged form above pH 6.

3.1.4 Aromatic amino acids

The aromatic amino acids tryptophan (Trp, W), tyrosine (Tyr; Y) and phenylalanine (Phe; F) all contain a conjugated ring system. There is an indole group in tryptophan, a phenol group in tyrosine and a benzene ring in phenylalanine. Since the indole and phenol groups contain polar groups (NH and OH), the side chains of tryptophan and tyrosine have polar properties as well as nonpolar properties.

The aromatic ring systems are planar and the π-electrons are delocalized, which give the ring system particular properties. The π-electrons will create a partial negative charge above or below the ring whereas the ring itself will be partially positive. If two aromatic side chains are in proximity, either "face-to-face" or "edge-to-face" an attractive interaction, a π–π interaction, can be formed between the rings. The force is highly dependent on distance (it decays with r^{-6}) and geometry (it is stronger if the rings are not perfectly face-to-face but rather shifted slightly). The perhaps best example of π–π interactions is the stacking of base pairs in DNA. In the DNA double helix, the distance between bases is only ~ 3.4 Å and each base pair is rotated ~ 30°.

The side chain of histidine is not only polar and ionizable but also aromatic. Thus, also histidine can participate in π–π interactions. The partially negative charge of the aromatic ring system can also form interactions with cations. In proteins, the side chains of arginine and lysine can form cation–π interactions. It should also be noted that the electronegative nitrogen atoms in the aromatic rings can function as acceptors in hydrogen bonds.

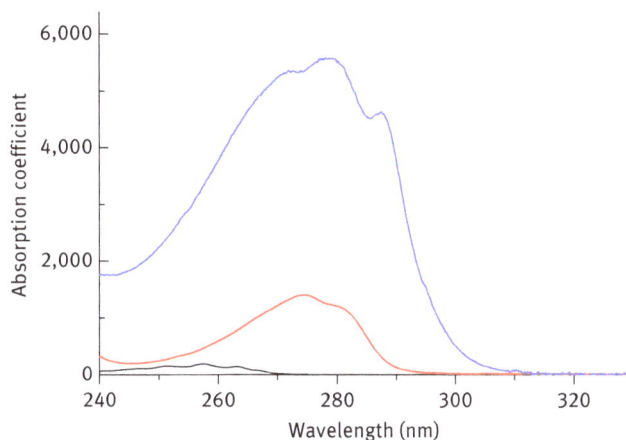

Figure 3.10: Absorption spectra of L-tryptophan (blue), L-tyrosine (red) and L-phenylalanine (black). Data from M. Taniguchi and J.S. Lindsey (2018). Database of absorption and fluorescence spectra of > 300 common compounds for use in photochemCAD. *Photochem. Photobiol.* 94:290–327).

Tryptophan, tyrosine and phenylalanine are important from a practical point of view. Since they absorb light in the near UV-region, their presence in proteins is very useful to measure the concentration of proteins (Figure 3.10). The spectral properties differ between the aromatic amino acids. Tryptophan absorbs ca. 5 times more light than tyrosine and ca. 50 times more than phenylalanine and the wavelengths of their absorbance maxima are different. At 280 nm, the absorbance of histidine is insignificant (Table 3.2).

3.1.5 Substituents

In some proteins, certain amino acids are irreversibly modified after synthesis. In collagen, a major component of fibrous tissues, such as cartilage, ligaments and tendons, proline residues are modified to hydroxyl-proline in a reaction that requires vitamin C and oxygen. If vitamin C is deficient, proline cannot be converted to hydroxyl-proline, which ultimately leads to scurvy. Carboxylation of the γ-carbon in the side chain of glutamate

Table 3.2: Spectral properties of the aromatic amino acids[a].

Amino acid	λ_{max}nm	Molar attenuation coefficient[b] $M^{-1} \cdot cm^{-1}$	Relative absorbance at 280 nm
L-Tryptophan	279	5579 ± 14	1.000
L-Tyrosine	278	1260 ± 2	0.224
L-Phenylalanine	257.6	195.1 ± 1.5	0.019

[a]Data from E. Mihalyi (1968). Numerical values of the absorbance of the aromatic amino acids in acid, neutral and alkaline solutions. *J Chem Eng Data* 13:179–182.
[b]IUPAC recommends to use attenuation coefficient instead of extinction coefficient or molar absorptivity.

is important for the function of osteocalcin (bone γ-carboxyglutamic acid–containing protein) in bone tissue and of several proteins involved in blood coagulation. The formation of carboxy-glutamate is dependent on vitamin K (Figure 3.11).

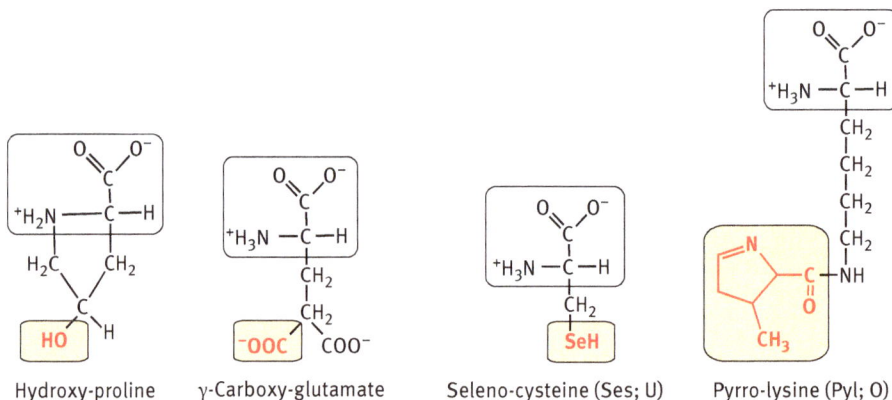

Figure 3.11: Common modified amino acids found in proteins.

Seleno-cysteine and pyrro-lysine are sometimes called the 21st and 22nd amino acids as they are inserted in a few proteins during synthesis. However, neither is coded directly by the genetic code; there is no codon that always translates to seleno-cysteine. Instead, seleno-cysteine is encoded by the UGA codon that normally translates to a stop signal but due to a seleno-cysteine insertion sequence in the mRNA and presence of selenium in the growth media, UGA is translated to seleno-cysteine. In the absence of selenium, UGA translates as a stop and leads to a premature end of the protein synthesis. The pKa (5.47) and reduction potential of seleno-cysteine are lower than that of cysteine.

There are 25 seleno-cystein-containing proteins in the human proteome. Several of these are involved in protection against oxidative stress, such as glutathione peroxidases and thioredoxin reductases.

Like seleno-cysteine, pyrro-lysine is also coded by a stop codon. In this case it is the "amber" stop codon UAG that codes for pyrro-lysine. Pyrro-lysine appears not to be present in eukaryotic proteins, as it has been found only in some archaea and bacteria.

Further reading

Barrett, G.C. and Elmore, D.T. (2008). Amino acids and peptides. Cambridge University Press. Cambridge.

Gonzalez-Flores, J.N., Shetty, S.P., Dubey, A. and Copeland, P.R. (2013). The molecular biology of selenocysteine. *Biomol Concepts* 4:349–365.

Hunter, S.A. and Sanders, J.K.M. (1990). The nature of pi-pi interactions. *J Amer Chem Soc* 112:5525–5543.

Vickery, H.B. and Schmidt, C.L.A. (1931). The history of the discovery of the amino acids. *Chem Rev* 9:169–318.

Wu, G. (2010). Amino acids. Biochemistry and nutrition. CRC Press. Boca Raton, FL.

4 The team

Amides are important substances in the chemical industry, particularly in the pharmaceutical area. An amide can be formed by a condensation reaction between an amine and a carboxylic acid. The conventional reaction occurs via an active ester. Although the reaction appears simple, it is a rather complicated process, requiring several steps of protecting and deprotecting reactive groups as well as removal of unwanted by-products (Figure 4.1).

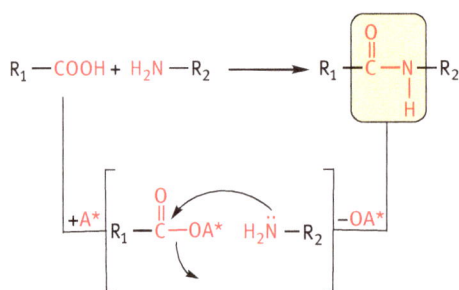

Figure 4.1: Formation of an amide by condensation of a carboxylic acid and an amine. A nucleophilic attack by the amine on the activated carboxylic acid, in the presence of coupling reagents and basic solvent, generates a new amide bond. Adapted with permission from V.R. Pattabiraman and J.W. Bode. (2010). Rethinking the amide bond synthesis. *Nature* 480:471–479.

Nature has devised a similar but superior route for the formation of amide bonds or peptide bonds, as it usually is called in proteins, between amino acids. The condensation reaction occurs on the ribosome, assisted by a template messenger RNA (mRNA) and a transfer RNA (tRNA) in addition to several other factors (Figure 4.2). The order amino acids should condensate is coded into the mRNA and tRNA carries an activated amino acid (attached by an ester coupling) to the site of reaction on the ribosome. The synthesis is error-free, does not produce unwanted products and can produce very large protein containing several thousands of amino acid residues. Although it is possible to chemical synthesize proteins (by solid-phase processes), it is not feasible to produce long (>150 residues) error-free proteins.

4.1 The peptide bond

The delocalization of the lone pair electrons of the nitrogen to the carbonyl, gives the peptide bond ca. 40% double bond character (Figure 4.2B). The bond length between the peptide bond's carbonyl carbon and amide nitrogen (1.33 Å) is therefore shorter than a single-carbon nitrogen bond (1.47 Å) but longer than a double-carbon nitrogen

https://doi.org/10.1515/9783111350684-004

A

B

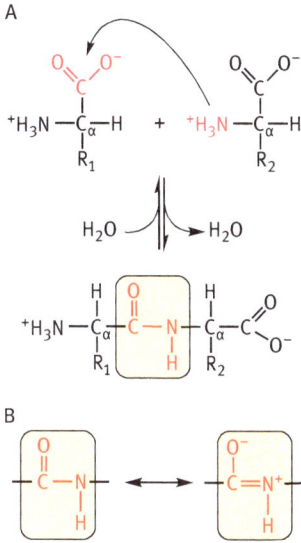

Figure 4.2: Peptide bond formation. (A) Schematic reaction mechanism of condensation of two amino acids and the loss of a water molecule. The peptide bond can be hydrolyzed by certain enzyme, proteases. (B) The peptide bond is stabilized by delocalization of the lone pair of the nitrogen to the carbonyl, giving the bond between the nitrogen and carbonyl carbon a partial double bond character.

bond (1.27 Å). A planar conformation of the six atoms (carbonyl carbon and oxygen, and amide nitrogen and hydrogen as well as the two α-carbons) involved in the peptide group is most stable (Figure 4.3).

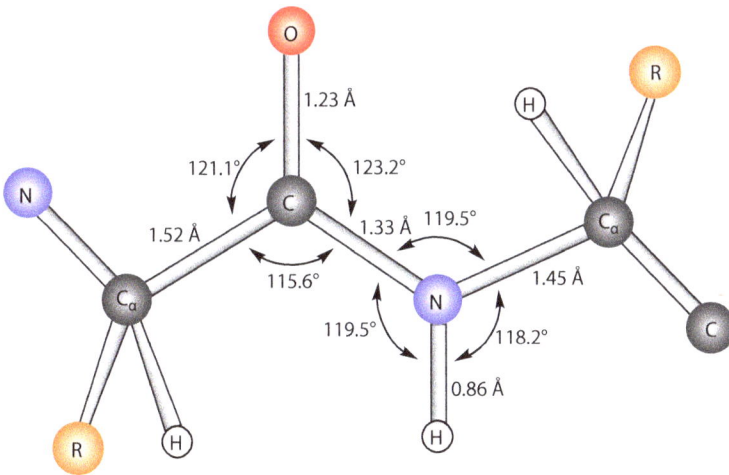

Figure 4.3: A *trans* peptide bond. Structural dimensions of average bond lengths and angles are based on X-ray crystal data from C. Ramakrishnan and G.N. Ramachandran (1965). Stereochemical criteria for polypeptide and protein chain conformations. II. Allowed conformations for a pair of peptide units. *Biophys. J.* 5:909–933, R.A. Engh and R. Huber (1991). Accurate bond and angle parameters for x-ray protein-structure refinement. *Acta Cryst. A* 47:392–400 and A. Jabs, M.S. Weiss and R. Hilgenfeld (1999). Non-proline cis peptide bonds in proteins. *J. Mol. Biol.* 286:291–304.

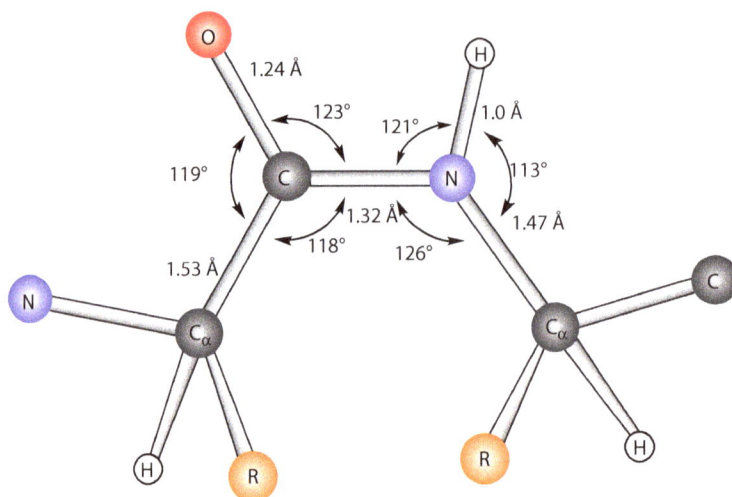

Figure 4.4: A *cis* peptide bond. Data from G.N. Ramachandran and C.M. Venkatachalam (1968). Stereochemical criteria for polypeptides and proteins 4. Standard dimensions for cis-peptide unit and conformation of cis-polypeptides. *Biopolymers* 6:1255–1262.

The partial double bond character of the peptide has a very important structural role. There are two possible conformations of the peptide bond; the carbonyl oxygen and the amide proton can be in a *trans* or *cis* conformation (Figures 4.3 and 4.4). As it turns out, nearly all peptide bonds are in the *trans* conformation. It is estimated that only around 0.05% of nonproline peptide bonds are *cis*, while around 6% of X-Pro peptide bonds are *cis*. Due to the double-bond character, there is a rather high energy cost to change the conformation of a *cis* peptide bond, and particularly of a nonproline *cis* peptide bond, to *trans* and vice versa.

In contrast to the peptide bond, there is free rotation (within certain limits) of the bonds between α-carbon and carbonyl carbon and between amide nitrogen and α-carbon. Since the amide planes are planar, the rotation around each of these bonds is enough to describe the structure of the chain of amino acid residues or the protein backbone. For this purpose, two torsion or dihedral angles are defined; Φ (phi) and ψ (psi). Φ is the rotation of the amide plane around the N–Cα bond and ψ is the rotation of the amide plane around the Cα–C bond (Figure 4.5). The dihedral angles Φ and ψ are relative to a plane defined by the amide nitrogen, α-carbon and carbonyl carbon. As the peptide group is planar, the dihedral angle ω is close to +180° (−180°) or 0° for a *trans* or *cis* peptide bond, respectively.

Although a single bond in principle allows free rotation, some conformations are energetically preferred. A staggered conformation is more stable than an eclipse conformation (Figure 4.6). The energy difference between a staggered conformation and an eclipse conformation of ethane is ~ 12 kJ/mol. Thus, there is an energy barrier to free

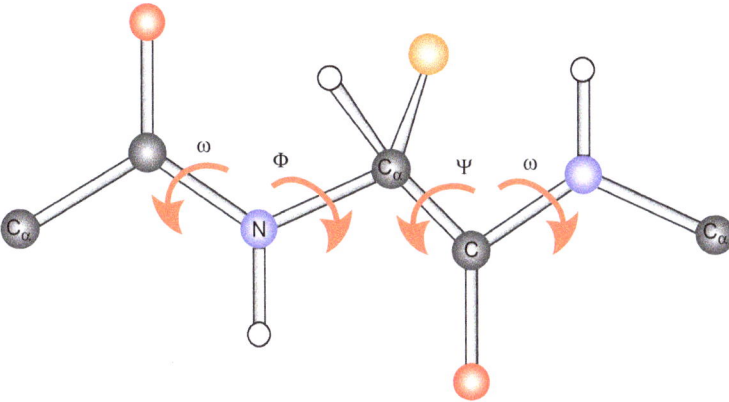

Figure 4.5: Dihedral angles. The dihedral angles Φ, ψ and ω are defined as +180° (or −180°) for a fully extended chain of amino acid residues.

rotation around the single C–C bond in ethane. For any molecule, the most stable conformation is the one with the lowest energy, which implies that whenever possible a molecule would prefer a staggered conformation rather than an eclipse conformation.

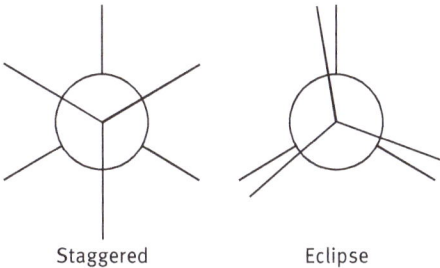

Staggered   Eclipse

Figure 4.6: Newman projection of staggered and eclipse conformations.

4.2 The Ramachandran graph

When both Φ and ψ are +180° (or −180°) the protein backbone is fully extended and has a staggered (or staggered-like) conformation. Dihedral angles close to zero give rise to a forbidden eclipse-like conformation as some atoms get too close to each other and clashes. Obviously, all dihedral angles that lead to clashes between side chains and/or atoms of the backbone are forbidden. Using "normally allowed" and "outer limit" van der Waals contact distances Ramachandran and colleagues were able to define permitted dihedral angles that were supported by analysis of known structures of di- and tripeptides. The result indicated that all sterically allowed conformations fall within two discrete areas, centered around $\Phi = -60°$ and $\psi = -40°$ and around $\Phi = -120°$ and $\psi = +130°$

except for dihedral angles involving glycine or proline. The original conformation space has increased as more and more protein structures have been solved and the dihedral angles have been determined. The Ramachandran graphs in Figure 4.7, are based on dihedral angles from structures of a very large number of actual proteins. The fundamental work by Ramachandran is still used as a control for determined protein structures, as all dihedral angles must be within permitted areas; otherwise, the structure probably contains errors. When validating a protein structure, it is expected that 98% or more of the dihedral angles fall within the favorable Φ and ψ angles.

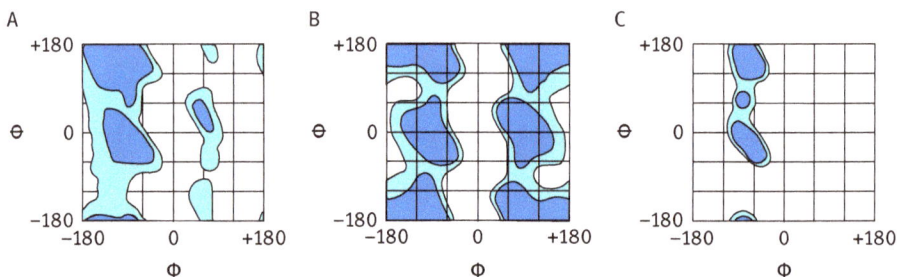

Figure 4.7: Ramachandran graphs of favorable (dark blue) and allowed (light blue) dihedral angles from structures of actual proteins. Conformational space of (A) all residues except for glycine and proline, (B) glycine and (C) proline. Data from UCSF Chimera. (E.F. Pettersen, T.D. Goddard, C.C. Huang, G.S. Couch, D.M. Greenblatt, E.C. Meng and T.E. Ferrin (2004). UCSF Chimera–a visualization system for exploratory research and analysis. *J. Comput. Chem.* 25:1605–1612).

4.3 The primary structure

Protein synthesis starts by condensation between the carboxyl group of one amino acid with the amino group of another amino acid, as shown in Figure 4.2. The resulting molecule now contains two amino acid residues, a dipeptide. The addition of another amino acid results in a tripeptide, and so on until the final polypeptide or protein is synthesized. This linear sequence of amino acid residues represents the primary structure of the polypeptide.

The polypeptide has different ends; there is a free amino group at one end and a carboxyl group at the other end. The end with an amino group is termed the N-terminal (or amino terminal) whereas the other end is the C-terminal (or carboxyl terminal) as shown in Figure 4.8. Therefore, all polypeptides have a direction, from N- to C-termini.

Since the transcription process has the same directionality (from the 5'-end to 3'-end) as the protein synthesis, translation of the template can even begin before the template is completed.

The backbone of all polypeptides and proteins is identical, a number of amino acid residues connected by peptide bonds, as shown in Figure 4.8. What distinguishes one

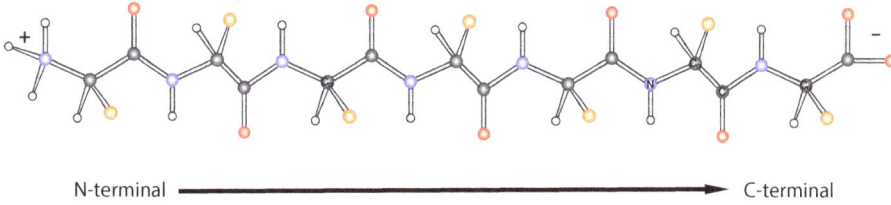

N-terminal ━━━━━━━━━━━━━━━━━━━━━━━▶ C-terminal

Figure 4.8: A polypeptide in its fully extended conformation, showing the directionality from the N-terminal to the C-terminal.

polypeptide from another one is the side chains. Thus, the function of any polypeptide or protein depends on the properties of the side chains.

Depending on the amino acids composition, the polypeptide may be charged. In most proteins, the termini as well as other ionizable groups are ionized at neutral pH. Since the ionization depends on the pH of the solvent, the ionization changes when pH is altered. At a certain pH, the total number of positive charges equals the total number of negative charges and the net charge is zero. This particular pH is called the proteins isoelectric point, pI. At any pH below the pI, the protein will have a positive net charge and at any pH above the pI it will have a negative net charge.

As charged groups may be involved in ion–ion interactions that are of great importance for the structure of the protein, altered ionization may affect such interactions, which could have devastating consequences for the protein structure.

Although most proteins contain a few hundred amino acid residues, there are proteins that only contain a few tens of residues. The largest human protein is the muscle protein titin that contains around 36,000 residues and has a molecular mass of 4,000,000 Dalton. Thus, there is an enormous range of sizes among proteins found in any organism, whether it is of bacterial or human origin.

Before the polypeptide becomes a functional protein, it must be folded correctly. During the folding process, nonpolar residues are placed away from contact with the surrounding water and hydrogen bonds and other noncovalent interactions are formed. Since the conformation space is limited by the allowed ranges of Φ and ψ angles, the backbone folds into two major conformations: one helical and the other extended in addition to loops.

4.4 The secondary structure

The folding process requires not only a driving force to initiate folding but also that the folded version of the protein is more stable than the unfolded. The driving force of folding is the hydrophobic effect that leads to minimal contacts between nonpolar groups and water. During the folding, several noncovalent interactions form: van der Waals interactions between nonpolar side chains as well as hydrogen bonds and ion–ion interactions. These interactions stabilize the folded conformation of the protein.

In 1951, Linus Pauling and Robert Corey proposed in a series of papers that 3.7 residue α-helices, 5.1 residue γ-helices and pleated-sheets are present in proteins. They deduced these fundamental structure elements from crystal structures of small molecules and Pauling's theoretical framework on chemical bonding. They also suggested that these structures are stabilized by hydrogen bonds between the carbonyl oxygen in one residue and the amide hydrogen in another residue. Later numerous examples of solved protein structures have shown that most of these predictions were correct, with the exception of the γ-helix that has not been found in any protein. Today we know that α-helices and pleated sheets (usually called β-sheets) together with turns and loops are common structural elements in all proteins. These structural elements are commonly called secondary structures. When looking at secondary structures, only the backbone is considered, and the side chains are not involved in the stabilization of α-helices and β-sheets as well as all other secondary structure elements.

4.4.1 α-Helix

The α-helix is a right-handed coil, with 3.6 amino acid residues per turn (not 3.7 as suggested by Pauling and Corey) or 13 backbone atoms. The rise of each turn or the pitch of the α-helix is 5.41 Å. Since there is no space within the helix, all side chains point to the periphery with a lateral spacing of 100°. The α-helix is stabilized by hydrogen bonds between the amide C=O group in residue i and the amide N—H group in residue $i + 4$, as illustrated in Figure 4.9. The hydrogen bond length is around 2.8 Å and the dihedral angles Φ and ψ for α-helix are −57° and −47°, respectively. An α-helix does not have a specific length but usually contains between 5 and 40 residues. There are also examples where the helical length is much longer; the muscle protein myosin is such an example.

The propensity to find a certain amino acid residue in an α-helix varies. A residue with a straight side chain, such as alanine or leucine, occurs more often in an α-helix than a residue with a bulky or branched side chain, like tyrosine or valine. As expected, proline has a very low propensity for α-helices due to its lack of amide hydrogen and the constrained structure caused by the cyclization. It is perhaps more surprising that glycine also has a low propensity for α-helices. However, this is probably because the small side chain (only a hydrogen atom) allows more backbone flexibility and the inability to participate in side chain interactions.

A proline residue has no hydrogen atom bound to the amide nitrogen atom due to cyclization of the side chain. Therefore, the carbonyl oxygen (C=O) of the residue four residues before the proline has no partner (amide proton) to form a hydrogen bond with. A proline residue in an α-helical segment is therefore generally considered as a helix breaker; an α-helix may begin after or end before a proline residue. This in

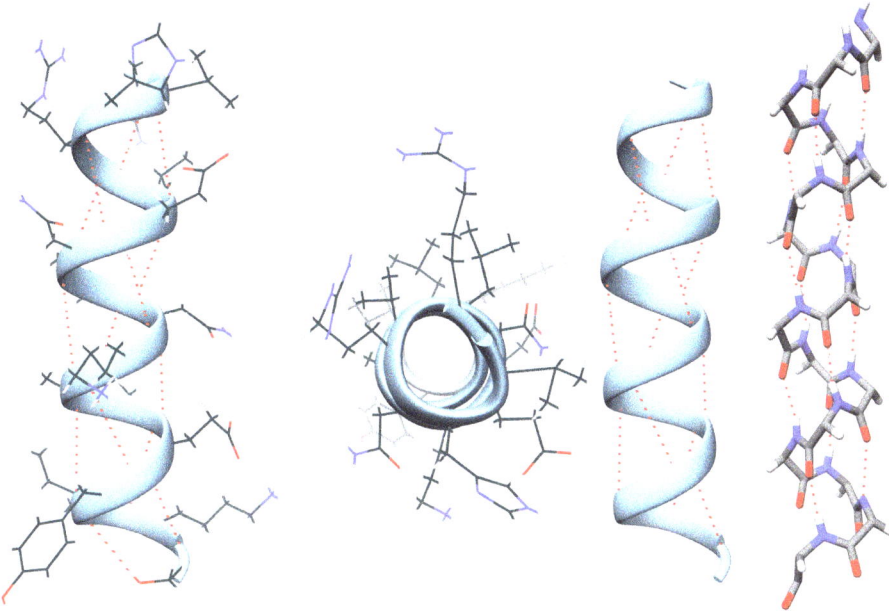

Figure 4.9: The α-helix. All side chains are pointing out from the helix. The hydrogen bonds between the carbonyl oxygens and the amide proton is indicated by the red dotted line.

not entirely correct as there are amino acid sequences with a proline residue that form noninterrupted α-helices but the presence of the proline creates a kink in the helix instead of ending the helix (Figure 4.10).

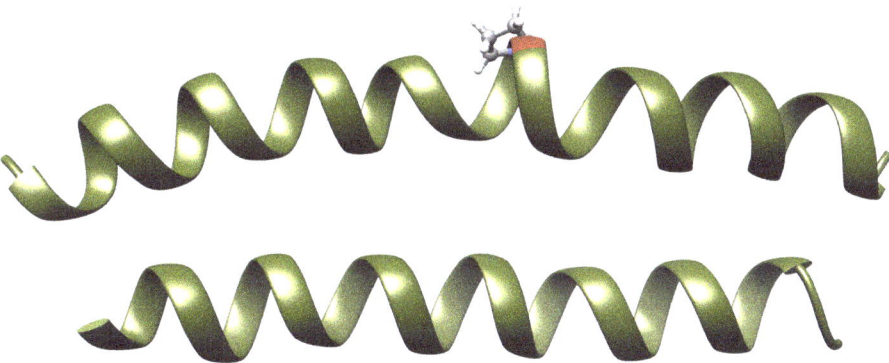

Figure 4.10: The presence of a proline residue in the sequence creates a kink in the helix. In the upper α-helix, the proline residue is marked in red and its side chain visible. Below a typical α-helix without proline residues.

As the peptide bond has a trans conformation the amide N—H and C=O groups in the same peptide unit are pointing in opposite directions. As both the amide hydrogen and the carbonyl oxygen both have partial but opposite charges, the peptide unit is polarized and has a weak dipole moment. At the N-termini, the amide N—H groups in the four most N-terminal residues have no amide C=O group to form hydrogen bonds with. Analogous at the C-terminal, the amide C=O groups in the four most C-terminal residues have no amide N—H group to interact with. This makes the whole α-helix a somewhat stronger dipole. The presence of acidic residues at the N-terminal and basic residues at the C-terminal stabilizes the helix.

The chemical properties of the α-helix are determined by the side chains since these are in contact with the surrounding solvent. Thus, in water it can be expected that most of the exposed side chains are polar. Likewise, in a nonpolar medium, such as the interior of a membrane, the α-helix would contain mostly nonpolar amino acid residues. Often one side of the helix has polar properties, whereas the other side is mostly nonpolar. Such an amphiphatic helix can interact with other amphiphatic helices and form a structure with a nonpolar outside and a polar channel, as illustrated in Figure 4.11.

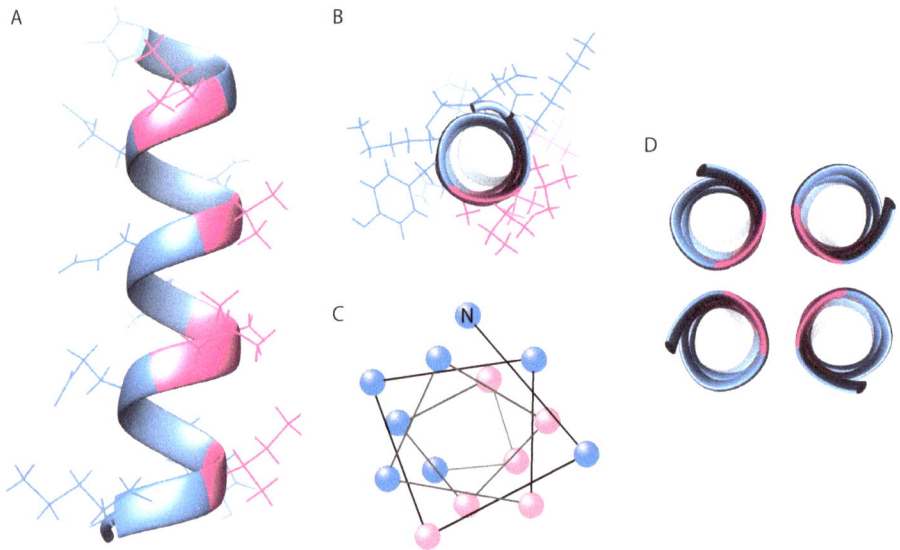

Figure 4.11: An amphiphatic α-helix, with a polar side (pink) and a nonpolar side (blue) viewed from the side (A) or from the top (B). (C) Helical wheel representation, looking down from the N-terminal end toward the C-terminal as in B. (D) Four amphiphatic helices may form a structure with a nonpolar outside and a polar inside. The nonpolar outside allows the helices to be placed in a membrane where the polar inside (pink) may function as a channel through the membrane.

The right-handed α-helix is not the only helical structure in proteins. Distortion of the regular α-helix due to a proline residue or overall folding of the protein can create other helical structures, such as the right-handed 3_{10}-helix and π-helix. Both occurs only in short segments due to less favorable energies. In the 3_{10}-helix the amide C=O group in residue i is hydrogen bonded to the amide N—H group in residue $i + 3$. There are 10 atoms between the oxygen and hydrogen in the hydrogen bond, hence the name 3_{10}-helix. Therefore, the twist of the 3_{10}-helix is tighter and makes it narrower than the α-helix.

In the π-helix there are about 4.1 residues per turn. The π-helix is also stabilized by hydrogen bonds between the amide C=O group and the amide N—H group but in this case it is between residue i and $i + 5$. This translates to 16 atoms between the atoms in the hydrogen bond. π-helical segments are short, usually with seven to ten amino acid residues and two π-type hydrogen bonds. It is suggested that a segment of π-helical structure is created when an amino acid is inserted in an α-helix. Thus, π-helices are in general flanked by α-helices and often a proline residue marks the end of the π-helix. The dihedral angles Φ and ψ for the π-helix are centered around −57° and −70°, respectively, precisely on the border to the allowed region of the Ramachandran graph (Table 4.1).

Table 4.1: Average dihedral angles for regular secondary structures.

Secondary structure	Φ (pfi)	ψ (psi)	ω (omega)	Residue per turn	Rise per residue
Right-handed α-helix	−57	−47	+180	3.6	1.50
Left-handed α-helix	+57	+47	+180	3.6	1.50
3_{10}-Helix	−49	−26	+180	3.0	2.00
π-Helix	−57	−70	+180	4.4	1.15
Poly-proline helix I	−83	+158	0	3.33	1.9
Poly-proline helix II	−79	+150	+180	3.0	3.12
Parallel β-sheet	−119	+113	+180		3.2
Anti-parallel β-sheet	−139	+135	−178		3.4

Adapted from G.N. Ramachandran and V. Sasisekharan (1968). Conformation of polypeptides and proteins. *Adv. Prot. Chem.* 23:283–438 and IUPAC-IUB Commission on Biochemical Nomenclature (1970). Abbreviations and symbols for the description of the conformation of polypeptide chains. Tentative rules. *Biochemistry* 9: 3471–3479.

Short sequences with several proline residues can fold into a helical arrangement, called poly-proline helix. The peptide bonds have a cis conformation in the right-handed type I poly-proline helix (PPI), whereas in the left-handed type II poly-proline helix (PPII) the conformation is trans. PPI helices have so far not been found in any protein. There are three residues per turn in the PPII helix and no hydrogen bonds. PPII helices are important in signal transduction, transcription and movement (Figure 4.12).

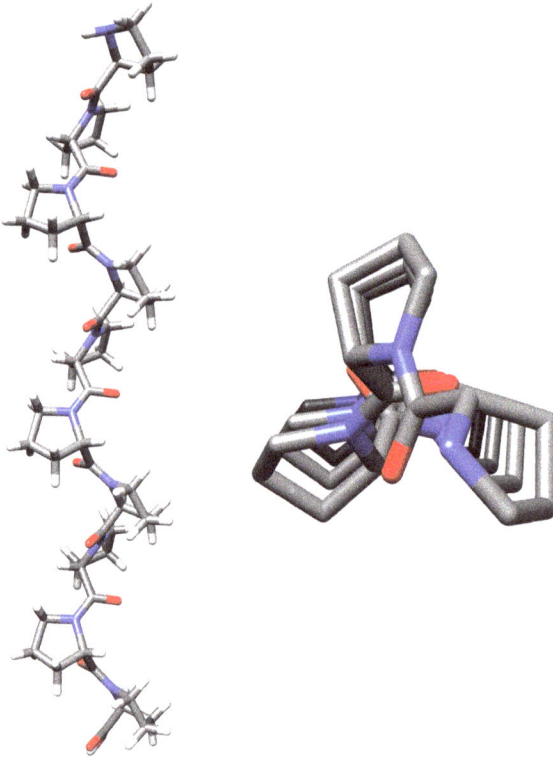

Figure 4.12: Type II poly-proline helix. When viewed parallel to the helix axis, it is clearly seen that a poly-proline in PPII conformation has a threefold symmetry and that every fourth residue is in the same position.

4.4.2 β-Strands and β-sheet

β-Strands are extended structures with 3.2–3.4 Å between the amino acid residues. The favorable dihedral angles Φ and ψ are ca. −130° and +120°, respectively, slightly less than for a fully extended polypeptide chain ($\Phi = \psi = \pm180°$). Two or more β-strands can form a β-sheet, where hydrogen bonds between the amide C=O group of one β-strand and the amide N—H group of another β-strand stabilizes the β-sheet. In a β-sheet, the strands are either parallel and point in the same direction or antiparallel and point in opposite directions (Figure 4.13). The repeat length or pitch of a parallel β-sheet is 6.5 Å compared to 7.0 Å in an antiparallel β-sheet. When viewed from the side, the β-sheet is not flat but pleated, with the α-carbons at the kinks and neighboring side chains alternately protrude above or below the sheet. Since steric hindrance is less important, residues with bulky or branched side chains are often found in β-sheets.

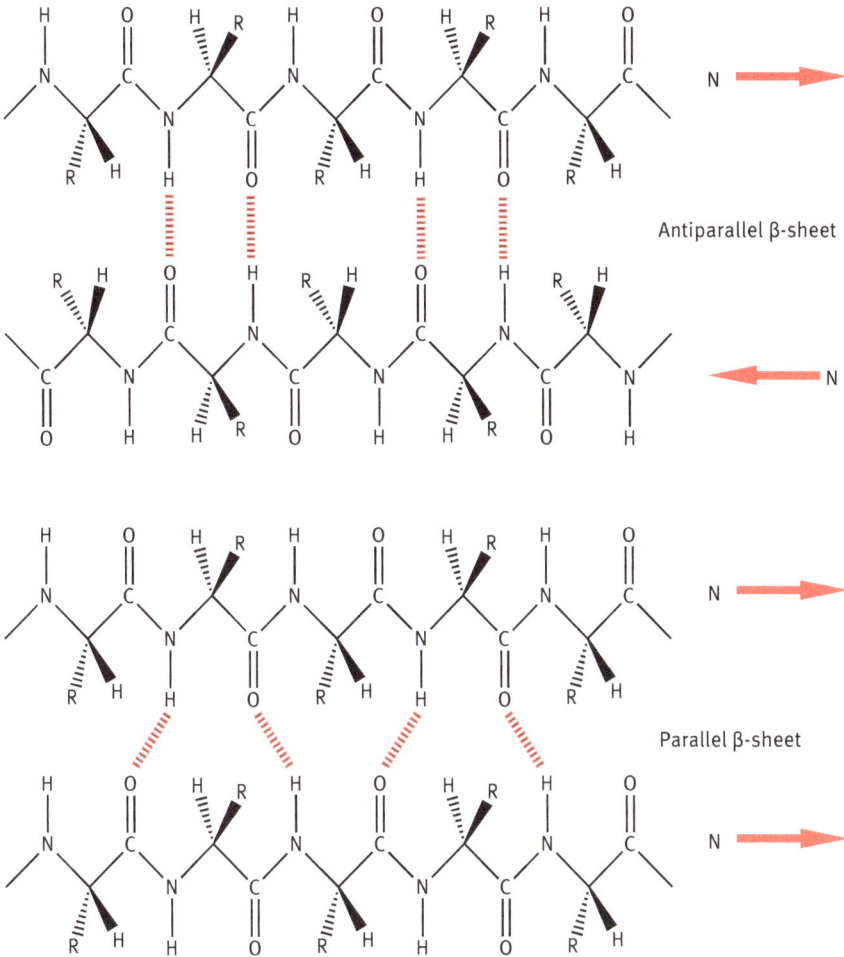

Figure 4.13: Parallel and anti-parallel β-sheets. Hydrogen bonds between the strands are indicated by dashed lines. The direction of each strand is indicated by arrows.

The hydrogen bonds in parallel β-sheets are slightly distorted from the optimal orientation as the C=O and N—H groups do not line up perfectly due to steric reasons. This makes the hydrogen bonds in parallel β-sheets a little weaker than those in antiparallel β-sheets and the parallel β-sheet energetically less stable.

Like α-helices, β-sheets are common structure elements in all proteins. Usually β-sheets contain between 2 and 5 β-strands, although there are β-sheets with many more strands such as the green fluorescent protein that contains 11 strands that form a cylinder. The length of a β-strand varies from only a few residues to more than 10 residues.

The β-strands are not straight but have a slight right-handed twist that gives the β-sheet a bent appearance (Figure 4.14).

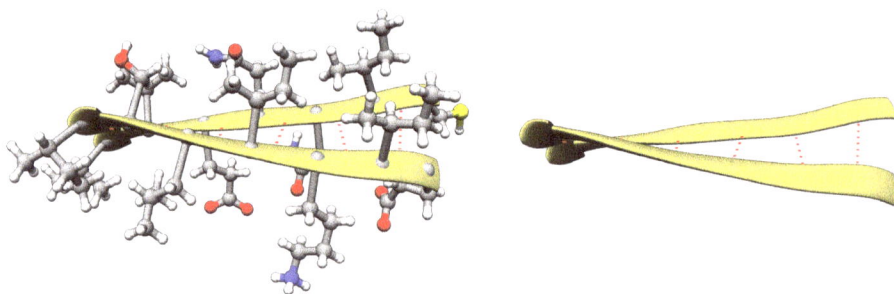

Figure 4.14: The right-handed twist of a parallel β-sheet. The ball-and-stick representation shows that every second side chain protrudes either above the β-sheet or below. Dashed lines indicate hydrogen bonds.

Since the backbone is continuous, β-strands are connected to each other by stretches with other structures. Anti-parallel β-strands can be connected by short loops or β-turns, creating a so called β-hairpin. This is not possible for parallel β-strands. Parallel β-strands are instead connected by right-handed or left-handed crossover connections. Most crossover connections in proteins are right-handed (Figure 4.15).

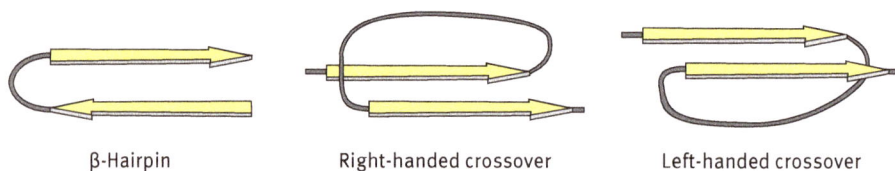

| β-Hairpin | Right-handed crossover | Left-handed crossover |

Figure 4.15: Connectivity of anti-parallel and parallel β-strands.

Often the crossover connection contains an α-helix, creating a β–α–β structure motif. In the enzyme triosephosphate isomerase eight parallel β-strands form a barrel-like structure where all strands are connected by a right-handed crossover containing an α-helix (Figure 4.16).

4.4.3 Turns and loops

In order for a protein to fold into a compact globular structure α-helices and β-strands must be connected with reverse turns and loops. Reverse and inverse turns are structure elements defined by their backbone dihedral angles (Table 4.2).

Even though there are several types of reverse turns that occur in proteins, β-turns are the most common type. The β-turn is defined by four residues that create a tight reverse turn. The structure is stabilized by a hydrogen bond between the carbonyl oxygen of residue i with the amide hydrogen of residue $i + 3$. The peptide bond between residue

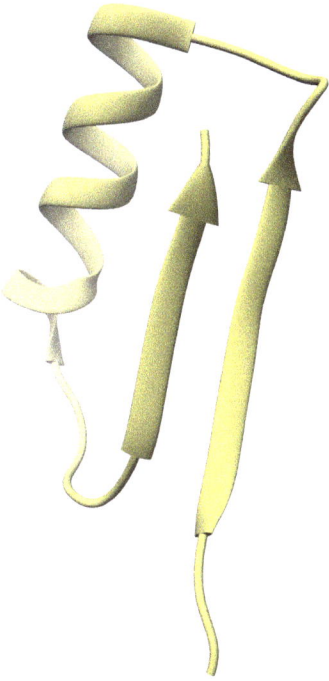

Figure 4.16: A typical β–α–β motif. The structure is extracted from the enzyme triosephosphate isomerase (pdb: 1YPI) that contains 8 parallel β-strands that all are connected by an α-helix in a right-handed crossover conformation.

Table 4.2: Average dihedral angles for reverse and inverse turns.

Type	Φ_{i+1}	ψ_{i+1}	Φ_{i+2}	ψ_{i+2}
Reverse type I β-turn	−60°	−30°	−90°	0°
Reverse II	−60°	+120°	+80°	0°
Inverse I′	+60°	+30°	+90°	0°
Inverse II′	+60°	−120°	−80°	0°
Reverse γ-turn	+70°	−60°		
Inverse γ-turn	−70°	+60°		

Adapted from G.N. Ramachandran and V. Sasisekharan (1968). Conformation of polypeptides and proteins. *Adv. Prot. Chem.* 23:283–438 and IUPAC-IUB Commission on Biochemical Nomenclature (1970). Abbreviations and symbols for the description of the conformation of polypeptide chains. Tentative rules. *Biochemistry* 9: 3471–3479.

$i + 2$ and $i + 3$ in type I is flipped by 180° compared to a type II β-turn. Position $i + 1$ in both types is often occupied by a proline residue as it can easily attain the proper conformation. In type II β-turns the carbonyl oxygen would clash with the side chain β-carbon of residue $i + 2$; therefore, this position is occupied by glycine. Inverse β-turns are mirror images of reverse β-turns in which the signs on the dihedral angles are the opposite (Figure 4.17).

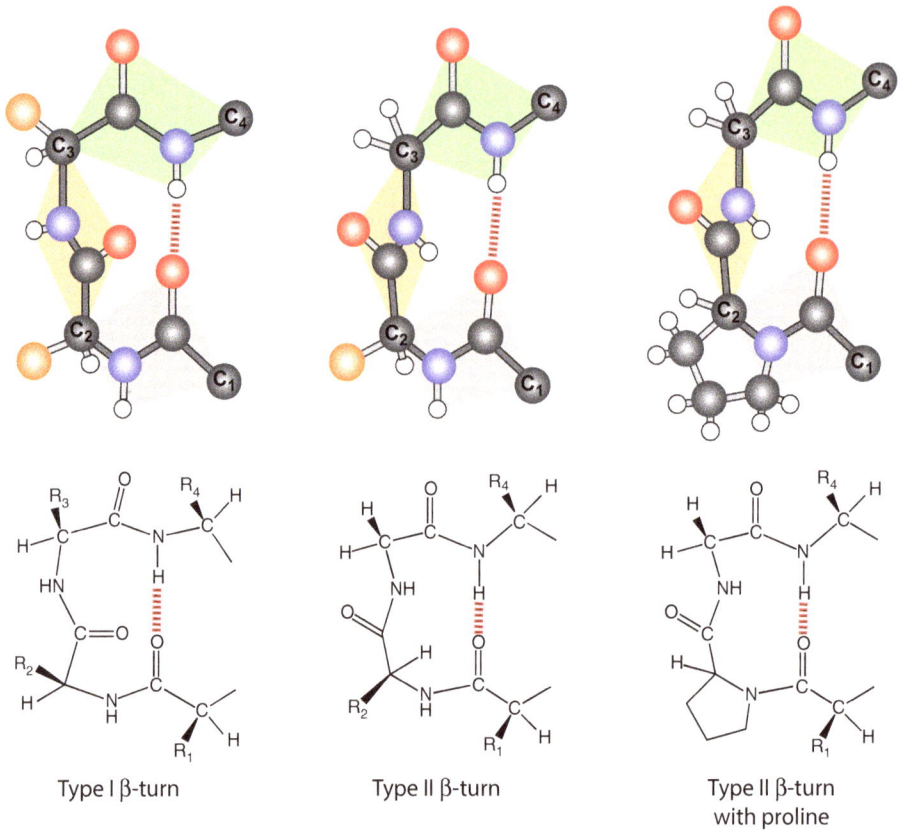

Figure 4.17: Type I and II β-turns. Dashed lines indicate hydrogen bonds.

The tightest defined turn is the γ-turn, that contains only three residues and is stabilized by a hydrogen bond between residue i and $i + 2$. Other occurring turns are α-turns (hydrogen bonds between residue i and $i + 4$) and π-turns (hydrogen bonds between residue i and $i + 5$).

A backbone stretch of six or more amino acid residues, with any amino acid sequence, can form a nonregular structure called a Ω-loop. Neither the number of residues nor the hydrogen bonding pattern is fixed in a Ω-loop. The name comes from its structural resemblance to the Greek letter Ω (omega).

Both turns and loops are usually located to the surface of globular proteins; therefore, they may also be directly involved in signal recognition. Since a loop generally is flexible, it is also possible that the loop structure can undergo a conformational change and thus adapt its structure to an interacting partner.

4.5 Secondary structure prediction

Peter Chou and Gerald Fasman developed an empirical method to predict the secondary structure from a given amino acid sequence. The method is based on the propensity of an amino acid residue to be in an α-helix or β-strand conformation and a few simple rules. The propensity values in Table 4.3 were calculated from 2168 protein structures. The initial propensity values were based on a very small dataset, containing only 15 proteins.

An α-helix is predicted if four of six residues in a row has propensity for helix formation. The helix is extended in both directions until the average helix propensity falls below 1 calculated for four consecutive residues. A similar rule applies for β-strands. A strand is predicted if three of five residues have propensity for β-strands formation and the strand is extended in both directions until the average strand propensity falls below 1 calculated for four consecutive residues.

Table 4.3: Secondary structure propensity of the amino acids.

Amino acid	α-Helix	β-Strand	Coil
Alanine	**1.39**	0.75	0.80
Glutamate	**1.35**	0.72	0.86
Leucine	**1.32**	1.10	0.68
Glutamine	**1.29**	0.76	0.89
Methionine	**1.21**	0.99	0.83
Arginine	**1.17**	0.91	0.91
Lysine	**1.11**	0.83	1.00
Valine	0.89	**1.86**	0.64
Isoleucine	1.04	**1.71**	0.59
Tyrosine	0.95	**1.50**	0.78
Phenylalanine	1.10	**1.43**	0.76
Cysteine	0.74	**1.31**	1.05
Tryptophan	1.06	**1.30**	0.79
Threonine	0.76	**1.23**	1.07
Proline	0.50	0.44	**1.72**
Glycine	0.47	0.65	**1.62**
Asparagine	0.77	0.62	**1.39**
Aspartate	0.89	0.55	**1.33**
Serine	0.82	0.85	**1.24**
Histidine	0.92	0.99	1.07

Data from S. Constantini, G. Colonna and A.M. Facchiano (2006). Amino acid propensies for secondary structures are influenced by the protein structural class. *Biochem. Biophys. Res. Communic.* 342:441–451.

4.6 The tertiary structure

The tertiary structure is defined as the functional conformation of a polypeptide chain. When the unfolded polypeptide chain folds into a highly ordered conformation, secondary structure elements, such as helices, pleated sheets, turns and loops, come together to establish the native conformation. Residues far away from each other in the primary structure end up close to each other when the polypeptide chain folds and may even form bonds with each other. During the folding process, hydrogen bonds break, particularly those involving water molecules, as bonds instead form between groups in the side chains or between groups in a side chain and the backbone amide groups. Juxtaposition of charged side chains allow for ion–ion interactions. Thus, the whole folding process seeks to find the most stable conformation of the protein, that is, the conformation with minimal energy.

Although all bonds and interactions are important for the stability of the protein, the driving force of folding is the hydrophobic effect. Hence, a polypeptide must contain nonpolar amino acid residues that together create a hydrophobic core in order to fold; without a hydrophobic core, there will be no folding. When nonpolar residues are "hidden" in the interior due to folding order is increased but at the same time caged water molecules are "released" which increase their disorder. Taken together, these opposing effects on order lead to increased entropy, as required for a spontaneous process.

Proteins and particular globular proteins are very compact as the folding tends to leave as little empty space as possible. Nonpolar side chains are preferably placed in the interior, avoiding the surrounding water, and close to each other that allows for van der Waals interactions or hydrophobic interactions. Aromatic side chains may be close enough to induce π–π interactions. However, polar groups also end up in the nonpolar interior.

When a polar group transfers from a polar to a nonpolar milieu, any hydrogen bonded water molecules are released. Since it is energetic unfavorable to omit hydrogen bonding groups from hydrogen bonds, these groups need to form new hydrogen bonds. This can be done by forming α-helices or β-sheets as these secondary structures are stabilized by hydrogen bonds between the amide groups in the backbone. Similarly, polar side chains need to find "new" hydrogen bonding partners.

Residues with charged side chains are usually found on the surface, where they can interact with other charged groups or salt ions. When a charged group is transferred to the nonpolar interior, it is more difficult to foresee the consequence as the chemical environment can influence the pK_a-value of an ionizable group considerably. For instance, the side chain carboxyl group of glutamate has a pK_a-value around 2 in water, but in the interior of a protein the pK_a may increase to 7 or even more.

As illustrated in Figure 4.18, the native structure of the protein is stabilized by many different and usually weak bonds and interactions. As mentioned above, the hydrophobic effect is the driving force in protein folding, but it is the bonds that form that stabilize the protein structure. The weak but numerous van der Waals interactions contribute considerably to the stability. Even though the intracellular environment is reducing, covalent disulfide bonds can persist in the interior of a protein and thereby contribute to the stability. When intermolecular hydrogen bonds between polar groups and water molecules exchange for intramolecular hydrogen bonds between polar groups of the polypeptide chain, the energy gain is not very large. In fact, the energy difference between a folded and unfolded (or denatured) protein is in most cases not large. A protein can be unfolded by breaking a few hydrogen bonds or ion–ion interactions. This can happen due to a small change in pH that would affect the ionization of a certain group, which could lead to the loss of an ion–ion interaction or a hydrogen bond.

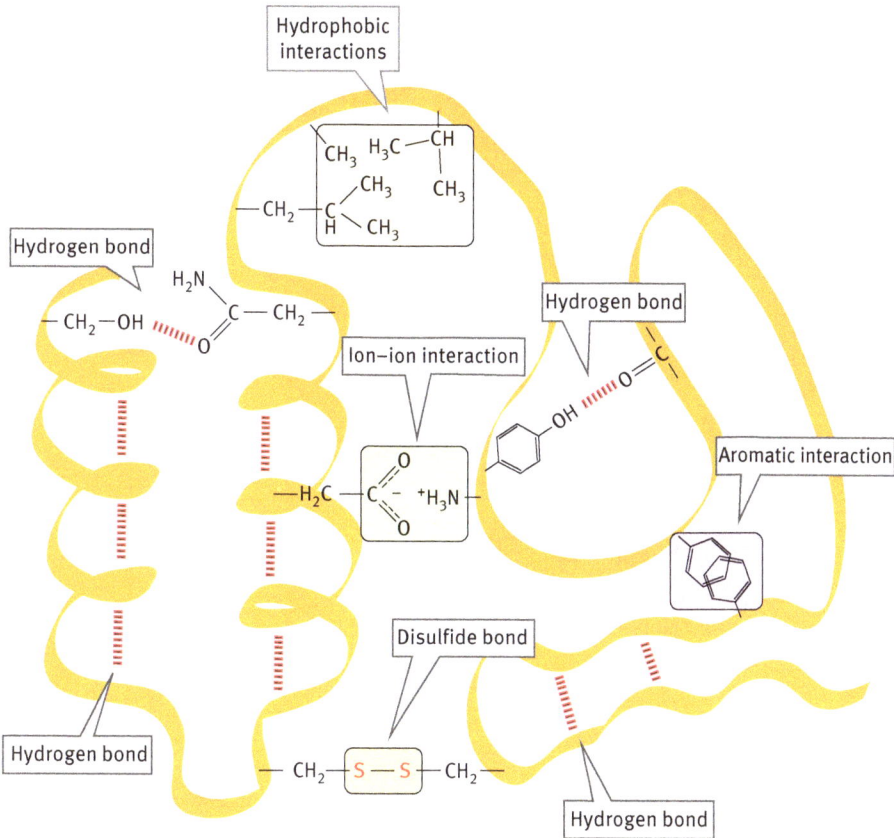

Figure 4.18: Stabilizing bonds and interactions in a native protein.

4.7 The quaternary structure

Some proteins contain two or more polypeptide chains. If the polypeptide chains are the result of translation of different genes, each polypeptide chain is usually called a subunit. The oxygen transporter hemoglobin is composed of two αβ-subunits, where each αβsubunit in turn contains one α-subunit and one β-subunit (Figure 4.19). Then there are proteins that are synthesized as a single polypeptide chain but before being functional one or more peptide bonds must be hydrolyzed. The proteolytic enzyme chymotrypsin that catalyzes hydrolysis of peptide bonds is synthesized as an inactive chymotrypsinogen. Activation requires hydrolysis of four peptide bonds and formation of several disulfide bonds, thus creating three polypeptide chains.

The quaternary structure relates the position of each subunit or polypeptide chain in space to all other subunits or polypeptides.

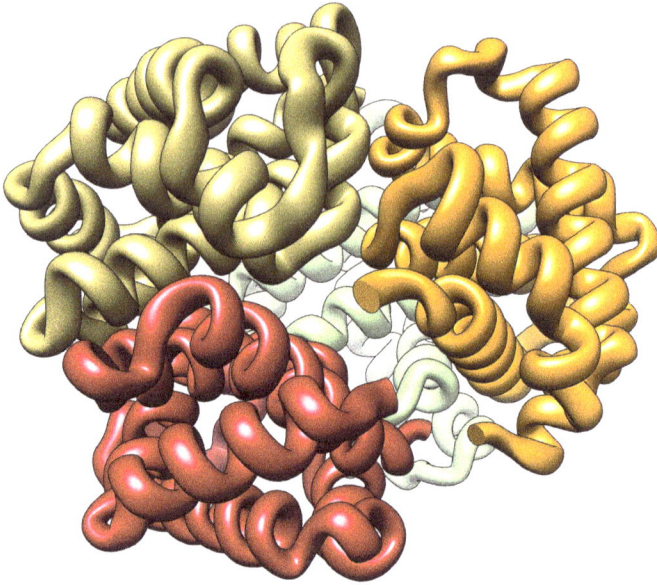

Figure 4.19: A tetrameric protein composed of four subunits.

Further reading

IUPAC-IUB commission on biochemical nomenclature. Abbreviations and symbols for the description of the conformation of polypeptide chains. Tentative rules (1969). *Biochemistry* 9:3471–3479.

Chou, P.Y. and Fasman, G.D. (1978). Empirical predictions of protein conformation. *Annu Rev Biochem* 47:251–276.

Cooley, R.B., Arp, D.J. and Karplus, P.A. (2010). Evolutionary origin of a secondary structure: Pi-helices as cryptic but widespread insertional variations of alpha-helices that enhance protein functionality. *J Mol Biol* 404:232–246.

Engh, R.A. and Huber, R. (1991). Accurate bond and angle parameters for x-ray protein-structure refinement. *Acta Cryst A* 47:392–400.

Jabs, A., Weiss, M.S. and Hilgenfeld, R. (1999). Non-proline cis peptide bonds in proteins. *J Mol Biol* 286:291–304.

Pauling, L., Corey, R.B. and Branson, H.R. (1951). The structure of proteins; two hydrogen-bonded helical configurations of the polypeptide chain. *Proc Natl Acad Sci U S A* 37:205–211.

Pettersen, E.F., Goddard, T.D., Huang, C.C., Couch, G.S., Greenblatt, D.M., Meng, E.C. and Ferrin, T.E. (2004). Ucsf chimera–a visualization system for exploratory research and analysis. *J Comput Chem* 25:1605–1612.

Ramachandran, G.N. and Sasisekharan, V. (1968). Conformation of polypeptides and proteins. *Adv Protein Chem* 23:283–438.

Ramachandran, G.N. and Venkatachalam, C.M. (1968). Stereochemical criteria for polypeptides and proteins .4. Standard dimensions for cis-peptide unit and conformation of cis-polypeptides. *Biopolymers* 6:1255–1262.

Ramakrishnan, C. and Ramachandran, G.N. (1965). Stereochemical criteria for polypeptide and protein chain conformations. Ii. Allowed conformations for a pair of peptide units. *Biophys J* 5:909–933.

Stewart, D.E., Sarkar, A. and Wampler, J.E. (1990). Occurrence and role of cis peptide bonds in protein structures. *J Mol Biol* 214:253–260.

5 The league

The conformations an unfolded protein can attain is in principle limitless, in contrast to the limited conformations of a folded protein. Although the folded state often is assumed to be a single conformation, it is in fact an ensemble of conformations due to the inherent dynamic properties of proteins. However, it is much smaller than the ensemble of the unfolded state. Thus, the entropy change of folding a protein is negative. At the same time, it is important to realize that the total entropy change of the folding process is positive when the entropy change of water is included.

For obvious reasons, the folding of a polypeptide chain is a spontaneous reaction; thus, the change in Gibbs free energy must be less than zero for the reaction:

$$\Delta G = \Delta G_{folded} - \Delta G_{unfolded} < 0$$

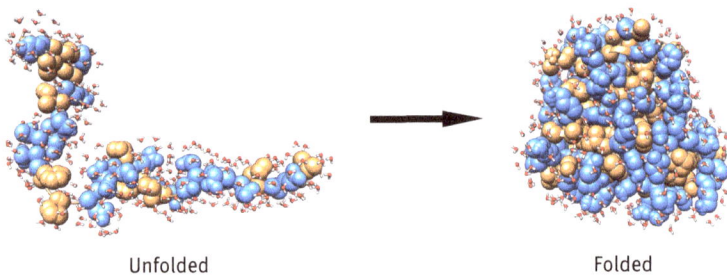

Unfolded   Folded

Figure 5.1: Gibbs free energy of folding.

As discussed before, the driving force for protein folding is the hydrophobic effect. Therefore, any folded protein has a hydrophobic core. The hydrophobic core consists of a set of nonpolar residues that through the folding avoids the unfavorable energy contact with water molecules. This leads to less exposed nonpolar groups and at the same time less water molecules that are "caged" around these nonpolar groups (Figure 5.1). During the folding process, new hydrogen bonds, salt bridges and van der Waals interactions as well as disulfides are formed. Together these forces contribute to the final stability of the folded protein.

In an unfolded protein, it can be assumed that most of the polar groups form hydrogen bonds with water molecules. In the folded state some, if not most, of these hydrogen bonds are broken and new ones are formed, with both water molecules and other polar groups.

The gain in free energy will depend on where the hydrogen bonds are formed. A hydrogen bond between a polar group and a water molecule that is exchanged for a bond between two polar groups on the surface of the protein does not change the free energy significantly. Since most water molecules are excluded from the interior of the folded protein, a polar group that ends up in the interior of the folded protein need to

https://doi.org/10.1515/9783111350684-005

find a polar group (or a water molecule) to bind to; otherwise it will destabilize the folded state. A hydrogen bond in a nonpolar environment is much stronger than the one in water due to the lower dielectric constant.

Charged side chains are predominantly localized to the surface of soluble proteins, where they form salt bridges with other charged groups. Water dipoles may form layers around the charges, thereby effectively screen the attractive (or repulsive) forces. Although the force is reduced by the water molecules, these ion-ion interactions contribute to the free energy gain and stabilize the folded structure. Similar to interior hydrogen bonds, an interior salt bridge is much stronger than a surface exposed one, due to the much lower dielectric constant. Transferring a salt bridge from water to a nonpolar surrounding with a dielectric constant around 2–3 may make the bond up to 40 times stronger.

In the folded state, most nonpolar residues end up in the interior of the protein, excluded from contact with the surrounding water. This increases the entropy of the water molecules and contributes to the energy gain. The folding process brings residues that can be far away from each other in the primary structure close enough to form van der Waals interactions. Although such an interaction is weak, the large number of van der Waals interactions may contribute significantly to the stability of the folded protein.

Two cysteine residues may form a covalent disulfide bridge (S-S-bond) if the conditions are right. However, since the intracellular milieu is reductive, a disulfide bridge that is exposed on the surface of a folded protein most probably will not survive and will be reduced to free thiols (-SH). A disulfide bridge inside the folded protein without contact with the reducing milieu has a greater possibility to survive and contribute to the free energy gain and the stability of the protein.

In the process of protein folding, secondary structure elements form that places side chains as well as amide groups in the backbone in such a way that as many groups as possible are involved in some type of bond or interaction. It is worth repeating that a nonbonded polar or charged group is a strong destabilizing factor.

5.1 More water

A fully folded soluble and globular protein has a very compact structure, with a mostly polar surface and a predominant nonpolar interior. All polar groups exposed on the surface, form bonds with either other polar groups on the surface or with water molecules. In fact, a globular protein is surrounded by at least two hydration layers. It also appears that the stereochemistry of the water molecules in these hydration layers are distorted (Figure 5.2).

There is usually no or very little void space in a folded protein. If a polar internal cavity is formed, it usually contains one or a few water molecules. An internal cavity without buried water is most probably lined with nonpolar side chains. These "lake-

Hydration layer

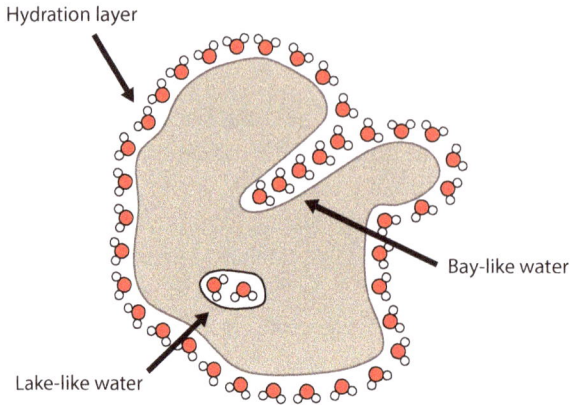

Bay-like water

Lake-like water

Figure 5.2: Hydration layer and buried water.

like" water molecules do not exchange with the bulk water in contrast to "bay-like" water. However, "bay-like" water molecules exchange with the bulk water on a much slower time scale than water molecules in the hydration layers.

Loops and other connecting structures are often present on the surface of proteins. In contrast to the rather rigid interior of a protein, these structures can be very dynamic. Structural changes in a loop due to binding of a ligand or a mutation can easily be accommodated without affecting the overall structure.

5.2 Domains

In the early 1970s, a "domain hypothesis" was proposed based on the observation that there are structural separate regions in immunoglobulins and that these regions are under genetic control. A few years later, Donald Wetlaufer suggested that not only immunoglobulins but also many other proteins contain compact and continuous intra-chain regions of 40–150 residues. The presence of such structural entities or domains in many, if not most, proteins is nowadays fully established. It is also perfectly clear that domains constitute the basic structural, functional or evolutionary unit of proteins.

A domain is defined as a continuous sequence profile of around 40–250 residues. Any domain always has a hydrophobic core and therefore it can fold and be expressed independently of the rest of the protein. Although the amino acid sequence of a particular domain can differ from protein to protein, the overall structure and function are generally the same; however, there are exceptions from this rule.

The src homology domain 3 (SH3) binds proline-rich sequences, usually in a left-handed type II poly-proline helix, with a minimal consensus Pro-X-X-Pro sequence. The aliphatic prolines are bound to a hydrophobic pocket on the SH3 domain. Additional nonproline residues can also interact with the SH3 domain and thereby influence the affinity. Although the structures of SH3 domains in different proteins are very similar, the amino acid sequences need not to be very well conserved. Only 17 of 60 residues of the SH3 domains of myosin 1e and α-spectrin are identical; still the structures overlap very well (Figure 5.3).

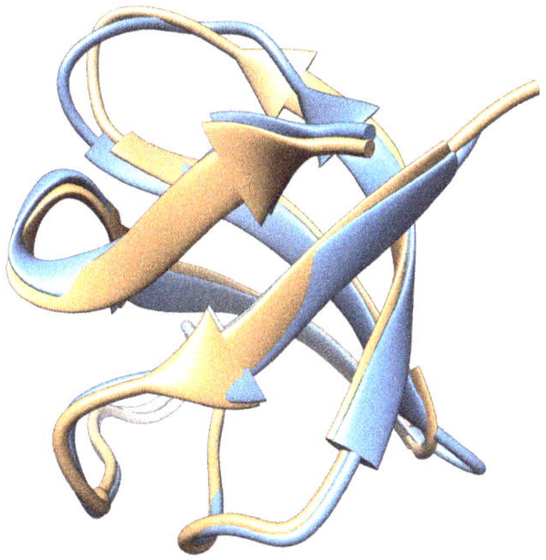

```
                1                               30
                |       |           |           |
Myosin 1e      GPLGSPQCKALYAYDAQDTDELSFNANDII
α-spectrin     DETGKELVLALYDYQEKSPREVTMKKGDIL

                31                              60
                |       |           |           |
Myosin 1e      DIIKEDPSGWWTGRLRGKQGLFPNNYVTKI
α-spectrin     TLLNSTNKDWWKVEVNDRQGFVPAAYVKKL
```

Figure 5.3: Comparing SH3 domains. (Top) The myosin 1e (pdb: 2XMF) SH3 domain structure (sandy brown) was superimposed on the α-spectrin SH3 (pdb: 1U06) domain structure (blue). (Bottom) Alignment of the two amino acid sequences indicates a large sequence discrepancy; only 17 (red) of 60 residues are conserved.

The list of structurally identified domains is long but still limited. Some domains are widespread, such as the ANK (ankyrin) repeat domain, and found in more than 1,500 proteins whereas others, like the TRAF (tumor necrosis factor receptor-associated factors) domain, are only present in a single protein family. As mentioned before, a certain domain in this or that protein usually has the same (or similar) function. This assumption has been useful to infer functions of newly sequenced proteins as the identification of a domain rests on the similarity of the amino acid sequence. However, it should be noted that not all domains have a specific function. Some common domains are shown in Figure 5.4.

(a)

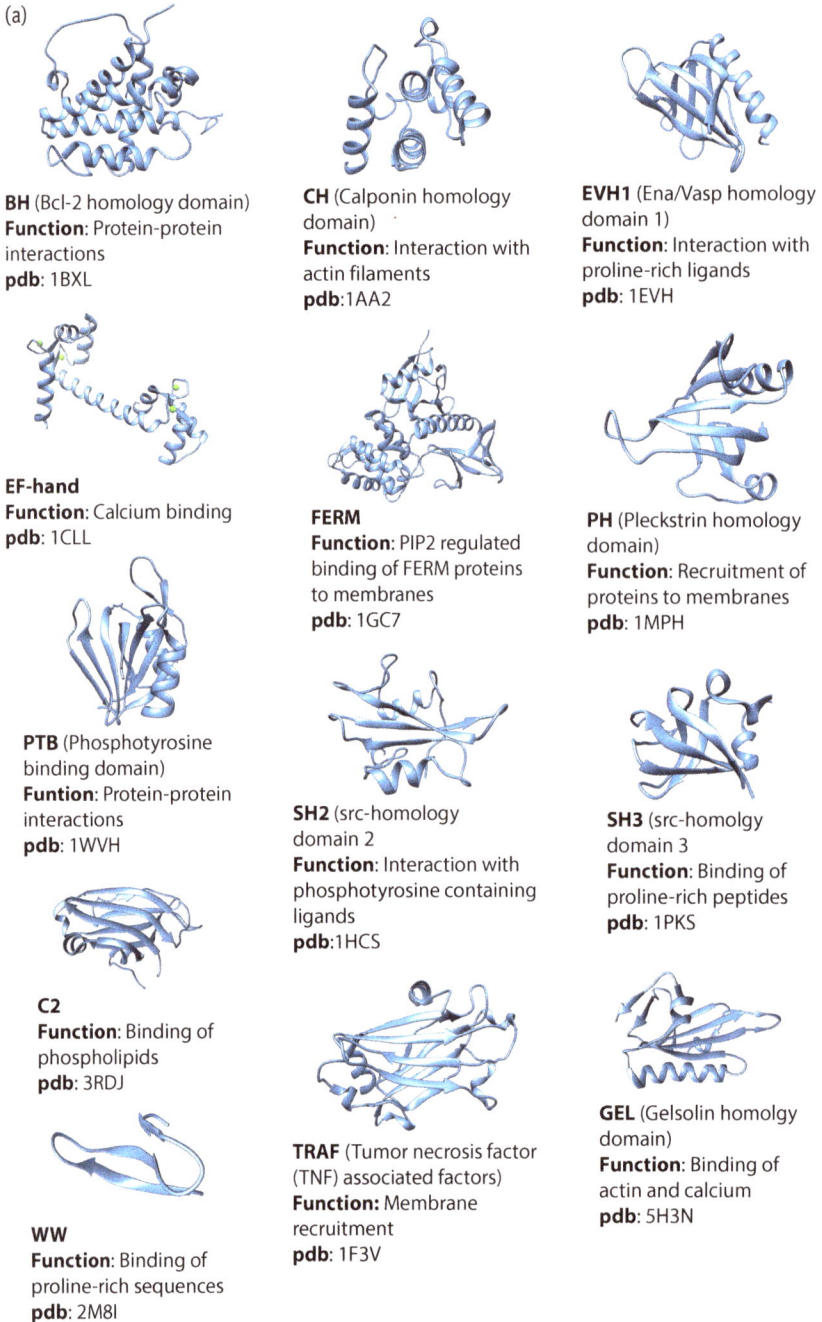

BH (Bcl-2 homology domain)
Function: Protein-protein interactions
pdb: 1BXL

CH (Calponin homology domain)
Function: Interaction with actin filaments
pdb:1AA2

EVH1 (Ena/Vasp homology domain 1)
Function: Interaction with proline-rich ligands
pdb: 1EVH

EF-hand
Function: Calcium binding
pdb: 1CLL

FERM
Function: PIP2 regulated binding of FERM proteins to membranes
pdb: 1GC7

PH (Pleckstrin homology domain)
Function: Recruitment of proteins to membranes
pdb: 1MPH

PTB (Phosphotyrosine binding domain)
Funtion: Protein-protein interactions
pdb: 1WVH

SH2 (src-homology domain 2
Function: Interaction with phosphotyrosine containing ligands
pdb:1HCS

SH3 (src-homolgy domain 3
Function: Binding of proline-rich peptides
pdb: 1PKS

C2
Function: Binding of phospholipids
pdb: 3RDJ

TRAF (Tumor necrosis factor (TNF) associated factors)
Function: Membrane recruitment
pdb: 1F3V

GEL (Gelsolin homolgy domain)
Function: Binding of actin and calcium
pdb: 5H3N

WW
Function: Binding of proline-rich sequences
pdb: 2M8I

Figure 5.4: Some common protein domains. The domain structures are from solved structures of representative proteins.

(b)

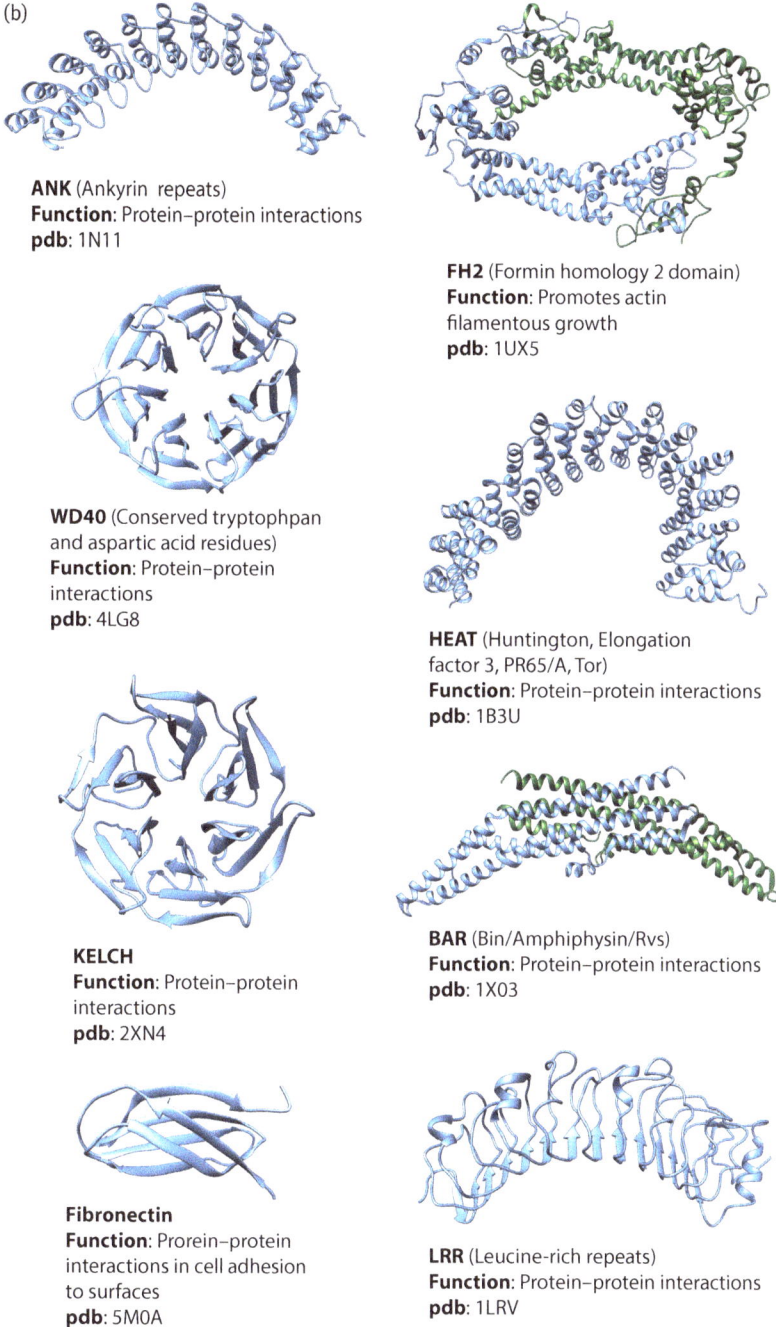

ANK (Ankyrin repeats)
Function: Protein–protein interactions
pdb: 1N11

FH2 (Formin homology 2 domain)
Function: Promotes actin
filamentous growth
pdb: 1UX5

WD40 (Conserved tryptophpan
and aspartic acid residues)
Function: Protein–protein
interactions
pdb: 4LG8

HEAT (Huntington, Elongation
factor 3, PR65/A, Tor)
Function: Protein–protein interactions
pdb: 1B3U

KELCH
Function: Protein–protein
interactions
pdb: 2XN4

BAR (Bin/Amphiphysin/Rvs)
Function: Protein–protein interactions
pdb: 1X03

Fibronectin
Function: Prorein–protein
interactions in cell adhesion
to surfaces
pdb: 5M0A

LRR (Leucine-rich repeats)
Function: Protein–protein interactions
pdb: 1LRV

Figure 5.4 (continued)

5.2.1 How to acquire a domain

Genetic material is not stable but undergoes continues changes. It is estimated that there are around 100 mutations in the human genome per generation. On an evolutionary time scale (rather millions of years), more profound changes have occurred, through whole genome duplications and rearrangements as well as due to gene duplications and mutations. Genes or gene fragments have also been transferred between species (horizontal gene transfer). Thus, with time an organism's genome evolves by modification of inherent genetic material and by acquiring genetic material from other organisms. The modification can be advantageous or deleterious for the organism. An advantageous mutation would enhance the survival of the organism and would be retained, whereas a deleterious mutation eventually would be deleted. A mutation can also be neutral, in that it alters the gene and changes the genotype but not the gene product or the phenotype of the organism.

Early on in evolution, it can be assumed that most (if not all) proteins were rather small and the whole protein could more or less be considered as a domain. Whole genome duplications as well as gene duplication increased the genome size and complexity. At the same time mutations in the genes gave rise to proteins with new functions, thereby expanding the protein domain space.

With time, gene fusions and other gene rearrangements created larger proteins. When genes coding for two small or single domain proteins fused, the new gene gave rise to a larger protein with two distinct domains, a so-called multidomain protein. Although present in the same protein, each domain would fold independent of each other and the global folded structure would resemble that of the two single domain proteins.

In a way, domains can be considered as building blocks or modules that can be put together like bricks, in any order and any numbers as illustrated in Figure 5.5.

Around half of all proteins, independent of organisms, contain more than one domain. Multidomain proteins are also more common among eukaryotes than among bacterial. It is estimated that up to 70% of eukaryotic proteins are multidomain proteins.

Since a domain is an independent entity, it is generally assumed that neither the structure nor the function is affected by where in the protein the domain is placed or by the presence of other domains. The ankyrin domain (ANK) is an illustrative example that this is both right and wrong. The ANK domain is a common scaffold for protein-protein interactions but at the same time the domain is found in proteins with very divergent functions. The ANK domain does not interact with a specific sequence or structure such as an SH3 or a CH domain does.

In a folded multidomain protein, it is possible that the separate domains come close to each other and may form intramolecular bonds that might influence the domain fold. It appears that the overall domain fold is independent of the presence of other folded regions. Secondary structure elements, α-helices and β-strands, are in principle unaffected. However, less rigid regions of the domain, such as a loop, may well have a different structure in the folded protein (Figure 5.6).

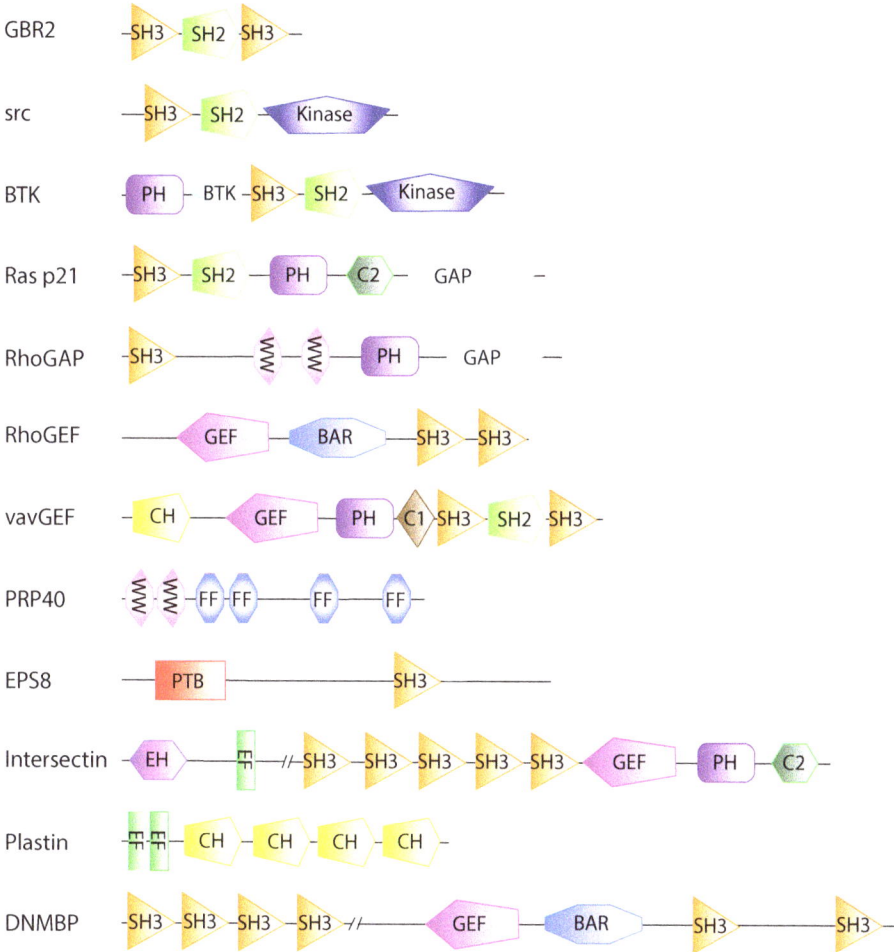

Figure 5.5: Domain arrangements in some proteins involved in protein-protein interactions. SH3: src homology domain 3 binds proline-rich regions; SH2: scr homology domain 2 interacts with phosphotyrosine regions; Kinase: protein kinase domain that catalyzes phosphorylation of tyrosine resides; PH: pleckstrin homology domain binds to membranes; BTK: Bruton's tyrosine kinase cystein-rich domain; C2: protein kinase C conserved region is a Ca^{2+}-binding domain; GAP: G-protein activating domain; WW: domain with two conserved tryptophan residues that binds proline-rich sequences; GEF: guanine nucleotide exchange factor for small G-proteins; BAR: Bin/Amphysin/Rvs domain is involved in endocytosis and actin reorganization; C1: cysteine-rich domain involved in recruitment of proteins to the membrane; FF: domain with two conserved phenylalanine residues; PTB: phosphotyrosine-binding domain; EH: Eps15-homology domain is involved in vesicle transport and regulation of endocytosis; EF: EF-hand domain binds intracellular calcium; CH: calponin homology domain interacts with filamentous actin.

The domain concept has been pivotal in acquiring functional cognizance of newly discovered proteins, without experimental evidence. Thus, by analyzing the amino acid sequence, the presence of one or more domains may predict a certain function of the

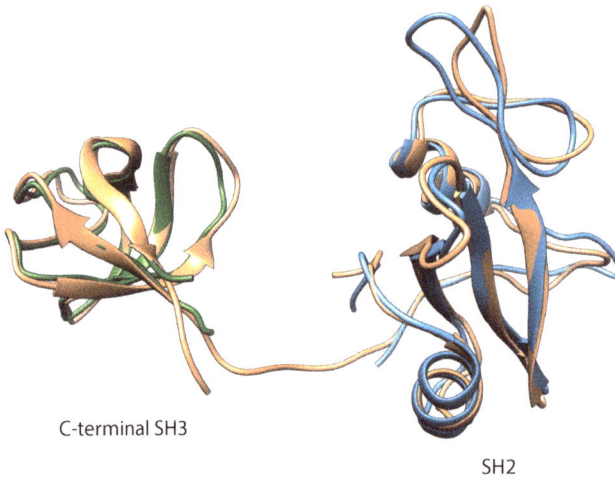

C-terminal SH3

SH2

Figure 5.6: The domain structure is retained in the global structure. The determined domain structures of the SH2 (pdb: 1GHU, blue) and SH3 (pdb: 1IO6, green) superimposed on the structure of full length GBR2 (pdb: 1GRI, sandy brown). The structure of the N-terminal SH3 has not been determined and was therefore removed from the structure of GBR2 for clarity.

protein. Whether it is right or wrong, the prediction may help to determine how to proceed, for example, by indicating suitable assays to do in order to establish the function.

There are several web resources available that analyze a raw nucleotide or amino acid sequence in order to identify domains. These resources (e.g., SMART, Superfamily, InterPro and Prosite) all rely on sequence similarity; if a sequence is similar enough to a domain's consensus sequence, it is considered to comprise that particular domain. Although these resources use algorithms that give slightly different results, they usually identify the same domains but the localization of the domain may differ. The domain analysis of human GRB2 is shown in Table 5.1. From the determined structure, it is known that the GRB2 comprise two SH3 and one SH2 domains and these domains are correctly identified by the web resources, even though the localization differs.

5.2.2 Classifying domains

There are reasons to group proteins. It can give information on structure and biochemical function on a newly sequenced or isolated protein as well as give insight into the evolution of a group of proteins with similar function. There are two major databases that classify proteins into groups with related structures: SCOPe (Structural Classification of Proteins – extended) and CATH (Class, Architecture, Topology, Homol-

Table 5.1: Domain analysis of human GRB2.

	N-terminal SH3		SH2		C-terminal SH3	
	Start	**End**	**Start**	**End**	**Start**	**End**
Superfamily[a]	5	52	60	159	160	209
SMART[b]	1	57	58	141	159	214
InterPro[c]	1	58	58	152	156	215
Prosite[d]	1	58	60	152	156	215

[a]A.P. Pandurangan, J. Stahlhacke, M.E. Oates, B. Smithers and J. Gough (2019). The superfamily 2.0 database: A significant proteome update and a new webserver. *Nucleic Acids Res.* 47:D490–D494.
[b]I. Letunic, S. Khedkar and P. Bork (2021). SMART: Recent updates, new developments and status in 2020. *Nucleic Acids Res.* 49:D458–D460.
[c]T. Paysan-Lafosse, M. Blum, S. Chuguransky, T. Grego, B.L. Pinto, G.A. Salazar, M.L. Bileschi, P. Bork, A. Bridge, L. Colwell, J. Gough, D.H. Haft, I. Letunic, A. Marchler-Bauer, H. Mi, D.A. Natale, C.A.Orengo, A.P. Pandurangan, C. Rivoire, C.J.A. Sigrist, I. Sillitoe, N. Thanki, P.D. Thomas, S.C.E Tosatto, C.H. Wu and A. Bateman (2023). Interpro in 2022. *Nucleic Acids Res.* 51:D418–D427.
[d]C.J. Sigrist, E. de Castro, L. Cerutti, B.A. Cuche, N. Hulo, A. Bridge, L. Bougueleret and I. Xenarios (2013). New and continuing developments at PROSITE. *Nucleic Acids Res.* 41:D344–D347.

ogous superfamily). The classification in both cases is based on experimentally determined domain structures and has hierarchical organizations. It should be noted that SCOPe and CATH do not always give the same result.

At the top level of hierarchy, SCOPe defines classes that describe the content of secondary structure elements. There are four major classes: all α-proteins, all β-proteins, α/β-proteins and α + β-proteins as well as several other minor classes, like small proteins and coil-coiled proteins. Folds, at the second level, group domains with similar overall tertiary structure and connectivity. For example, the all α-proteins class contains around 290 distinct folds. At the next hierarchy level, each fold is classified into superfamilies. Domains of a certain superfamily not only have structurally similarity but are also evolutionary related. A superfamily contains a number of protein families, each with similar sequences but different functions. Domains that are placed in the same family are more closely related and their sequence identity is much higher. Each protein family comprises groups of proteins with similar sequences and functions that originate from different species (Figure 5.7).

CATH only defines three classes at the top level, mainly α, mainly β and α/β, that are equivalent to SCOPe's classes but also includes a fourth class, few secondary structures, that has no analogy in SCOPe. Architecture corresponds to folds in SCOPe and topology group structures into fold families. Homologous superfamilies in CATH are similar to protein families in SCOPe. Within each superfamily, proteins with similar structure and function are organized in functional families (Figure 5.8).

Due to alternative splicing, the ca. 20,000 protein-coding genes of the human genome are translated into ca. 100,000 different proteins and post-translational modifications may even increase this number. However, this does not mean that there are

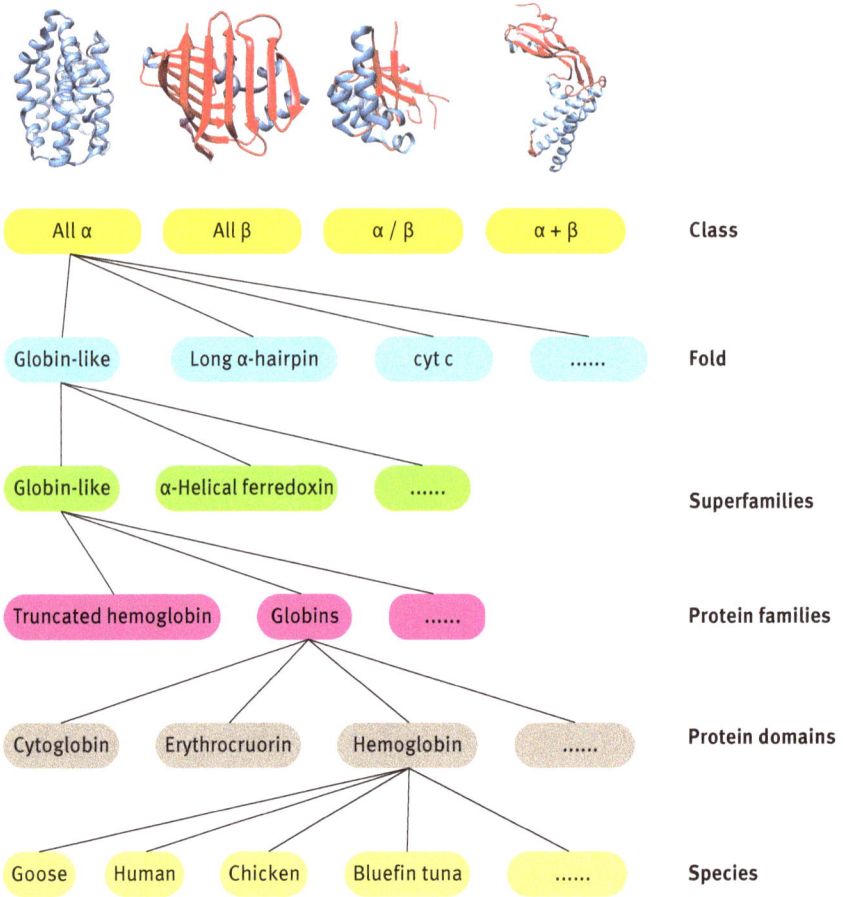

Figure 5.7: The SCOPe classification. The example shows the hierarchy for globin domains.

100,000 completely different folded proteins. Both SCOPe and CATH identify that the protein universe contains less than 1,400 different folds or around 5,000 protein families. This implies that same structure themes reoccur in several proteins.

5.3 Motifs

The distinction between motifs and domains can be confusing; a domain has a motif and a motif can be a domain. A motif is a combination of α-helices and β-strands connected by loops or turns that forms a supersecondary structure that can be found in several proteins. The structure elements need not to be continuous and can be far away from each other in the peptide chain. Upon folding the secondary structure elements are placed close to each other, forming the motif as illustrated in Figure 5.9.

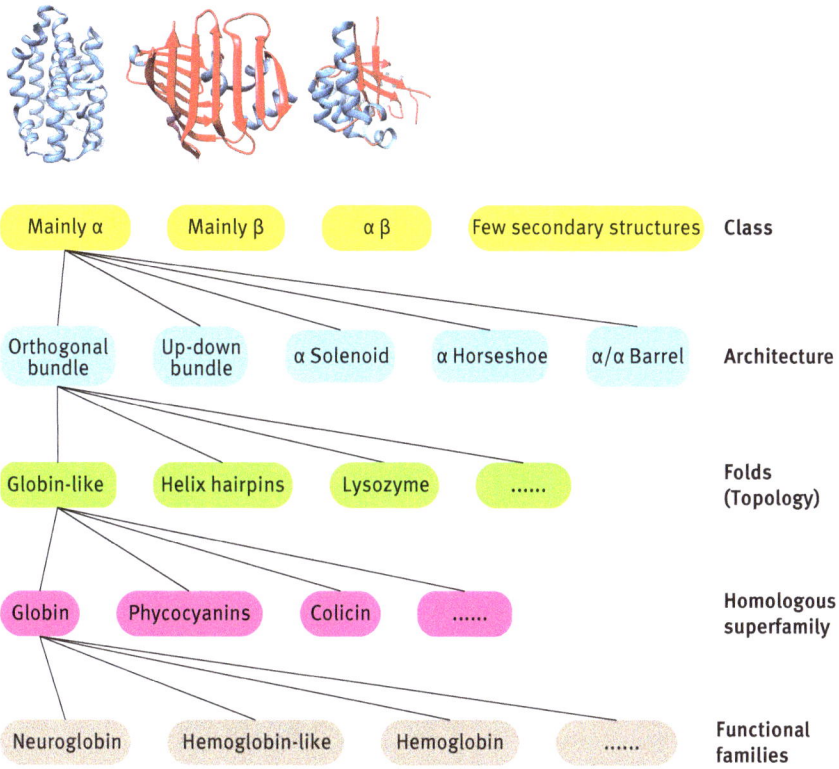

Figure 5.8: The CATH classification. The example shows the hierarchy for globin domains.

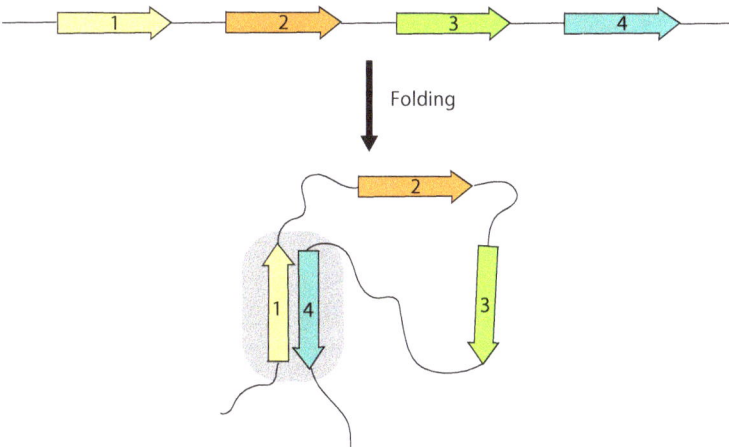

Figure 5.9: When the peptide chain folds, region 1 and 4 are placed close to each other and form a structural motif.

The difference between a motif and a domain is that the amino acid sequence of the domain is always continuous and that the domain is an independent entity. The sequence of a motif may but need not be continuous. In addition, motifs or rather the consensus sequence of the motifs are used to recognize domains!

One of the first structural motifs discovered was the Rossman fold (Figure 5.10). The fold was initially found in lactate dehydrogenase, an enzyme involved in anaerobic glycolysis. Later it was found to be present in dehydrogenases and other enzymes utilizing the nucleotide coenzymes NAD^+, $NADP^+$ and FAD. The classical Rossman β-α-β fold is composed of 6 alternating β-strands and α-helices where the β-strands form an extended β-sheet and the α-helices are placed on each side of the β-sheet. The coenzyme specificity appears to be determined by residues in the loop region between the first β-strand and the first α-helix. A Gly-X-Gly-X-X-Gly sequence in this region seems to indicate a NAD^+ or FAD-dependent enzyme whereas a Gly-X-Gly-X-X-Ala sequence is often present in $NADP^+$-dependent enzymes.

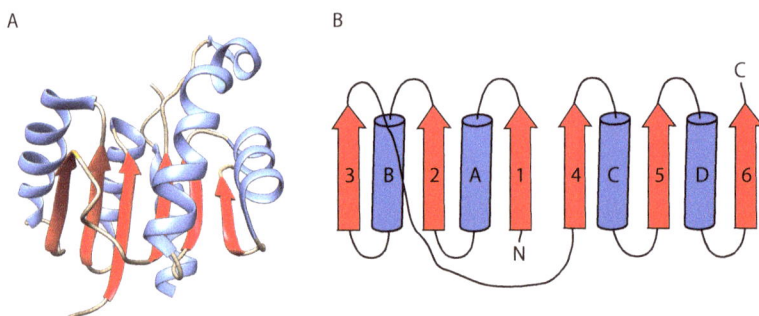

Figure 5.10: The Rossman fold. (A) The extended β-sheet (red) is surrounded by α-helices on both sides (blue). The Rossman fold is taken from dogfish lactate dehydrogenase (pdb: 3LDH). (B) Schematic representation of typical β-α-β connectivity of a Rossman fold.

There are two types of motifs: structural motifs, as illustrated above, and sequence motifs. A sequence motif is an amino acid (or nucleotide) pattern that is common and thus present in several different proteins. It usually has a distinct biological function. An illustrative example is the zinc finger motif. It was first identified in a transcription factor as a zinc-binding domain required for binding to DNA. The "classic" zinc finger motif contains a short β-hairpin and a α-helix, in a β-β-α structure (Figure 5.11). The zinc ion is coordinated by four amino acid residues; either two cysteines and two histidines (C2H2) or three cysteines and a single histidine (C2HC or CCCH). Transcription factors usually contain several zinc fingers that determine both the affinity of DNA binding and where along the DNA the protein binds.

Later it has turned out that there are many superfamilies with zinc finger motifs. They differ both in sequence and structure; some do not even bind zinc but other metals or no metal at all. Some interact with DNA or RNA and others with proteins.

The classical C2H2 zinc finger motif is identified by a sequence where there are two or four residues between the two cysteines and three or five residues between the two histidines. The zinc coordinating residues are in turn separated by 12 residues, where the third residue usually is nonpolar; or in the short form C-X(2,4)-C-X (12)-H-X(3,5)-H. In other types of zinc fingers the spacing is different. For example, CCCH motifs are recognized by the consensus sequence C-X(8)-C-X(5)-C-X(3)-H.

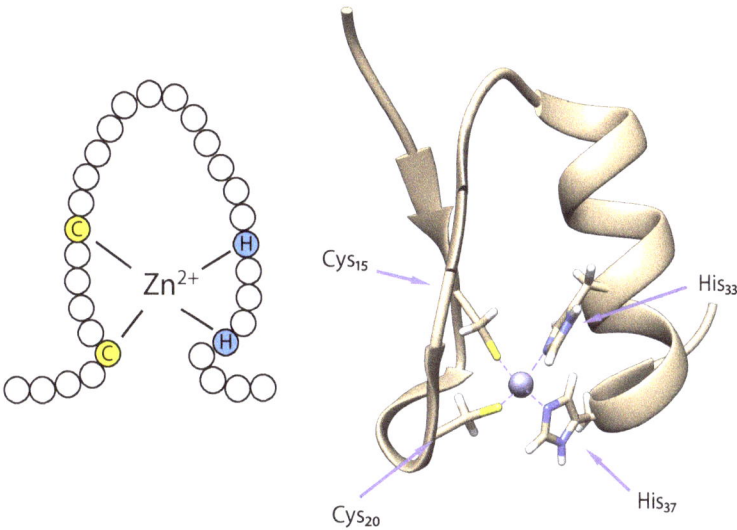

Figure 5.11: The classic zinc finger motif. The structure shows the N-terminal and first zinc finger in *Xenopus laevis* transcription factor IIIA. The zinc ion is coordinated by cysteines and histidines. Depending on the type of the motif, both the coordinating residues and the distance between residues varies.

5.4 Topology and connectivity

If we consider four β-strands, there are several ways they can be connected; they might be parallel, antiparallel or a mix of both. Independent of how the secondary structure elements are arranged in relation to each other, all arrangements will have the same topology.[1] However, the connectivity is different, as illustrated in Figure 5.12.

---

1 *Encyclopedia Britannica* defines topology as "branch of mathematics, sometimes referred to as 'rubber sheet geometry,' in which two objects are considered equivalent if they can be continuously deformed into one another through such motions in space as bending, twisting, stretching, and shrinking while disallowing tearing apart or gluing together parts. The main topics of interest in topology are the properties that remain unchanged by such continuous deformations. Topology, while similar to geometry, differs from geometry in that geometrically equivalent objects often share numerically measured quantities, such as lengths or angles, while topologically equivalent objects resemble each other in a more qualitative sense."

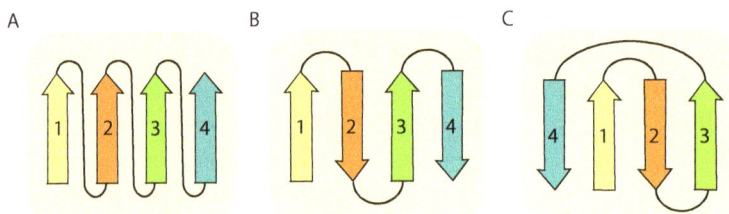

Figure 5.12: Different arrangement of β-strands. β-Strands can be parallel (A) and antiparallel (B) as well as a mix of both parallel and antiparallel. An antiparallel up-down arrangement, as in (B) is also called a β-meander. The arrangement of β-strands in (C) is called a Greek key due to the similarity to patterns seen on old Greek pottery.

In a parallel β-sheet (Figure 5.12A), each β-strand is connected to the next through a right-handed crossover. Antiparallel β-strands (up-and-down) can be connected by short loops or β-turns (Figure 5.12B), creating β-hairpins or a combination of short and long loops. The connectivity shown in Figure 5.12C is called a Greek key, due to its resemblance to the decorative patterns used on old pottery. In a Greek key, β-strand 1 is often hydrogen bonded to β-strands 2 and 4.

A β-sheet with many (often eight or more) β-strands forms a cylindrical structure, called a β-barrel, to form as many hydrogen bonds as possible. Through the cylindric shape of the β-barrel, the first β-strand can form hydrogen bond to the last β-strand in the sheet (edge-to-edge). In the up-and-down arrangement, the first β-strand is adjacent to the second in an antiparallel manner and so on until the barrel is closed.

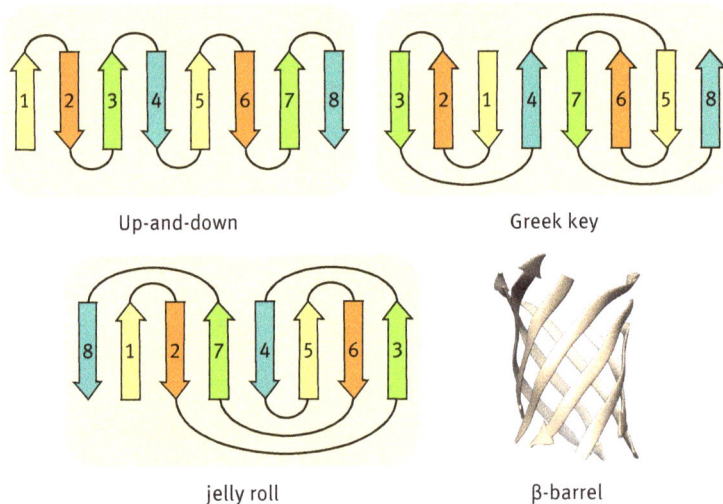

Figure 5.13: Arrangement of β-strands. Although the connectivity differs, all arrangements form a β-barrel. The β-barrel structure is adapted from pdb: 6CZG.

Many proteins contain β-barrels where sequence adjacent β-strands are not adjacent in the final structure. In one type of arrangement the connectivity follows the Greek key pattern. Another common pattern is the so-called jelly roll (Figure 5.13).

What about motif consisting of mostly α-helices? A common motif is the helix-turn-helix (HTH) motif, particularly found in many proteins that regulate gene expression but also as well as in many other protein groups. The HTH motif contains an α-helix connected by a short turn to a second α-helix, which fits snuggly into the major groove of the DNA and recognizes a certain nucleotide base sequence. Bacterial HTH proteins usually bind DNA as dimers whereas eukaryotic HTH proteins can interact as monomers (Figure 5.14A). Another common motif is the leucine zipper, created by two α-helices that form a homodimer. In the α-helices, every seventh residue is a leucine. Since there are 3.6 residues per turn in an α-helix, all leucines will line up on the same side of the helix and in the dimer leucines in one helix interdigitate with the

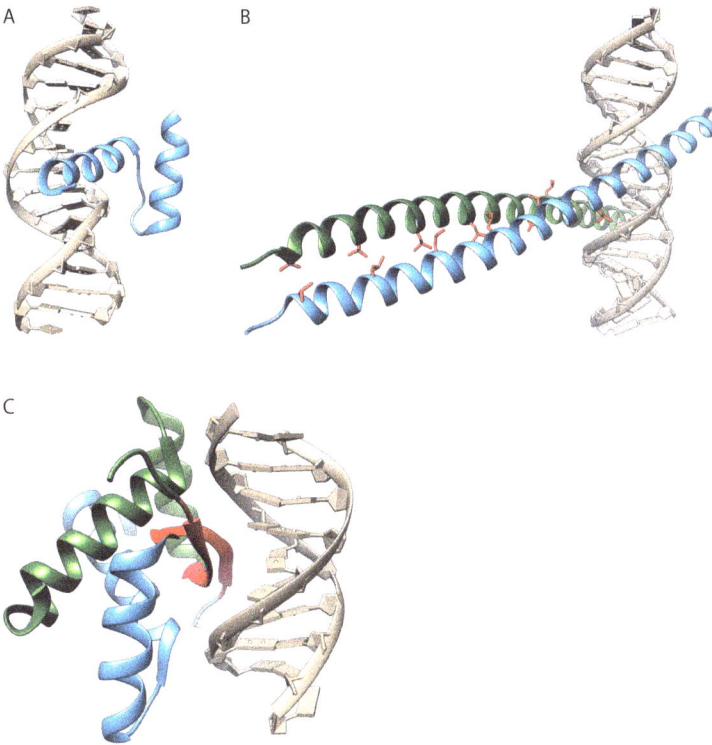

Figure 5.14: DNA-binding motifs. (A) In the helix-turn-helix motif the first α-helix is connected to the second α-helix, the so-called recognition helix that interacts with the major groove of DNA. (B) In the leucine zipper, luecine residues (red) in one α-helix interdigitate with leucines on the other α-helix, burying nonpolar residues from the surrounding solvent. The C-terminal region of the α-helices contains basic residues that interact with the DNA. (C) In the ribbon-helix-helix motif, an antiparallel β-sheet with two β-strands (red) is created that interacts with the major groove of DNA.

leucines in the other helix, thereby burying nonpolar residues. In the C-terminal of the α-helices there are basic residues (lysine and arginine residues) that interact sequence-specific with the major groove of DNA (Figure 5.14B).

Ribbon-helix-helix is another motif found in proteins interacting with nucleic acids. In this motif, two monomers interact in such a way that a β-strand from each monomer is placed adjacent to each other, forming an antiparallel β-sheet that interacts with the major groove of DNA (Figure 5.14C).

Although very similar, the helix-loop-helix (HLH) motif is considered different from the helix-turn-helix (HTH) motif. Again, the HLH motif is found not only in proteins interacting with nucleic acids but also in many other proteins, such as calcium-binding proteins like calmodulin and parvalbumin.

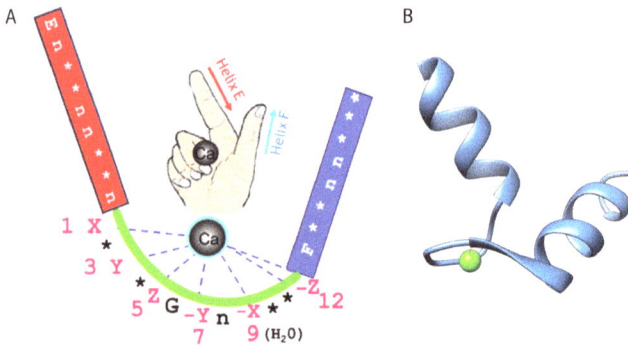

Figure 5.15: The EF-hand motif. (A) Six residues in the loop connecting the two α-helices coordinate the calcium ion in a pentagonal bipyramidal arrangement. The first (X) and last (–Z) residues in the binding loop are invariant aspartate or glutamate. The other residues, with the exception of –Y, bind the ion through a side chain oxygen. –Y binds the ion through the carbonyl oxygen. A water molecule bridges the side chain oxygen of residue –X and the calcium ion. G: glycine, E: glutamate, n: nonpolar residue, *: any residue. Reproduced with permission from Y. Zhou, T.K. Frey and J.J. Yang (2009) Viral calciomics: interplays between Ca^{2+} and virus. *Cell Calcium* 46:1–17. (B) The structure shows one of the four EF-hand motifs of human calmodulin (pdb: 1CLL). The coordinated calcium ion is indicated by the green ball.

In many calcium-binding proteins, the calcium ion is coordinated in a pentagonal bipyramidal arrangement where the six residues in the loop region coordinate the calcium ion. The motif is also known as an EF-hand since initially α-helices E and F of parvalbumin were used to illustrate the structure of the motif (Figure 5.15). The residues that coordinate the ion are often labeled X, Y, Z, –Y, –X and –Z, referring to their locations in the coordination sphere. The residues in position X and -Z are invariant aspartate or glutamate. The residues in all other positions, except –Y, usually ligate the calcium ion through their oxygen bearing side chains. The residue in position –Y coordinates the calcium ion through the carbonyl oxygen.

The EF-hand motif is usually present in pairs and there are usually two pairs in most proteins, with the possibility to bind four calcium ions. However, in many proteins some of the residues in one or more EF-hand motifs that are involved in the co-ordination of the calcium ions are not able to coordinate the ion. Such sequences are in many cases recognized as EF-hand motif unable to bind calcium ions; a calcium ion insensitive EF-hand.

EF-hand-containing proteins can be divided into two major groups; calcium sensors and calcium buffers. Calcium sensors, like calmodulin, troponin C and most other EF-hand proteins, sense a change in calcium ion concentration and translate this into various responses. Calcium binding often induces a large conformational change that frequently exposes a hydrophobic region that allows interaction with the target protein. Calcium buffers, like parvalbumin and calbindin, modulate the intracellular free calcium ion level.

5.5 Why should proteins be grouped?

In order to group proteins or domains with similar functions, we need to consider both topology and connectivity. Although the secondary structure elements may be the same, the order they are connected can differ substantially and this will most probably give rise to different functions. This is illustrated in Figure 5.16, with the C-terminal domain of arginine repressor (pdb: 1XXA) and microphage migration inhibition factor (pdb: 1UIZ). Both are classified as α + β proteins with a two-layer sandwich architecture. There are four β-strands forming a β-sheet with two α-helices layered on one side, giving a very similar topology. However, the connectivity is very different. In the repressor protein, the β-strands are antiparallel whereas in the migration inhibition factor, the β-strands in each pair are parallel but the two pair of β-strands are antiparallel. The order of β-strands in the β-sheets is also different; in the repressor the strand order is 1243 and in the inhibition factor it is 1234.

These two proteins do indeed have different functions, showing the importance not only to look at the overall structure but also consider how secondary structure elements are connected to each other in the final native structure.

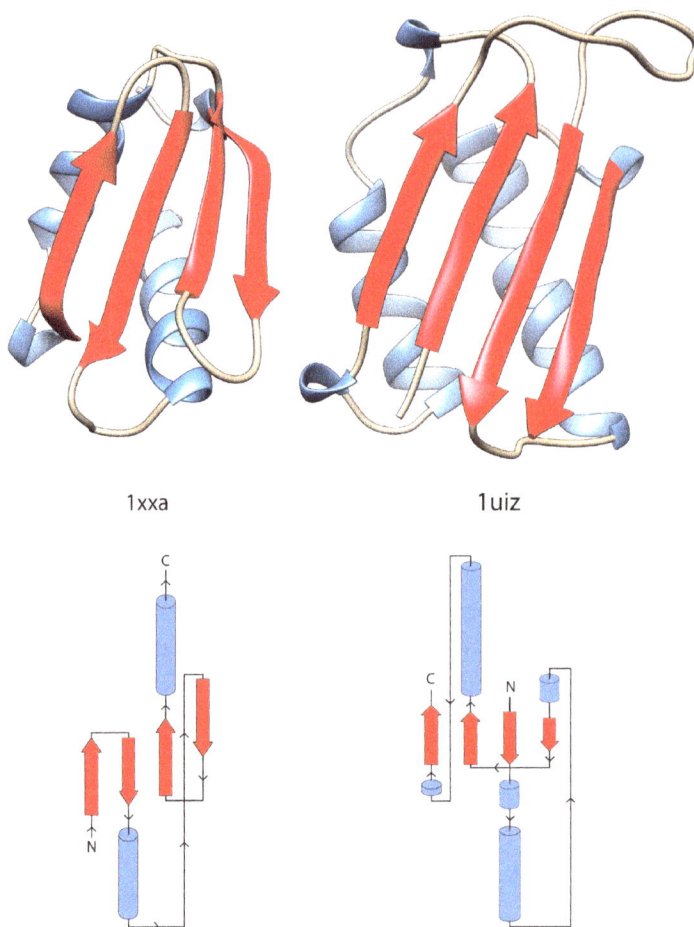

Figure 5.16: Topology and connectivity. The β-sheet of both the C-terminal domain of arginine repressor (pdb: 1XXA) and microphage migration inhibition factor (pdb: 1UIZ) contains four β-strands (red). A two-layered sandwich is formed with the two α-helices (blue). PDBSum was used to determine and draw the connectivity. The last 14 residues of the migration factor were removed from the figure to increase clarity.

Further reading

Aravind, L., Anantharaman, V., Balaji, S., Babu, M.M. and Iyer, L.M. (2005). The many faces of the helix-turn-helix domain: Transcription regulation and beyond. *FEMS Microbiol Rev* 29:231–262.

Carugo, O. (2016). Statistical survey of the buried waters in the protein data bank. *Amino Acids* 48:193–202.

Carugo, O. (2017). Protein hydration: Investigation of globular protein crystal structures. *Int J Biol Macromol* 99:160–165.

Dawson, N.L., Lewis, T.E., Das, S., Lees, J.G., Lee, D., Ashford, P., Orengo, C.A. and Sillitoe, I. (2017). Cath: An expanded resource to predict protein function through structure and sequence. *Nucleic Acids Res* 45:D289–D295.

Fox, N.K., Brenner, S.E. and Chandonia, J.M. (2014). Scope: Structural classification of proteins–extended, integrating scop and astral data and classification of new structures. *Nucleic Acids Res* 42:D304–309.

Han, J.H., Batey, S., Nickson, A.A., Teichmann, S.A. and Clarke, J. (2007). The folding and evolution of multidomain proteins. *Nat Rev Mol Cell Biol* 8:319–330.

Harrison, S.C. (1991). A structural taxonomy of DNA-binding domains. *Nature* 353:715–719.

Kawasaki, H. and Kretsinger, R.H. (2017). Structural and functional diversity of ef-hand proteins: Evolutionary perspectives. *Protein Sci* 26:1898–1920.

Wetlaufer, D.B. (1973). Nucleation, rapid folding, and globular intrachain regions in proteins. *Proc Natl Acad Sci U S A* 70:697–701.

6 El clásico

Animals and many other organisms require a constant supply of molecular oxygen for survival. In animals, molecular oxygen is needed for the energy conversion in the mitochondria, more precisely to allow electron transport to fuel oxidative phosphorylation to produce ATP. Molecular oxygen is transported from the lungs to the tissue by the blood system. Since the solubility of molecular oxygen is very low, the amount of dissolved molecular oxygen would not satisfy the demand in tissues. Therefore, molecular oxygen is transported bound to hemoglobin, the major protein in red blood cells.

Hemoglobin and myoglobin are two proteins intimately associated with oxygen transport. Both proteins contain a nonprotein moiety, a heme group to which molecular oxygen binds. Myoglobin is a small globular protein present in all animals. It functions as a temporarily storage for molecular oxygen, particularly important in diving animals such as whales and seals. Independent of source, myoglobins contain around 150 amino acid residues and have a molecular mass of ~17 kDa. Hemoglobin is larger, around four times as large as myoglobin. It is composed of four subunits (two α-chains and two β-chains), where each subunit is very similar to myoglobin. When venous blood passes the alveolars in the lungs, molecular oxygen is taken up by hemoglobin, transported by the blood to the tissues where it is released.

6.1 Myoglobin

In 1958, John Kendrew and coworkers presented the first three-dimensional structure of a protein, namely, of sperm whale myoglobin, albeit at low resolution (6 Å). In this structure, it was not possible to discern any details – only that myoglobin appeared to be composed of a single polypeptide chain folded into a compact globular form. A few years later, Kendrew had improved the resolution to 2 Å, and it was possible to see that myoglobin contained a number of right-handed α-helices separated by short irregular structures. However, even at this resolution, some ambiguities remained that would have been possible to solve if the amino acid sequence of myoglobin would have been known. Finally, in 1961 Kendrew's group had increased the resolution to 1.5 Å, which made it possible to discern nearly all side chains.

The high-resolution tertiary structure showed, as expected, that myoglobin is a very compact globular molecule composed of eight right-handed α-helices (labeled A-H) connected by turns and loops. Nearly 78% of all residues are located in helices. Mainly polar residues are exposed on the surface of myoglobin, whereas the majority of the nonpolar residues are located to the interior of the protein and the four prolines are found at either end of an α-helix (Figure 6.1). Independent of the source of myoglobin,

https://doi.org/10.1515/9783111350684-006

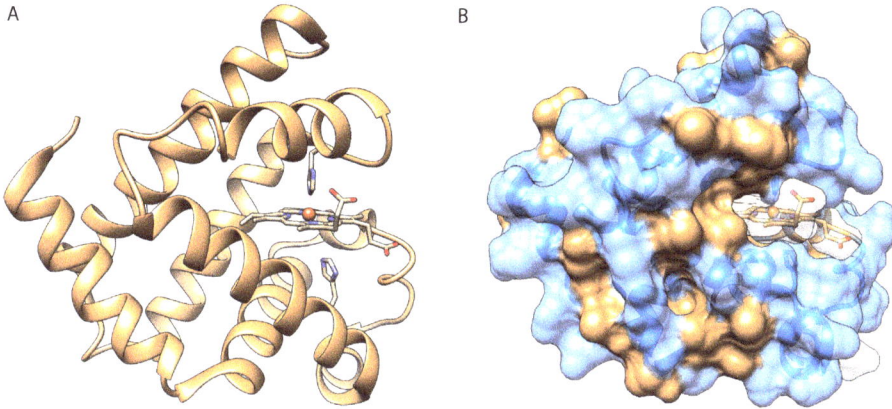

Figure 6.1: Myoglobin. (A) High-resolution structure of sperm whale myoglobin (pdb: 1mbd) showing the location of the heme group and the two histidine residues in the hydrophobic cleft in the protein. (B) Molecular surface colored to show exposed polar (blue) and nonpolar (sandy brown) amino acid residues.

the tertiary structures are nearly identical despite differences in the primary structures; in sperm whale and human myoglobins, 16 residues out of 153 are different.

The functional part of myoglobin is not the protein but rather the heme group and its coordinated Fe(II). The heme group is attached to the protein in a deep hydrophobic cleft by several nonpolar contacts and by iron coordination to the side chain of a histidine residue. The polar propionic groups of heme lie on the outside and the nonpolar vinyl groups are buried in the hydrophobic cleft, surrounded by nonpolar residues (Figure 6.2). The iron is coordinated to four nitrogens of the protoporphyrin IX ring and to histidine (F8) in helix F. On the distal side of the heme, another histidine is located, but this histidine is too far away to coordinate Fe(II). Thus, in deoxymyoglobin the iron is penta-coordinated by the four nitrogens in heme and one in histidine F8, whereas in oxymyoglobin Fe(II) is octahedrally coordinated, with molecular oxygen as the sixth ligand. The binding of molecular oxygen changes the electron configuration around Fe(II), which leads to a change in color, from dark red to pale red.

Since it is the nonprotein moiety and not the protein that is directly involved in binding of oxygen, one may wonder why the protein is needed. There are two main reasons. First, in solution, free heme Fe(II) oxidizes rapidly to Fe(III), which is incapable of reversible binding of molecular oxygen. Second, carbon monoxide (CO) would bind much stronger to heme than O_2 if it were not for the distal histidine E7. Thus, the protein provides an environment suitable for reversible oxygen binding.

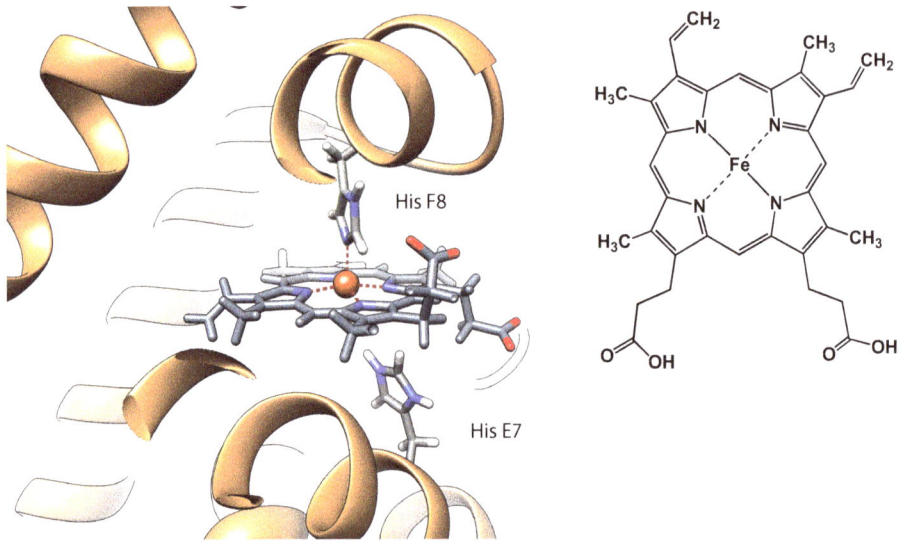

Figure 6.2: The function of the two histidines F8 and E7. The heme group is attached to myoglobin in a hydrophobic cleft. The Fe(II) in heme is coordinated to histidine F8. His E7 is located on the distal side of the heme group, reducing the affinity of carbon monoxide (CO) for the iron about 200 times.

6.2 Hemoglobin

In 1960 a structure at 6.5 Å resolution of horse hemoglobin was presented by Max Perutz and coworkers. The structure showed that each of the four subunits resembled myoglobin closely. But it was not until 1970 that the structures of both oxyhemoglobin and deoxyhemoglin were solved and allowed detailed analysis of the oxygen binding.

Although hemoglobin of most organisms is tetrameric, there are examples showing that all hemoglobins are not always tetrameric. The single polypeptide chain of lamprey and hogfish hemoglobin is monomeric when oxygenated but forms dimers and tetramers upon deoxygenation. Even so, the overall fold of each polypeptide chain is very similar (Figure 6.3). As in myoglobin, the heme group is attached in a hydrophobic cleft. Fe(II) in the heme group is coordinated to histidine F8 and histidine E7 is placed on the other side of the heme group. Thus, independent of the source of hemoglobin or myoglobin, the surrounding of the heme group is more or less the same.

Tetrameric hemoglobin is better described as a dimer of dimers. An α-chain interacts with a β-chain forming an αβ-dimer, which in turn interacts with another αβ-dimer, forming the $(\alpha\beta)_2$ tetramer. This conglomerate of four subunits has a major impact on the oxygen affinity. Although molecular oxygen is bound to the heme group in both myoglobin and hemoglobin, the affinities differ (Figure 6.4). At the oxygen partial pressure in the lungs (~100 mmHg) both would be saturated with molecular oxygen. In tissue where the pressure drops to 20–40 mmHg (depending on the tissue), only a few percent of loaded

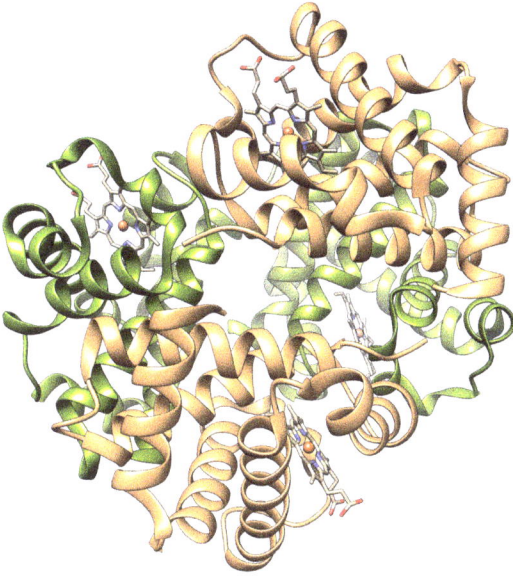

Figure 6.3: Oxygenated human hemoglobin. The tetrameric human hemoglobin consists of two α-chains (sand) and two β-chains (green), forming $(\alpha\beta)_2$ tetramer (pdb: 2DN1).

oxygen would be released from myoglobin. Hemoglobin, on the other hand would release up to 40% of bound oxygen. The propensity of hemoglobin to unload oxygen much more effectively than myoglobin depends on the tetrameric structure. Comparing the structures of deoxygenated and oxygenated hemoglobin, it is clear that there is a structural difference.

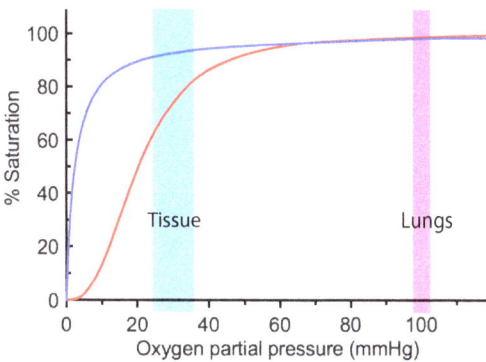

Figure 6.4: Oxygen binding to myoglobin and hemoglobin. At the oxygen partial pressure of lungs, both myoglobin (blue) and hemoglobin (red) would be saturated with bound molecular oxygen. In the tissues, when the oxygen partial pressure is much lower, only hemoglobin would unload a substantial fraction of bound oxygen.

The sigmoid shape of the oxygen saturation curve of hemoglobin implies that the affinity for molecular oxygen is dependent on the oxygen partial pressure and that the binding of oxygen is cooperative. Thus, at high levels of oxygen, as in the lungs, hemoglobin binds molecular oxygen more strongly, whereas in tissue, with its lower level of oxygen, the binding is much weaker. This change in binding affinity is caused by a conformational change in the heme group that is transmitted to the whole protein when molecular oxygen binds to Fe(II). The deoxygenated and low affinity state is the so called "tense" or T-state, while the oxygenated and high affinity state is the so-called relaxed or R-state.

6.2.1 The role of heme

In deoxygenated hemoglobin, the heme group is slightly domed and the Fe(II) protrudes from the ring plane. When molecular oxygen binds, the Fe(II) is converted from a high-spin state to a low-spin state that has a smaller radius, which moves the Fe(II) ~0.6 Å into the heme plane and flattens the heme group. Since the iron atom is coordinated to His F8, this residue is pulled toward the heme group together with whole helix F (Figure 6.5). This movement causes a major conformation change of this subunit from a "tense" to a "relaxed" state, which effectively induces a "tense" to "relaxed" transition in the other subunits. Thus, the binding of the first molecule of oxygen converts hemoglobin from a low-affinity state to a high-affinity state.

His E7 and other residues on the distal side of the heme group are much less affected by oxygenation, and appears not to change to any larger degree.

Initially it was believed that the conformation change was a two-state transition $(T \rightarrow R)$, but further analysis has indicated that the "classical" R-state is an intermediate in the transition to the end-state $(T \rightarrow R \rightarrow R2)$.

Comparison of deoxyhemoglobin with oxygenated hemoglobin shows that the major structural changes occur between the two $\alpha\beta$-dimers, or more specifically between the interfaces between subunits α_1 and β_2 and between subunits α_2 and β_1. The interface between the α- and β-subunits in each $\alpha\beta$-dimer is much less affected, probably due to stronger association. The oxygen induced $T \rightarrow R \rightarrow R2$ transition causes the $\alpha_1\beta_1$-dimer to rotate ~23° relative to the $\alpha_2\beta_2$-dimer (Figure 6.6). This rotation brings the β-subunits closer to each other and reduces the opening of the central cavity. Simultaneously several hydrogen bonds, salt bridges and van der Waals contacts break and new ones are formed. This is particularly noticeable at the interface between the C-helix of α_1 and the FG-corner of β_2 and between the symmetrically related FG-corner of β_1 and the C-helix of α_2 (Figure 6.7). In the T-state, His 97 of the β_2-subunit is inserted between Thr41 and Pro44 of the α_1-subunit. Oxygenation causes His97 to slide pass Thr41 and get inserted between Thr38 and Thr41. Thus, Thr41 function as a barrier between the tense and relaxed states. Several other contacts between β_2 and α_1 change due to the transition. For instance, the T to R transition breaks the hydrogen bonds between Asp99 of β_2 and Thr38 and Tyr42 of α_1, but instead a hydrogen bond between Asn102 of β_2 and Asp92 of α_1 forms.

At the C-terminus, salt bridges involving Arg141 of the α-subunits and His146 of the β-subunits break when hemoglobin is oxygenated. In the deoxygenated state Arg141 in α_1 is hydrogen bonded to Val34 of β_2 and form salt bridges to Asp126 and Lys127 of α_2. Tyr140 has intramolecular hydrogen bond to Val93. The C-terminal residue of the β_2-subunit, His146 has a salt bridge to Lys40 of α_1. His146 also form an internal salt bridge to Asp94 and Tyr145 forms a hydrogen bond with Val98. All these bonds break upon oxygenation. In addition, the C-terminal His146 in $\beta2$ and the N-terminal Val1 in $\alpha2$ become partially deprotonated.

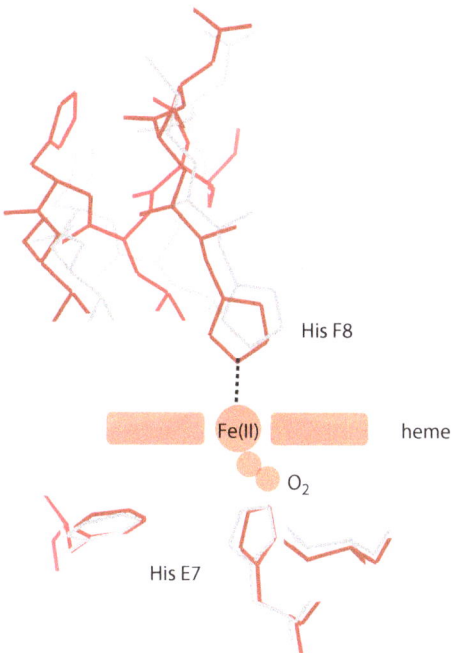

Figure 6.5: Oxygenation of heme induces a conformational change in the subunit. In the deoxygenated state (gray), the heme group is slightly domed and Fe(II) is placed above the heme plane. Oxygenation (red) flattens the heme group and the iron moves into the plane of the heme group. This pulls HisF8 and the whole F helix toward the heme group and leads to a major conformational change in the subunit that induces conformational changes in the other subunits.

6.2.2 The Bohr effect

Evolution has created an excellent transporter of molecular oxygen; it gets fully loaded during the passage through the capillaries in the lungs and unloads oxygen easily in the tissue where it is used in oxidation processes, such as ATP production in the

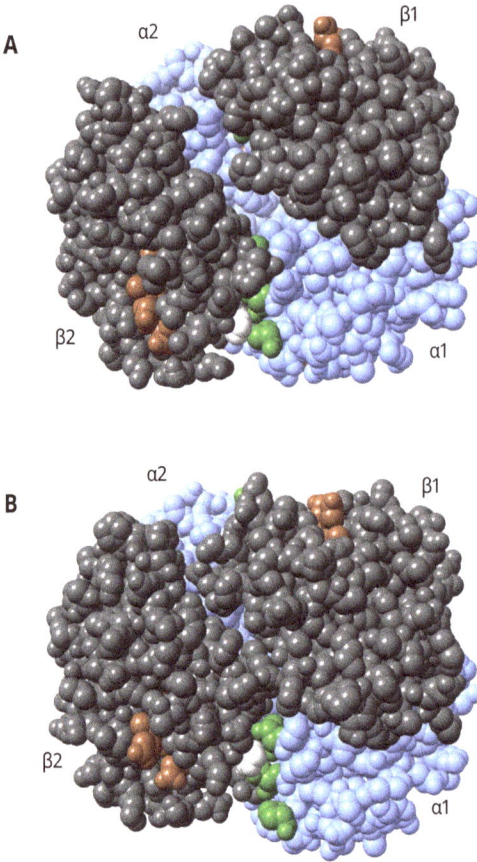

Figure 6.6: Hemoglobin. (A) Structure of deoxygenated hemoglobin (pdb: 1HGA). (B) Structure of oxygenated hemoglobin (pdb: 1HHO). The two αβ-dimers are related to each other by a twofold axis of symmetry. Oxygen-binding to heme (brown) induces conformational changes in both the α-chains (blue) and β-chains (gray). In the T → R → R2 transition $\alpha_1\beta_1$ is rotated ~23° relative to $\alpha_2\beta_2$ and the central cavity gets narrower. The rotation causes several hydrogen bonds, salt bridges and van der Waals contacts to break and new ones to form. See text for more details. His97 in β2 and Thr38, Thr41 and Pro44 in α1 are indicated in white and green, respectively.

mitochondria. A byproduct formed during the energy metabolism is carbon dioxide (CO_2). Thus, when oxygen is consumed in the tissue, the CO_2 level increases and is taken up by the red blood cells, and inside the cell carbonic anhydrase catalyzes its conversion to bicarbonate

$$CO_2 + H_2O \rightleftharpoons H^+ + HCO_3^-$$

Bicarbonate is then transported across the membrane into the plasma. It is estimated that that a major fraction (~60%) of blood CO_2 exists as bicarbonate in the plasma.

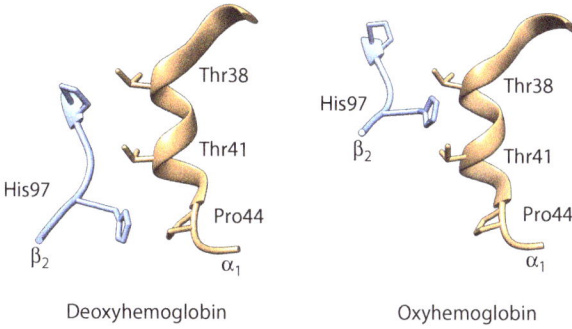

Figure 6.7: Transition from "tense" to "relaxed" state. Oxygenation causes His97 to move from a position between Thr41 and Pro44 in "tense" state to a position between Thr38 and Thr41 in the "relaxed" state. The α_1-subunit is colored sandy brown and the β_2-subunit is in blue.

This lowers the pH of the blood and as a consequence hemoglobin takes up protons, which in turn unloads even more oxygen. This is known as the Bohr effect.

In an actively respiring muscle, the oxygen's partial pressure may be as low as 20 mmHg. If the pH drops from 7.6 to 7.4, nearly 10% more oxygen is unloaded from hemoglobin (Figure 6.8). The proton uptake also stimulates bicarbonate formation; the Bohr effect facilitates CO_2 transport. The process is reversed in the lungs; the oxygen binding releases the Bohr protons, which causes carbonic anhydrase to convert bicarbonate to CO_2. CO_2 then diffuses out of the red blood cell and into the lungs and is expired.

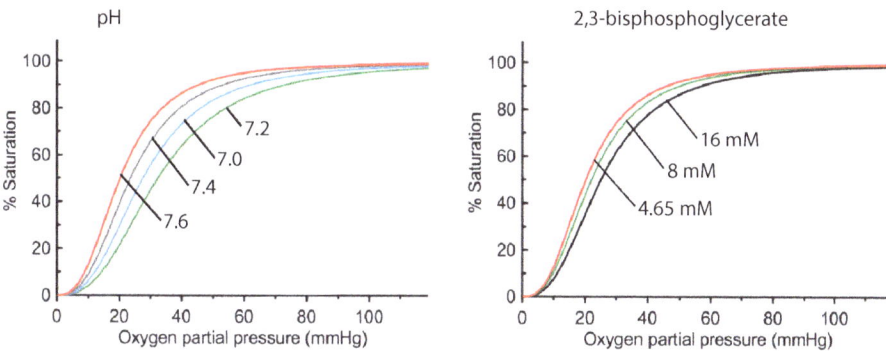

Figure 6.8: Effect of pH and 2,3-bisphosphoglycerate on oxygen binding to hemoglobin. The oxygen's partial pressure in actively respiring muscle tissue is indicated by the orange line.

Carbon dioxide in blood is present as bicarbonate (~85%), bound to hemoglobin (~10%) or dissolved as free CO_2 (~5%). The N-terminal amino acid residues in both deoxygenated and oxygenated hemoglobin can combine reversibly with CO_2 to form carbamates:

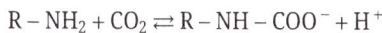

$$R - NH_2 + CO_2 \rightleftarrows R - NH - COO^- + H^+$$

Although the difference in carbamate formation between deoxygenated and oxygenated hemoglobin is rather small (~5%), this transport mechanism is still very important as it removes nearly half of the CO_2 that is expired during each respiration cycle.

6.2.3 2,3-Bisphoshoglycerate

Going from lowland to high altitudes above 1,800–2,000 meters leads to several prominent effects due to the lower oxygen levels. The immediate effect is probably a sensation that it is more difficult to breath and to perform physical activities. However, after a week or two, the body gets acclimatized to the lower level of molecular oxygen. Erythropoietin (EPO) is secreted from kidneys to the blood within a few hours, which triggers red blood cell synthesis in the bone marrow. More red blood cells can transport more oxygen, making it easier to breathe.

2,3-Bisphosphoglycerate is an allosteric regulator of hemoglobin affinity to molecular oxygen. It binds at a molar ratio of 1:1 to the central cavity between the β-subunits of deoxygenated hemoglobin. In oxygenated hemoglobin, the cavity is smaller and 2,3-bisphosphoglycerate cannot bind.

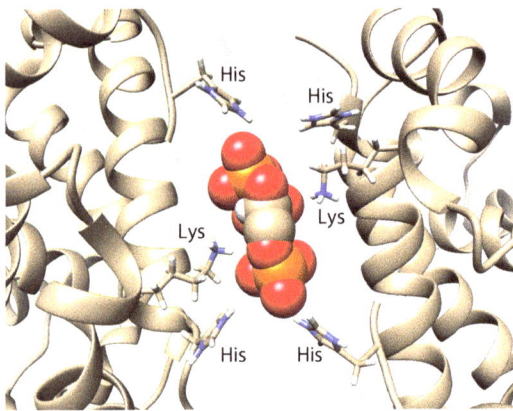

Figure 6.9: 2,3-Bisphosphoglycerate binds to the central cavity. His2, His143 and Lys82 of the β-subunits bind 2,3-bisphosphoglycerate.

2,3-Bisphosphoglycerate is bound by two histidines, one lysine and the N-terminal amino group of each β-subunit (Figure 6.9). Since 2,3-bisphosphoglycerate only binds to deoxygenated hemoglobin and stabilizes the T-state, the oxygen affinity decreases; less molecular oxygen will bind to heme when 2,3-bisphosphoglycerate is bound. At first sight, this seems contraproductive: would it not be better if more oxygen would bind?

High-altitude acclimatization induces conversion of 1,3-bisphosphoglycrate (from glycolysis) to 2,3-bisphosphoglycerate, increasing the normal concentration from 4.6 mM to more than 7–8 mM, and decreasing the oxygen affinity even more (Figure 6.8). The consequence of the lowered oxygen affinity is that somewhat less molecular oxygen will be loaded in lungs but more will be unloaded in the tissues. Therefore, the lower oxygen affinity will in fact be advantageous at high altitudes when the air contains less molecular oxygen.

Since the effect of high-altitude acclimatization lasts for some time after return to lowland, it has a certain interest among endurance athletes for obvious reasons.

6.2.4 Fetal hemoglobin

During development, from embryo to fetus and finally an adult, humans express different types of hemoglobin. During most of the fetus state, fetal hemoglobin (HbF) is expressed. Around birth and during the first year of life, HbF is replaced with adult hemoglobin (HbA). HbF consists, as HbA, also of four subunits, two α-subunits and two γ-subunits, and forms a $(\alpha\gamma)_2$ tetramer. The γ-subunit is similar but distinct from the β-subunit of HbA; 39 of its 146 amino acids are different. The altered functional properties of HbF are due to these substitutions.

Since the fetus blood does not pass the lungs, it must be oxygenated differently. HbF has higher affinity for molecular oxygen and binds 2,3-bisphosphoglycerate weaker than HbA. Therefore, oxygenated HbA of the pregnant woman can oxygenate the fetus HbF.

6.2.5 Sickle-cell diseases and sickle-cell anemia

Several mutations in the hemoglobin genes are known, leading to sickle-cell diseases. Some do not affect the carrier noticeably, such as the mutation Glu26 to Lys26 in the β-subunit of HbE, whereas others are extremely severe. It is estimated that around 5% of the world's population carries at least one mutated hemoglobin gene and that over 300,000 new-born each year are affected. The greater part of affected people is found in sub-Saharan Africa.

The most common disease linked to mutations in hemoglobin is sickle-cell anemia. The disease was discovered in the beginning of the twentieth century by the American physician James Herrick. He also found that the incidence of sickle-cell anemia was particularly high among African Americans.

Sickle-cell anemia is linked to a single mutation in the β-subunit, substituting thymine for adenine in the sixth codon of the β-subunit gene: GAG to GTG. The glutamate residue at position 6 is mutated to a valine in sickle-cell anemia hemoglobin (HbS); a highly polar and negatively charge residue is exchanged for a nonpolar one. This creates a nonpolar protrusion on the surface of helix A, independent of whether HbS is oxygenated or not.

In deoxygenated hemoglobin but not in oxygenated, there is a nonpolar pocket formed by a phenylalanine (Phe85) and a leucine (Leu88) residue in helix F of the β-subunit. At low oxygen levels, when deoxygenated HbS dominates, the nonpolar protrusion of β_2 fits snugly in the nonpolar pocket on β_1 (Figure 6.10) This leads to the formation of a fiber of hemoglobin that causes the red blood cell to change shape from the normal biconcave disc to a sickle-like shape, and hence the name of the disease – sickle-cell anemia.

The severity of the disease depends on whether both or only one of the β-genes are mutated. If only one of the genes inherited from the parents is mutated (heterozygote), the carrier may not even notice that she is affected as long as she does not embark on expedition to places with very low oxygen levels, like the Andes or other high mountains. As a homozygote carrier, the situation is much worse. Although treatments have improved, the disease leads to serious pathophysiological problems, sometimes lethal.

There is no coincidence that high incidence of sickle-cell anemia correlates with malaria endemic areas in the world as carrier of sickle-cell trait has some protection from malaria.

6.2.6 Allostery and cooperative binding

When the first molecule of oxygen binds to one of the hemoglobin subunits, this is communicated to the other subunits through the conformational change the initial binding induces. To explain the behavior of certain enzymes, that resembled the cooperative oxygen binding, Jacques Monod, Jeffries Wyman and Jean-Pierre Changeux proposed a model to describe the observed allosteric effects in 1965. The concerted MWC model could also be applied to analyze oxygen binding to hemoglobin. Around the same time another model, the sequential KNF model, was proposed by Daniel E. Koshland, George Némethy and David Filmer. In both models, it is assumed that a subunit in a multimeric protein can attain two states: "tense" or "relaxed" (Figure 6.11).

The concerted model assumes that the binding of a molecule of molecular oxygen to any of the subunits of hemoglobin induces the conformational change from the tense state to the relaxed state. It is "all or none" change; no "hybrid" quaternary structures occur with both tense and relaxed subunits.

In the sequential model, binding of the first molecule of molecular oxygen to one of the subunits induces a conformational change of that subunit, which induces a change in the neighboring subunits that increases the affinity for the next molecule of molecular oxygen. In contrast to the concerted model, the sequential model allows that some subunits are in the tense state and others in the relaxed state.

Although the concerted model describes the oxygen binding of hemoglobin rather well, there are discrepancies and some of the binding behavior is better described by the sequential model.

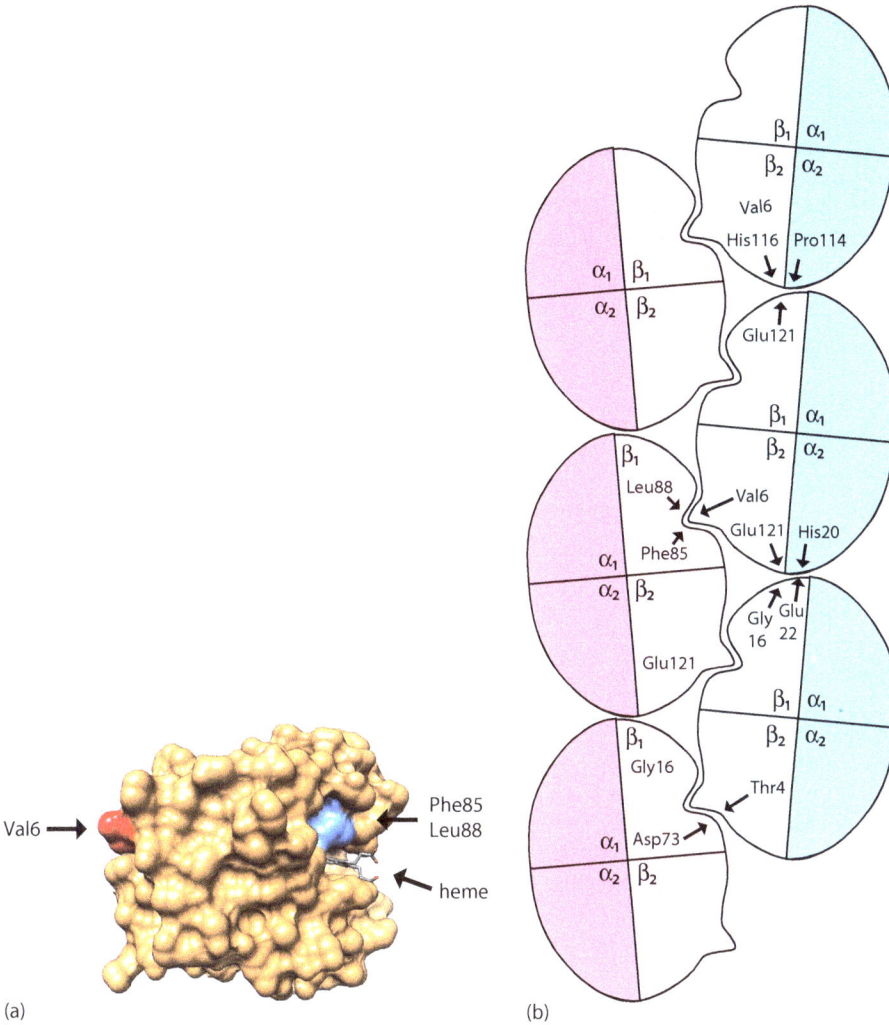

Figure 6.10: Formation of HbS fiber. (A) Beta-subunit of HbS with valine at position 6 (red) that protrudes creating a nonpolar sticky site. Phenylalanine and leucine at positions 85 and 88, respectively, are also indicated (blue). (B) Schematic structure of a hemoglobin fiber. The sticky protrusion of a β_2-subunit fits in the nonpolar pocket in β_1 of deoxygenated hemoglobin. Adapted with permission after B.C. Wisher, K.B. Ward, E.E. Lattman and W.E. Love (1975) Crystal structure of sickle-cell deoxyhemoglobin at 5 A resolution. *J Mol Biol* 98:179–194.

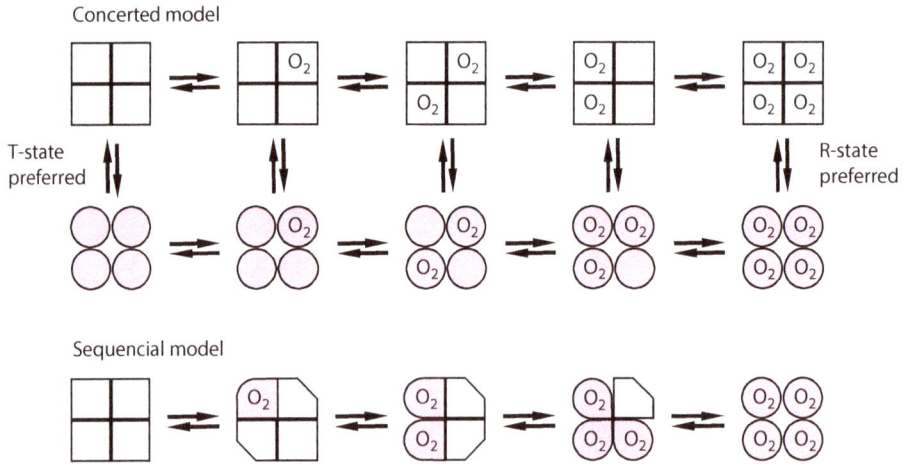

Figure 6.11: Illustration of the concerted and sequential models of cooperative binding.

Further reading

Ahmed, M.H., Ghatge, M.S. and Safo, M.K. (2020). Hemoglobin: Structure, function and allostery. *Subcell Biochem* 94:345–382.

De Santis, V. and Singer, M. (2015). Tissue oxygen tension monitoring of organ perfusion: Rationale, methodologies, and literature review. *Br J Anaesth* 115:357–365.

Eaton, W.A., Henry, E.R., Hofrichter, J. and Mozzarelli, A. (1999). Is cooperative oxygen binding by hemoglobin really understood? *Nature Struct Biol* 6:351–358.

Kendrew, J.C., Bodo, G., Dintzis, H.M., Parrish, R.G., Wyckoff, H. and Phillips, D.C. (1958). A three-dimensional model of the myoglobin molecule obtained by X-ray analysis. *Nature* 181:662–666.

Kendrew, J.C., Dickerson, R.E., Strandberg, B.E., Hart, R.G., Davies, D.R., Phillips, D.C. and Shore, V.C. (1960). Structure of myoglobin: A three-dimensional Fourier synthesis at 2 a. Resolution. *Nature* 185:422–427.

Kendrew, J.C., Watson, H.C., Strandberg, B.E., Dickerson, R.E., Phillips, D.C. and Shore, V.C. (1961). A partial determination by X-ray methods, and its correlation with chemical data. *Nature* 190:666–670.

Koshland, D.E. Jr, Nemethy, G. and Filmer, D. (1966). Comparison of experimental binding data and theoretical models in proteins containing subunits. *Biochemistry* 5:365–385.

Mairbaurl, H. and Weber, R.E. (2012). Oxygen transport by hemoglobin. *Compr Physiol* 2:1463–1489.

Manning, L.R., Russell, J.E., Padovan, J.C., Chait, B.T., Popowicz, A., Manning, R.S. and Manning, J.M. (2007). Human embryonic, fetal, and adult hemoglobins have different subunit interface strengths. Correlation with lifespan in the red cell. *Protein Sci* 16:1641–1658.

Monod, J., Wyman, J. and Changeux, J.P. (1965). On the nature of allosteric transitions: A plausible model. *J Mol Biol* 12:88–118.

Perutz, M.F., Rossmann, M.G., Cullis, A.F., Muirhead, H., Will, G. and North, A.C. (1960). Structure of haemoglobin: A three-dimensional Fourier synthesis at 5.5-a. Resolution, obtained by x-ray analysis. *Nature* 185:416–422.

Shibayama, N. (2020). Allosteric transitions in hemoglobin revisited. *Biochim Biophys Acta Gen Subj* 1864:129335.

Tanford, C. and Reynolds, J. (2001). Nature's robots. A history of proteins. Oxford University Press. Oxford.

7 The ball is round

Most proteins that are present in the soluble part of the cell have a globular form or at least a form that resembles a sphere. However, there are other forms. There are proteins that have a very extended structure and some even form thick fibers, which can be several micrometer long but very narrow. In this chapter we will look at some of these so-called fibrous proteins.

7.1 Collagen

Collagens make up some 25–35% of the total protein content in animals, thereby making it one, if not, of the most abundant protein. Collagens are found in connective tissues like skin, bone, cartilage and tendons where it provides mechanical strength. Depending on their functions, domain architecture and organization collagens are divided into different types. However, common for all collagen types is the ability to form a triple helix. Each of the 28 types is formed by three polypeptide chains, known as α-chains, that fold into longer or shorter stretches of a triple helical structure. In the fibril-forming collagens, most of the structure is triple helical, whereas other collagens only form short stretches with a triple helical structure. Other collagens form different types of networks, beaded filaments or anchoring fibrils.

Around 90% of the human collagens are made up by the main fibril-forming collagens (I, II, III, V and XI). Collagen I, mainly found in bone, consists of two α_1(I)-chains and one α_2(I)-chain whereas collagen II, predominate in cartilage, is a homotetramer of three α_1(II)-chains.

The human α_1(II) genes codes for a polypeptide chain of nearly 1500 amino acids and of these residues more than 800 are either glycine, proline or alanine. In all fibril-forming collagens around 50% of the residues are either glycine or proline. Each polypeptide has a repeating Gly-Xaa-Yaa-Gly-Xaa-Yaa sequence where Xaa and Yaa often are proline and hydroxyproline, respectively.

The three left-handed helical chains form a tight right-handed triple helix, held together by interchain hydrogen bonds. There are 3.3 residues per turn in the helix and the pitch is 0.95 nm, considerably longer than the 0.56 nm pitch in α-helices. Due to the repetitive nature, glycine is placed inside the triple helix, which allows for the much tighter twist (Figure 7.1).

Collagens are synthesized as precursor procollagen α-chains. These precursor molecules have long propeptides at both the N- and C-terminals and a signal recognition sequence that targets the polypeptide for the endoplasmatic reticulum. Hydroxylation and glycosylation in the endoplasmatic reticulum lead to the formation of hydroxylysine, hydroxyproline and O-linked glycosylation. After the post-translational modifications three α-chains associate to form procollagen. Procollagen is then transferred to

https://doi.org/10.1515/9783111350684-007

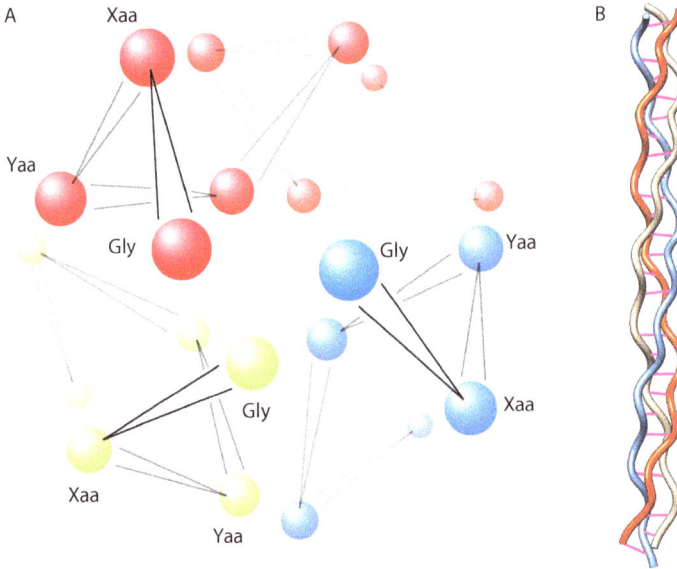

Figure 7.1: The collagen triple helix. (A) View along the molecular axis showing the location of the residues in the Gly-Xaa-Yaa triplet sequence. (B) The three α-chains coil around each other in the triple helical structure. Interchain hydrogen bonds (pink) stabilize the triple helix. Adapted from K. Beck and B. Brodsky (1998) Supercoiled protein motifs: the collagen triple-helix and the alpha-helical coiled coil. *J Struct Biol* 122:17–29.

the Golgi apparatus for packaging and export to the extracellular matrix. It is not fully clear where and how the last steps in the collagen fibril formation occur. It is believed that proteolytic processing of the propeptide occurs during secretion and that the mature collagen self-assembles into fibrils (Figure 7.2).

Collagen self-assembly creates microfibrils that grow both longitudinally and axially into mature collagen fibrils. As the collagen molecules assemble with staggered ends, a cross-striated pattern forms that can be observed in electron micrographs. Fibrils assemble into fibers that are stabilized by intramolecular and intermolecular cross-links catalyzed by lysyl oxidase. These fibers then bundle and form larger and larger structures, such as cartilage and tendons.

Fibronectin (FN) is another fibrous protein present in the extracellular matrix. FN is important in cellular processes such adhesion, migration, growth and differentiation. FN is usually present as a dimer, composed of two nearly identical ~250 kDa subunits. Although a single gene codes for FN, alternative splicing events give rise to 20 different variants of human FN. Disulfide bonds at the C-terminals link the monomers covalently. Around 90% of the FN sequence folds into three different types of FN repeats (12 type I, 2 type II and 15–17 type III repeats) connected by flexible regions. The type III FN repeats is one of the most common domains in vertebrate proteins.

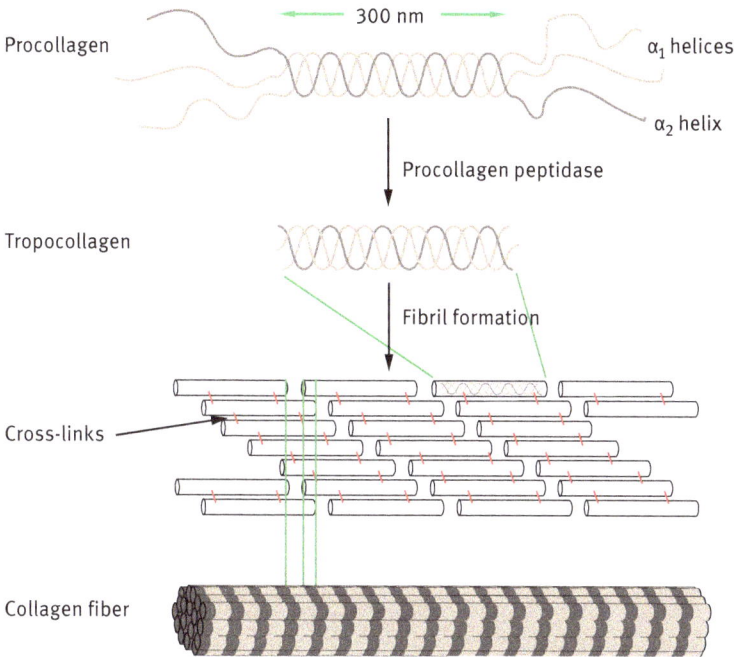

Figure 7.2: Collagen fibril formation. Collagen type I consists of two $\alpha_1(I)$-chains and one $\alpha_2(I)$. After synthesis and post-translational modifications in the endoplasmatic reticulum and export to the extracellular matrix, the collagen molecules self-assembly into fibrils. Adapted from S.P. Canelón and J.M. Wallace (2016) β-aminopropionitrile-induced reduction in enzymatic crosslinking causes in vitro changes in collagen morphology and molecular composition. *PloS One* 11:e0166392. Licensed by CC BY (creativecommons.org/licenses).

The different domains in FN have affinity for membrane-bound integrins, collagen and heparin. The interaction with integrins connects the extracellular matrix with the intracellular actin cytoskeleton. Thereby an external signal can be relayed across the membrane and exert an effect intracellularly. The structure of FN is also under tension control. When FN is stretched, new "cryptic" binding sites are exposed that causes FN to self-associate and form fibrils.

Another important component of the extracellular matrix is elastin. Elastin fibers provide elasticity to connective tissue in such a way that the tissue can recoil after being stretched or compressed. Elastin is assembled from tropoelastin, a soluble ~72 kDa precursor protein, that forms the core of elastin and is surrounded by a microfibril scaffold made from fibrillins (~300 kDa glycoproteins). The amino acid sequence of tropoelastin contains alternating hydrophobic and hydrophilic regions. The hydrophobic regions contain repetitive motifs, either proline-rich motifs: Val-Pro-Gly-Xaa-Gly (Xaa = any amino acid except Pro) or glycine-rich motifs: Yaa-Gly-Gly or Yaa-Gly-Gly-Zaa-Gly (Yaa, Zaa = Leu, Val or Ala). The proline-rich motifs are located to the center and the glycine-rich motifs to the ends of tropoelastin. The hydrophilic regions are rich in lysine and alanine and tend to form α-helices, whereas the hydrophobic regions are believed to be highly disordered.

Although tropoelastin is a very hydrophobic protein, around 90% of the amino acid residues are nonpolar, it is highly soluble in water and apparently hydrated. When stretched, the glycine-rich regions are believed to unfold, which will expose nonpolar side chains. Due to the hydrophobic effect this will reduce the entropy. As soon as the tension drops the polypeptide recoils, water molecules are released from the cage structures around the nonpolar side chains and the entropy increases. Thus, the recoil of elastin is entropy driven.

Due to the properties of collagen and elastase, the interest from the beauty industry is not surprising.

7.2 Fibroin – a valuable protein

Silk from the silkworm *Bombyx mori* has been used in textiles for more than 4,000 years. Its soft and lustrous appearance and good mechanical properties have made it a favorite fabric in clothing. More recently the biocompatibility and the possibility to chemical modifications, silk fibroin has been found useful in many areas.

Silk fibroin, which is used in fabrics, is obtained from *Bombyx mori* cocoons. The cocoon is made by a single continuous strand of silk about 900 meters long. Before the fibroin can be used, sericin (a group of soluble glycoproteins) must be removed. Sericin covers the surface of the fibroin fibril and acts as an adhesive to keep the structural integrity of the fibril.

In silk fibroin, a heavy chain (~ 390 kDa) and a light chain (~26 kDa) are linked by a disulfide at the C-terminal of the heavy chain. A glycoprotein P25 (~25 kDa) is noncovalently attached to the complex of the heavy and light chains. These proteins associate in a ratio of 6:6:1 in silkworm silk.

The silk fibroin heavy chain consists mainly of glycine (46%), alanine (30%), serine (12%) and tyrosine (5%). Together glycine, alanine, serine and tyrosine account for more than 93% of the amino acid content. The fibroin heavy chain contains a repetitive hexapeptide sequence of Gly-Ala-Gly-Ala-Gly-Ser and Gly-Ala/Ser/Tyr dipeptide repeats (Figure 7.3). In the amino acid sequence, every second residue in the heavy chain is a glycine, except at the N- and C-terminals. Since the heavy chain forms a stable anti-parallel β-sheet, all glycine side chains are on one side of the sheet and the side chains of alanine or serine are on the other side. Hydrogen bonds between amide nitrogens of one β-strand and carboxyl oxygens of the other anti-parallel β-strand stabilize the β-sheet.

In silk these β-sheets are layered on top of each other in such a way that a "glycine side" faces another "glycine side" and an "alanine/serine side" faces another "alanine/serine side." This allows for a tight packing of the β-sheets in layers, which contributes to the rigid structure and tensile strength of silk. The distance between "gly-

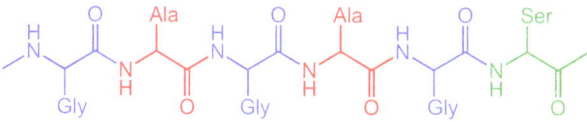

Figure 7.3: Hexapeptide repeat in fibroin heavy chain.

cine sides" (0.35 nm) is smaller than that between "alanine/serine sides" (0.57 nm) due to the size of the side chains; a glycine side chain (-H) is considerable smaller than the methyl group (-CH$_3$) of alanine (Figure 7.4).

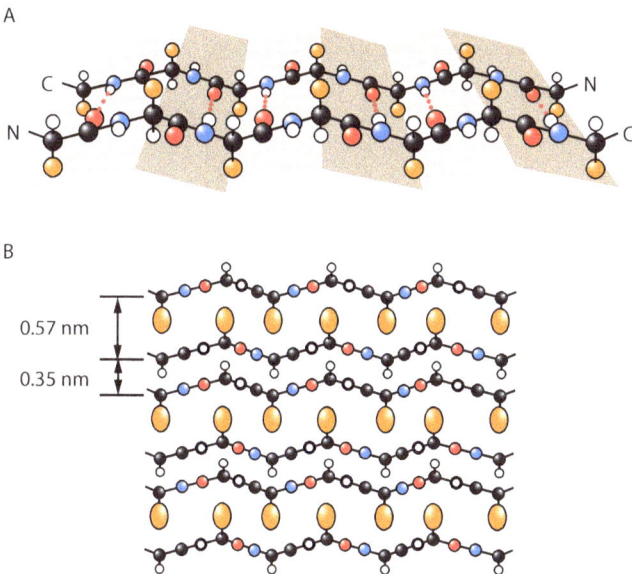

Figure 7.4: Silkworm fibroin. (A) In an antiparallel β-sheet, side chains (orange) are alternating on either side of the sheet. The β-sheet is stabilized by hydrogen bonds (red dots) between the β-strands. (B) Schematic illustration of the structure of silk. Since every other residue is a glycine, all glycine side chains are located to the same side of the β-sheet with all alanine/serine side chains on the other side of the β-sheet. The distance between the layers of β-sheet will therefore vary.

Spiders, like many other insects, also have silk-producing glands. The chemical and mechanical properties of spider silk differ from those of silkworm silk. The major protein in spider silk is spidroin, but each gland makes it own version of the protein, which leads to silk with different mechanical properties and uses. The major ampullate gland produces major ampullate spidroin (MaSp) 1 and 2, the main proteins in draglines (used in the outer rim of the web). The aciniform gland produces aciniform spidroin (AcSp) and a silk used to wrap captured prey. Although the amino acid sequences of the spidroins differ, they all shear a common structural feature. All spidroins contain a large repetitive core domain, making up around 90% of the amino acid residues. The core domain is flanked by nonrepeating N- and C-terminals (Figure 7.5A). The core domain of MaSp consists of highly repetitive glycine and alanine blocks. These blocks create very different structures. The alanine blocks form crystalline β-sheets, and the glycine blocks consist mainly of helical (β-spirals and 3_{10}-helices) and β-turn structures and are found in the so-called amorphous region (Figure 7.5B). The interplay between these two blocks, the hard crystalline segments and the elastic amorphous segments gives spider silk its extraordinary mechanical properties.

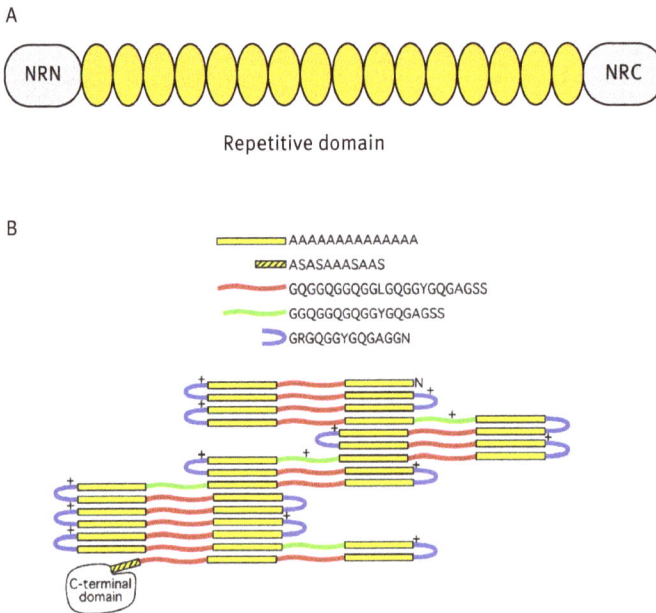

A

Repetitive domain

B

Figure 7.5: Spider silk. (A) Spidroins consist of a large repetitive core domain, flanked on both sides by nonrepetitive amino (NRN) and nonrepetitive carboxyl (NRC) terminals. Adapted from L. Eisoldt, C. Thamm and T. Scheibel (2001) Review the role of terminal domains during storage and assembly of spider silk proteins. *Biopolymers* 97:355–361. (B) Schematic arrangement of alanine (yellow) and glycine (red and green) blocks in spidroin. Reprinted with permission from M. Stark, S. Grip, A. Rising, M. Hedhammar, W. Engström, G. Hjälm and J. Johansson (2007) Macroscopic fibers self-assembled from recombinant miniature spider silk proteins. *Biomacromolecules* 8:1695–1701. Copyright (2007) American Chemical Society.

Insects belonging to Apoidea (bees) and Vespoidea (ants, wasps and hornets) also produce silk fibers. However, the silk produced by these insects has a different structure. It appears that these species have four genes that give rise to four distinct but structurally similar silk proteins. Bumblebee proteins are all around 32–35 kDa, whereas hornet proteins are a bit larger (36–47 kDa). The sequences are rich in alanine and serine residues; at least 30% of the amino acids are alanine. In the amino acid sequence alanine is predominantly found in the central region of the polypeptide chain and serine is located at the ends.

Analyses of the amino acid sequence predict that the central region of these proteins are α-helical and can form a coiled-coil structure. The secondary structures of the end regions vary depending on the source of the silk protein. Analysis has also indicated that the four silk proteins can associate and form a supercoil or four-helix bundle, where each polypeptide chain coils around the other chains (Figure 7.6).

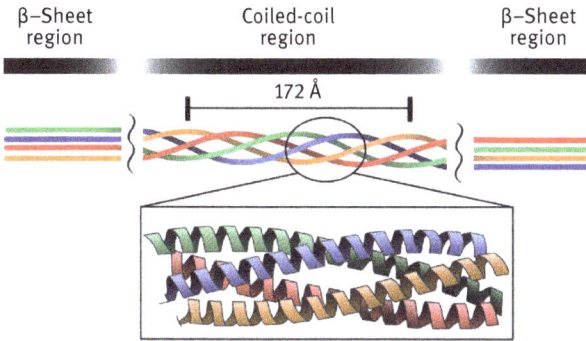

Figure 7.6: Hornet silk protein. The four distinct hornet silk proteins assemble and form a coiled-coil structure in the central region, surrounded by β-sheet regions. Reprinted with permission from T. Kameda, T. Nemoto, T. Ogawa, M. Tosaka, H. Kurata and A.K. Schaper (2014) Evidence of alpha-helical coiled coils and beta-sheets in hornet silk. *J Struct Biol* 185:303–308.

7.3 Coiled-coil and helical bundles

In the early 1950s, Francis Crick suggested that two right-handed α-helices that pack together are not fully parallel but rather twisted around each other into a coiled-coil structure. By placing a side chain in one helix between four residues in the other helix, like a "knob in the hole," the tightest packing and thus the most energetic favorable structure is obtained. As the pitch of an α-helix is 3.6 amino acid residues, the helices in the coiled-coil deform slightly, such that there are exactly seven residues per two turns. This induces a left-handed twist in the coiled-coil structure and a contact or "seam" between the two helices that also is twisted.

Crick also proposed that coiled-coils contain a seven-residue repeat pattern and that nonpolar residues preferentially are localized along the seam and therefore shielded from the surrounding polar milieu. Later high-resolution structure data has proven that Crick's proposal was correct.

The seven-residue or heptad repeat is usually denoted as *a-b-c-d-e-f-g* (Figure 7.7). In the heptad sequence, residues at positions *a* and *d* are usually occupied by nonpolar amino acids (the most common ones are Leu, Ile and Val). Positions *e* and *g* are

A

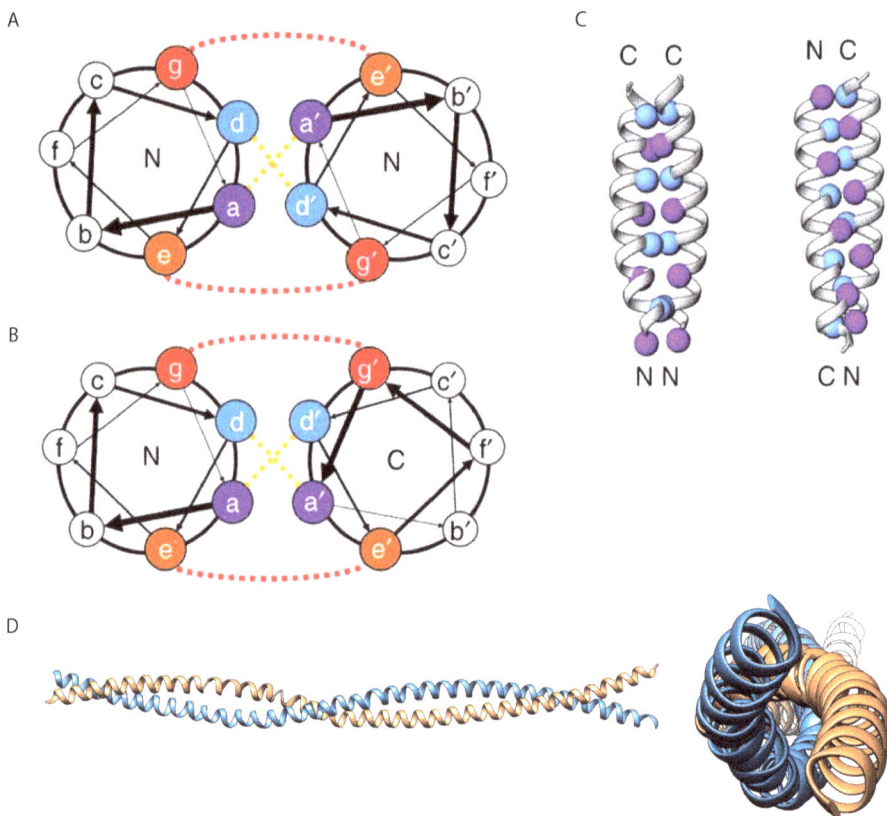

B

C

D

Figure 7.7: Schematic view of a two-helix bundle. Helical wheel representation of a heptad repeat in a parallel (A) and antiparallel (B) coiled-coil. (C) In a parallel coiled-coil residue at position *a* in one helix interacts with the residue at position *a* in the other helix. In antiparallel coiled-coils, residues at *a* and *d* interact. Residues at *a* and *d* are shown as spheres. Reprinted with permission from A. Ljubetič *et al.* (2017) Design of coiled-coil protein-origami cages that self-assemble in vitro and in vivo. *Nature Biotechnol* 35:1094–1101.(D) Sideview and top view of rabbit skeletal muscle troposmyosin (pdb: 2D3E), one of the first determined structures of a coiled-coil domain.

often occupied by charged residues, usually Glu and Lys. Due to this distribution of amino acid residues, the helices are amphiphatic, with a polar side and a nonpolar side. The nonpolar residues at position *a* and *d* in the heptad form the "seam" or the inter-helical hydrophobic core. The charged residues at *e* and *g* form inter-helical charged in-teractions. Depending on parallel or antiparallel packing of the helices, the placement of the hydrophobic core residues *a* and *d* differ. In a parallel arrangement of the helices, an *a* residue in one helix is placed close to an *a* residue in the other helix and vice versa for *d* residues (Figure 7.7A). When the helices are antiparallel, *a* and *d* residues in each helix are placed opposite to each other (Figure 7.7B).

The nonpolar residues in the seam pack tightly as a residue at position *a* or *d* in helix 1 can be placed in a "hole" created by four residues in helix 2, as illustrated in Figure 7.8.

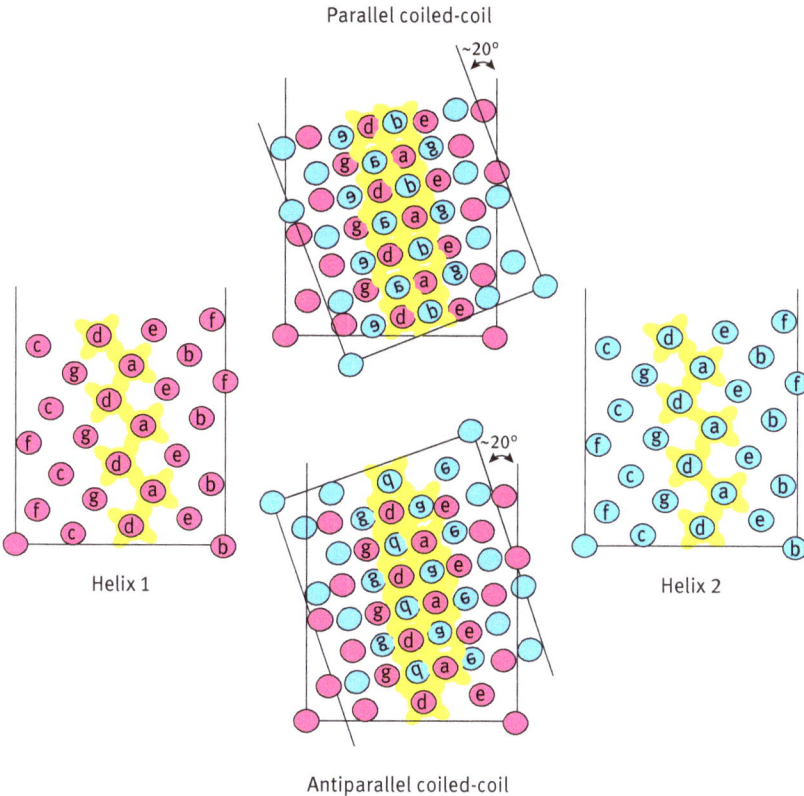

Figure 7.8: Helical net diagram. In this diagram the positions of the side chains along the surface of the α-helix are projected on a plane parallel with the helix axis (helices 1 and 2). By placing the projections face-to-face and tilt one of the projections 18°, a side chain in one helix fits precisely between four residues of the other helix. The hydrophobic core or "seam" between the two helices is colored yellow. Adapted after F. Crick (1953) The packing of α-helices – simple coiled-coils. *Acta Cryst* 6:689–697.

The nonpolar residue in one helix, "the knob," fits in the "hole" in the other helix, formed by four residues. In a parallel coiled-coil, a knob at position *d* in helix 1 is surrounded by *a*, *d* and *e* in one heptad and *a* in the next heptad (for short *adea*) of helix 2. An *a* knob in helix 1 docks into a hole formed by *dgad* in helix 2. When the helices are antiparallel, *a* and *d* knobs dock into holes formed by *adea* and *dgad*, respectively.

Coiled-coils are not restricted to be formed by two helices. There are numerous examples of proteins containing both three- and four-helix bundles. It is even possible to find proteins containing seven- and eight-helix bundles, such as PqqC, the enzyme catalyzing the final step in the biosynthesis of pyrroloquinoline quinone.

However, not all helices in helix bundles have the required heptad sequence to pack according to the "knob in the hole" model. The packing of many four-helix bundles is better explained by the "ridges in grooves" model, proposed by Cyrus Chothia, Michel Levitt and Douglas Richardson. In this model, it is assumed that the ridges are formed by the side chains whirling along the surface of the helix with short stretches of nonpolar residues. Two major types of ridges can be imagined: a right-handed ridge formed by rows of every forth residue and a left-handed ridge formed by rows of every third residue. The grooves are formed between the ridges, similar to the major and minor grooves observed in DNA. Tight packing of helices occurs when a nonpolar ridge of one helix intercalates with the nonpolar grooves on another helix. The inclination between the two helices depends on the type of ridges and grooves, but angles around 50° and 20° appear to predominate compared to an interhelical angle around 20° observed in "knob in hole" packing. Although helix packing in many bundles and globular proteins follow this rule, it is easy to find exceptions.

The interaxial distance between two helices depends on the nature of the side chains at the inter-helical interface. Long side chains, such as those in leucine or methionine, increase the distance whereas short side chains, such as those in glycine or alanine, allows a closer packing. The interaxial distance varies from 7 Å to 12 Å. The interpenetration of atoms at the interface is around 2.3 Å, which means that only the very last atoms in the side chain will make the contacts between the helices.

7.4 Intermediate filaments

Many properties of any eukaryotic cell rely on an intracellular network of proteins. This network or cytoskeleton determines shape, provides tensile strength and resilience, connects the cell physical and biochemical to the extracellular environment, generates forces that enable cell mobility and organizes the intracellular infrastructure. Although the word "skeleton" indicates something stable and nondeformable, the cytoskeleton is quite the opposite, it is highly dynamic and is continuously reshaping.

The cytoskeleton is not a single network but rather three distinct networks that cooperates, microfilaments, intermediate filaments and microtubules. Intermediate filaments are called so as their diameter (~10 nm) is between that of microfilaments (~7 nm) and microtubules (~25 nm).

Each network consists of a major protein component. Actin is the main constituent in microfilaments, which are also are called actin filaments. Microtubules are formed by tubulin. The major component of intermediate filaments depends on the cell type and where in the cell they are located (Table 7.1). Intermediate filaments in epithelial cell as well as skin and hair are built by keratins, while vimentin is the building block in fibroblasts, mesenchymal and endothelial cells. Intermediate filaments of lamins differ as these are present in the nucleus, while all other intermediate filaments are cytoplasmic.

Intermediate filaments form extensive networks in the cytoplasm of most cells from multicellular organisms. It is not clear whether intermediate filaments also are present in unicellular eukaryotes. Perhaps the most important function of intermediate filaments is to provide mechanical support for the plasma membranes, particularly at contacts with other cells and with the extracellular matrix. For instance, keratin intermediate filaments in epithelial cells interact with desmosomes at cell–cell contacts and with hemidesmosomes at contacts between cells and the underlaying basal membrane. The connection between intermediate filaments in one cell and intermediate filaments in neighbouring cells through the desmosome provides strength and rigidity to the entire epithelium. Similarly, the nuclear membrane is mechanically supported by a lattice of cross-linked lamin A and lamin C, which is connected to the inner nuclear membrane through interactions between lamin B and a lamin B receptor.

The structural role of intermediate filaments implies that they are rather static. However, intermediate filaments are dynamic, though much less so than microtubules and microfilaments. The assembly and disassembly of intermediate filaments in the cell are affected by phosphorylation of intermediate filament proteins. Phosphorylation of the nuclear lamins induces the required disassembly of the nuclear lamina and breakdown of the nuclear envelope during mitosis.

Table 7.1: Intermediate filament.

Type	Protein	Human genes	Size (kDa)	Location
I	Acidic keratins	32[a]	40–60	Epithelial cells, hair, skin
II	Neutral and basic keratins	35[b]	50–70	Epithelial cells, hair, skin
III	Vimentin	1	54	Fibroblast, white blood cells
	Desmin	1	53	Muscle cells
	Glial fibrillary acidic protein (GFAP)	1	51	Glial cells
	Peipherin	1	57	Peripheral neurons
IV	Neurofilament (L,M,H)	1 of each	67–200	Neurons
V	Lamins (A,B1,B2,C)	1 of each	60–75	Nucleus lamina
VI	Nestin	1	200	Stem cells of central nervous system

[a]4 pseudogenes.
[b]8 pseudogenes.

The structural organization of all intermediate filament proteins is similar even though they differ in size and amino acid sequence (Figure 7.9). They all have a ~45 nm long central α-helical rod-like domain of ~310 amino acids. The rod domain is flanked on the N- and C-terminals by nonhelical or disordered head and tail domains. The head and tail domains of the intermediate filament proteins vary in size, amino acid sequence and structure. The rod domain is highly conserved among all intermediate filament proteins and consists of four long helical regions separated by three short nonhelical regions.

During filament assembly, the rod domain of two polypeptide chains coil around each other forming a parallel coiled-coil dimer. Two dimers associate in an antiparallel manner forming a staggered tetramer. These tetramers associate laterally and form 58 nm long and 16 nm wide octamers of tetramers, also called unit length filaments. Unit length filaments then associate laterally to form very long filaments that undergo radial compaction, which reduces the diameter to 10 nm.

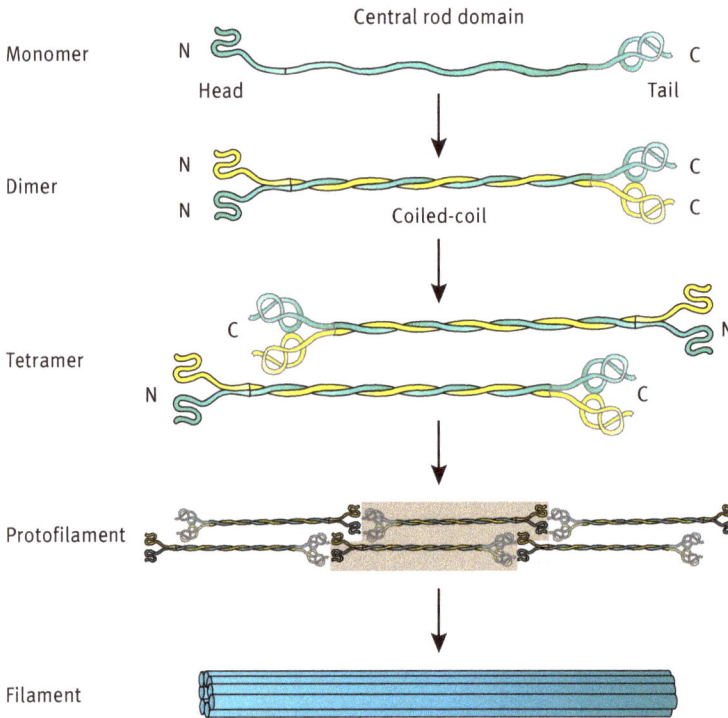

Figure 7.9: Assembly of intermediate filament. Assembly begins with the formation of a parallel coiled-coil dimer. Two dimers associate in an antiparallel and staggered manner forming a tetramer. Protofilaments are formed when tetramers associate in an end-to-end manner. In the final intermediate filament around eight protofilaments wound around each other in a rope-like fashion.

Due to the antiparallel tetramer, there is no directionality in intermediate filaments; both "ends" are equivalent. The distinct properties of each type of intermediate filaments are probably brought about by the varying head and tail domains of the intermediate filament proteins.

Keratins always form intermediate filaments based on coiled-coil heterodimers built by one type I keratin and one type II keratin. Vimentin and desmin, on the other hand, can form homodimers as well as heterodimers but never with keratin.

7.5 Microtubules

The beating of eukaryotic cilia and flagella and segregation of chromosomes during mitosis depend on microtubules. Microtubules also provide tracks for cargo transport from the cell center toward the periphery and back. In addition, microtubules organize the intracellular infrastructure by positioning organelles, such as the endoplasmic reticulum, the Golgi apparatus and mitochondria. It should be noted that bacterial flagella have a completely different structure and do not contain any microtubules.

The internal structure and function of motile cilia and flagella are more or less the same. Cells have many (>100) short cilia whereas there are usually only a few flagella (<5). The force generating structures in cilia and flagella are the axoneme. It consists of nine fused microtubule doublets that surround two single microtubules and several other proteins that link the microtubules to each other (Figure 7.10). Nexin connects the outer doublets to each other and radial spokes connect the outer ring of microtubules with the central singlets. The axoneme emanates from the basal body and pushes the plasma membrane outward. The basal body has a different structure as it lacks the central singlets and has nine triplet microtubules instead of doublets in the outer ring. The beating movement is caused by the motor protein dynein "climbing" on the doublet microtubules.

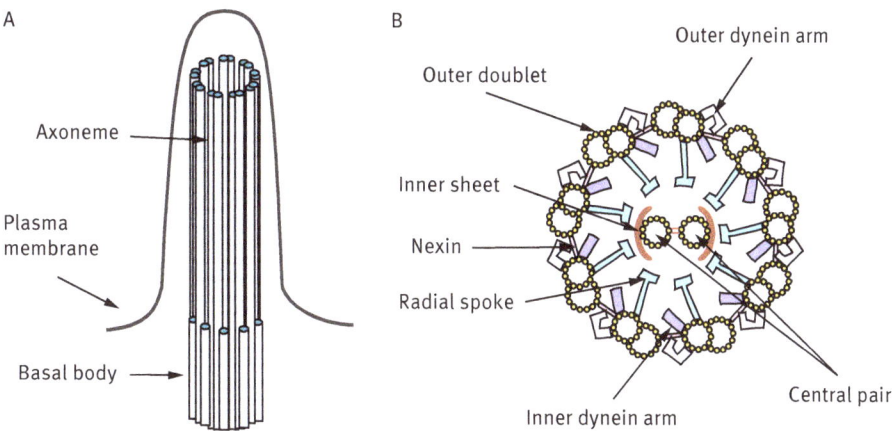

Figure 7.10: Schematic illustration of cilia. (A) A cilium surrounded by the plasma membrane. The axoneme emanates from the basal body. (B) Cross section of a 9 + 2 cilium showing the major proteins in the axoneme. Adapted with permission after C.B. Lindemann (2007) The geometric clutch as a working hypothesis for future research on cilia and flagella. Ann N Y Acad Sci 1101:477–493.

One of the most prominent functions of motile (secondary) cilia is to clean our airways. The constant beating of cilia on the surface of the epithelial cells lining the respiratory tract moves mucos from the lungs to the throat. The beating of cilia as well as flagella brings mobility to many unicellular organisms. However, cilia have other functions. Primary cilia lacking the central pair of microtubules are nonmotile and function as sensor receptors. It has also become evident that cilia are involved in development and have a role in some types of memory.

Cytoplasmic microtubules are organized differently. They usually emanate from a microtubule-organization center (MTOC). The MTOC contains a γ-tubulin ring complex and many other proteins. In animal cells the centrosome constitutes the MTOC and consists of 50 or so γ-tubulin ring complexes. The ring complex is believed to function as a nucleation template for the formation of a microtubule with 13 protofilaments.

Microtubules are assembled from globular heterodimers of α- and β-tubulins (Figure 7.11). The heterodimers associate or polymerize in a head-to-tail manner forming a protofilament. The protofilament has a polarity as at one end α-tubulin is exposed and the other end has a β-tubulin exposed; these ends are usually designated – (minus) and + (plus), respectively. The rigid and hollow 25 nm microtubule is then formed by

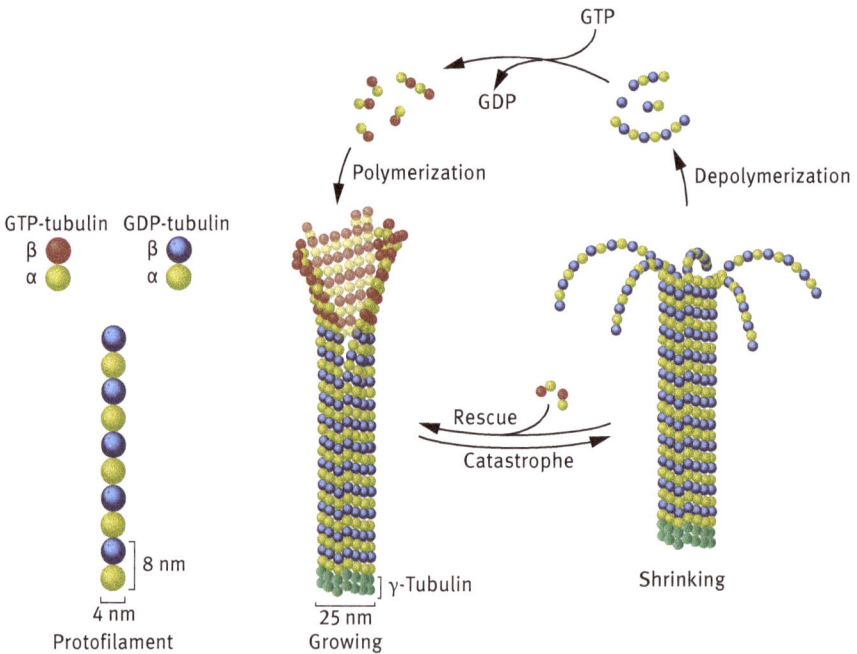

Figure 7.11: Dynamic instability of microtubules. Microtubules grow from microtubule-organization center (MTOC). MTOC contains several rings of γ-tubulin and other proteins. During growth, GTP-loaded α/β-tubulin dimers are added to the plus end of the growing microtubule. Eventually the terminal sheet closes and generates metastable blunt-ended microtubule that can pause, continue to grow or shrink.

lateral association of 13 protofilaments. The microtubule grows at the plus end by the addition of an α/β-dimer. Both α-tubulin and β-tubulin contain a bound guanine triphosphate (GTP) molecule. β-Tubulin with GTP forms a cap, which stabilizes the plus end from depolymerization and allows for continuous growth. However, with time the GTP bound to β-tubulin (but not the GTP bound to α-tubulin) hydrolyzes and forms guanine diphosphate (GDP). The hydrolysis induces a ~5° bend in the protofilaments and leads to depolymerization (also called catastrophe) of the microtubule. The shrinking microtubule continues to shrink until it is rescued by the addition of GTP-containing dimers. This process of alternating growth and shrinkage is called dynamic instability. After depolymerization, the bound GDP must be exchange for GTP before the dimer is ready to be added to the growing microtubule.

Microtubules in cilia and flagella are stable and do not undergo polymerization/depolymerization cycles as the ends of the axoneme are capped by the basal body at one end and by associated proteins at the tip of the cilia that prevent depolymerization.

The human genome contains 26 genes coding for tubulins. There are 12 α-tubulin genes (2 pseudogenes), 10 β-tubulin genes (1 pseudogene), 2 γ-tubulin genes and one gene each coding for δ- and ε-tubulin. There is also a zeta-tubulin present in some organisms but not in humans.

Tubulins are ~50–53 kDa proteins, capable of binding GTP and to catalyze hydrolysis of the bound GTP to GDP. The structure of the α/β-tubulin dimer has been inferred from crystallographic analysis of tubulin sheets in the presence of zinc ions and taxol. Due to the propensity of the α/β-tubulin dimer to polymerize and form microtubules, it has not been possible to determine the structure of the dimer directly neither of the individual α- and β-tubulins (Figure 7.12).

The structures of α- and β-tubulin are nearly identical, even though the sequence identity is only around 40% (Figure 7.12A). Certain regions are better conserved than others. For instance, the GTP-binding signature sequence, GGGTG(T/S)G, is conserved as well as other residues involved in the binding of GTP. CATH classifies tubulins as members of the α/β class. In the structure, two β-sheets, one with five parallel β-strands close to the N-terminal and one with four antiparallel β-strands close to the C-terminal, are surrounded by α-helices. The database Superfamily (as well as other web resources) identifies the N-terminal domain (the first ca. 245 residues) as a GTPase domain. The fold of the N-terminal half also resembles a Rossman fold, a motif known to bind dinucleotides, such as nicotine adenine dinucleotide (NAD^+). In both tubulins, the nucleotide is bound in a crevice, lined with several residues involved in the attachment of GTP or GDP.

The C-terminal part contains binding sites for taxol, a molecule that stabilizes microtubules from depolymerization, and sites for kinesin and other motor proteins.

The dynamic instability of microtubules is essential for correct cell division. Microtubule growth and shrinkage is involved in several of the steps a dividing cell undergoes. If the dynamic instability is prevented, cell division is also prevented. Therefore, drugs, such as taxol, that stabilize microtubules as well as drugs that cause

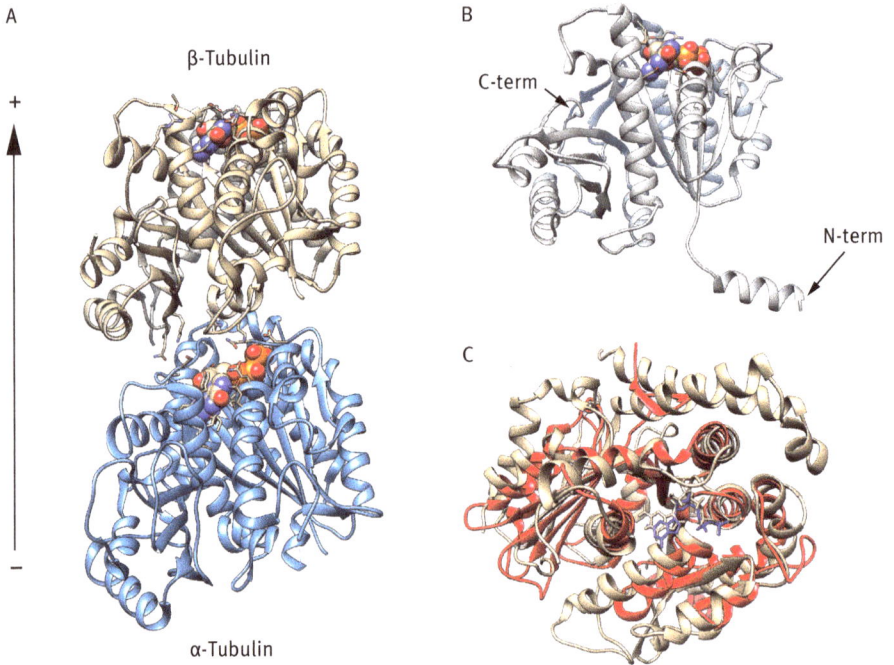

Figure 7.12: Structure of tubulin and FtsZ. (A) The α/β-tubulin dimer with GTP and GDP (pdb: 5EYP). α-Tubulin (blue) and β-tubulin (sand) with GTP and GDP, respectively. The arrow indicates direction of growth. Next α/β-tubulin dimer will be added to the plus end. B) *Methanocaldococcus jannaschii* FtsZ with GDP (pdb: 1FSZ). Notice the long extension at the N-terminal. (C) The structure of β-tubulin (sand) with GDP (sand) was superimposed with *M. jannaschii* FtsZ (red) with GDP (dark blue).

depolymerization have turned out to be efficient anticancer drugs by reducing growth of certain tumors.

Bacterial cell division, and in particular cytokinesis, is dependent on a protein called FtsZ. During cell division, FtsZ molecules assemble into a ring structure at the middle of the cell. After the newly replicated chromosomes have moved to opposite ends of the cell, the FtsZ ring constricts the cell and recruits several other proteins required for the final separation into two cells.

Structural analysis has shown that FtsZ and tubulins are homologous (Figure 7.12B). Although the sequence identity between FtsZ and tubulin is less than 20%, their folded structures are very similar. Like tubulin, the ~40 kDa FtsZ binds GTP, has GTPase activity and can assemble into structures that are similar to tubulin protofilaments, sheets and tubes. FtsZ and FtsZ-like proteins consist of four domains: a variable N-terminal domain, the core domain with a site for GTP-binding, a spacer domain (that is short in FtsZ but can be up to 300 residues in orthologous) and a C-terminal domain that interacts with other proteins. The most visible structural difference between tubulin and FtsZ is the N-terminal extension in FtsZ and the longer C-terminal domain in tubulin (Figure 7.12C).

7.6 Actin filaments

Similar to tubulin, actin is present in all eukaryotes and remarkably conserved; the identity between most sequences is more than 90%. One important function of actin in any organism is to create movement, either of the whole organism or of cells within the organism.

The human genome contains six genes for actin coding for three main groups of actin isoforms, α- β- and γ-actin. The α-actins are found in muscle tissue, whereas β- and γ-actins are present in most cell types. As in other organisms, all three isoforms are essential for all sorts of movement, whether it is due to muscular contraction or cell motility.

The structure of the ~42 kDa actin can be divided into four subdomains (Figure 7.13). Subdomains 1 and 3 are structurally related, which indicate that they may have emerged due to a gene duplication. There are two clefts between the subdomains. The upper cleft constitutes the binding site for adenosine triphosphate (ATP) and the ATP-bound divalent magnesium cation (Mg^{2+}). The nucleotide also provides a stabilizing linkage between the subdomains 2 and 4. The lower cleft between subdomains 1 and 3

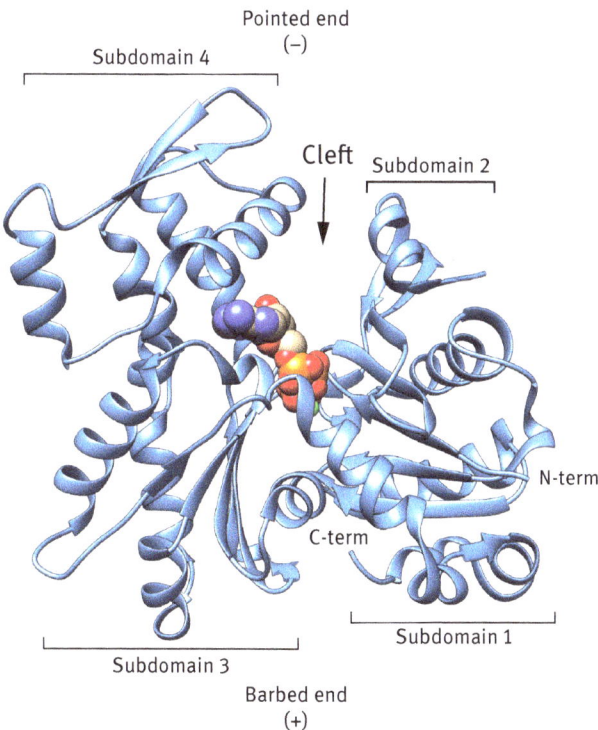

Figure 7.13: Structure of actin. The structure was determined with ATP and a calcium ion in the nucleotide-binding cleft (pdb: 3HBT). The space-filling model in the middle of the actin structure is ATP. The subdomains and the barbed and pointed ends are indicated.

is lined with mostly nonpolar residues and is important for interactions with several actin-binding proteins (ABPs). The lower cleft is also involved in the longitudinal contacts between actin monomers in the filaments.

In the presence of millimolar concentrations of divalent cations (usually Mg^{2+} and K^+), globular actin (G-actin) nucleates formation of actin filaments (F-actin). However, nucleation occurs only when the concentration is above a certain concentration. Below this concentration (the critical concentration) the actin concentration is too low to initiate polymerization. G-actin adds preferentially to the barbed or + (plus) end, although the filament is also elongated at the pointed or – (minus) end, albeit at a slower rate. At high elongation rates, the filament ends will contain actin monomers with bound ATP. However, with time the ATPase activity of actin will hydrolyze the bound ATP to adenosine diphosphate (ADP) and inorganic phosphate (P_i). When P_i is released from the actin monomer in the filament, it induces a conformational change that destabilizes the filament and the monomer dissociates from the end. At steady-state (when elongation and shortening rates are the same), the barbed end contains monomers with either ATP or ADP–P_i, which stabilizes the filament, whereas monomers at the pointed have released the P_i and therefore dissociate. Thus, at steady-state, the actin filament polymerizes at the barbed end and depolymerizes at the pointed end (Figure 7.14). This process is known as treadmilling.

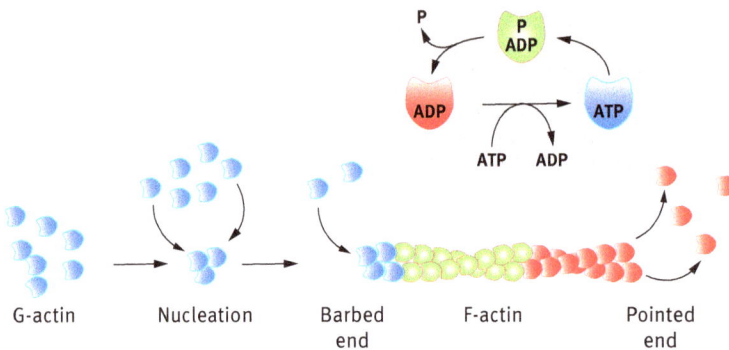

Figure 7.14: Actin polymerization. In the presence of Mg^{2+} and K^+, ATP-actin nucleates formation of actin filaments. The nucleus elongates by the addition of ATP-actin monomers. Elongation occurs preferentially at the barbed end, although actin monomers are added at the pointed end (though at a slower rate). The stability of the filament depends on the bound nucleotide; as long as ATP or ADP and P_i are bound the filament is stable. However, when P_i is released, a conformation change occurs that destabilizes the filament and ADP-actin dissociates from the pointed end. ADP needs to be exchanged for ATP before it can participate in the elongation again.

Actin and the actin cytoskeleton are essential for many, if not all, cellular processes. In addition to contractility and cell motility, the actin cytoskeleton is crucial for cell morphology and polarity, endocytosis, intracellular trafficking and cell division as well as many other functions. The state of actin and the actin cytoskeleton are regulated by numerous actin-binding proteins (ABPs). Some ABPs sequester actin from the pool of actin monomers available for filament formation, whereas others cap the end of the filament and inhibit further elongation (or dissociation). Other ABPs stabilize filamentous actin by binding along the length of the filament. Yet others nucleate formation of new filaments *de novo* or from the side of existing filaments. Some of the various functions of ABPs are depicted in Figure 7.15.

The actin cytoskeleton is highly dynamic and continuously remodeled, due to external as well as internal signals. A quick response to a signal requires access to polymerizable actin (ATP-G-actin). Profilin (~15 kDa) and thymosin (~5 kDa) are two proteins that bind actin and maintain around 50% of the cytoplasmic actin in a non-muscle cell in a monomeric but polymerization ready state. Without these sequestering proteins nearly all actin would be present as filamentous actin due to the intracellular salt concentration.

Profilin binds ATP- and ADP-actin with nearly equal affinity and also enhances the exchange of ADP for ATP. ATP-actin bound to profilin can elongate the barbed end, with nearly the same rate as free ATP-actin, but not the pointed end. In contrast to profilin, thymosin binds ATP-actin in preference to ADP-actin and cannot elongate either end of the actin filament.

Depolymerization of filamentous actin releases ADP-actin that binds preferably to profilin due to profilin's higher affinity. When bound to profilin, ADP is exchanged for ATP and actin becomes polymerization competent again. As the affinity for ATP-actin is similar, profilin and thymosin compete for ATP-actin. During active actin polymerization, the concentration of actin-free profilin is high and profil therefore snitch ATP-actin from thymosin and replenish the pool of polymerizable actin.

Cofilin (~19 kDa) and cyclase-associated protein (CAP, ~52 kDa) cooperates to increase actin dynamics. Cofilin increases the rate of ATP hydrolysis. At low concentrations, cofilin induces severing of actin filaments. This leads to dissociation of ADP-actin from the pointed end of the filament. Cofilin is then recycled when ADP-actin is transferred to CAP. Upon exchange of ADP for ATP, profilin binds the actin monomer and the profilin-actin complex can then elongate the barbed end of the actin filament. Together, cofilin, CAP and profilin as well as other ABPs regulate the process of treadmilling (Figure 7.15).

In order to nucleate filament formation, an actin dimer or, more likely a trimer must be formed. However, due to the instability of the dimer and trimer, this nucleation core must be stabilized. There are several proteins that are able to nucleate polymerization. One such group of proteins are the formins . It is believed that formin homodimers nucleate filament formation by stabilizing the dimer or trimer by binding to the barbed end of actin through the C-terminal formin homology 2 (FH2) domain. Although bound to the barbed end, the formin-stabilized nucleation core, can then be elongated by profilin-bound actin. Other proteins like spire and cordon blue (~150 kDa) contain tandem WASP-homology 2 (WH2) domains that bind actin monomers and thereby creates a nucleation core.

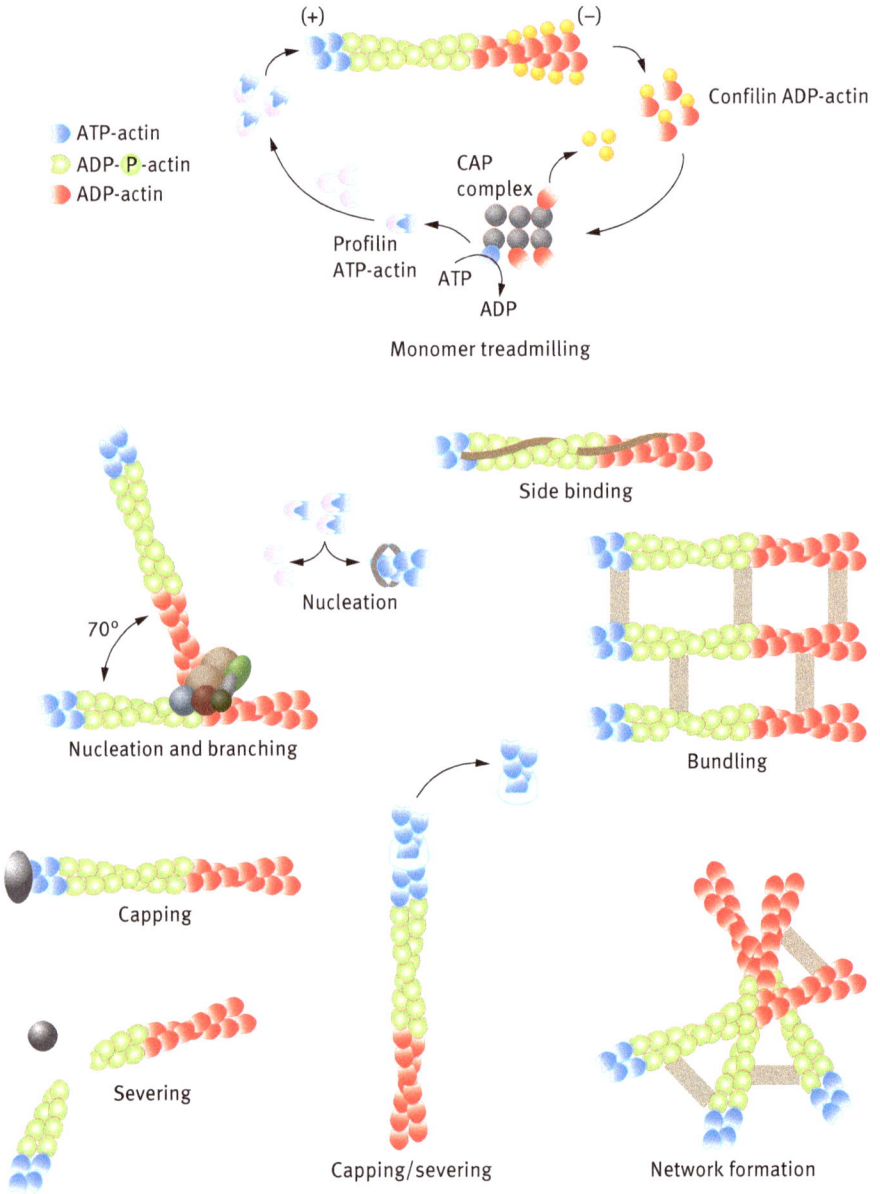

Figure 7.15: Actin-binding proteins (ABPs). Actin and the actin cytoskeleton are regulated by several ABPs with different functions. Adapted with permission after J. Baum, A.T. Papenfuss, B. Baum, T.P. Speed and A.F. Cowman (2006) Regulation of apicomplexan actin-based motility. *Nature Rev Microbiol* 4:621–628.

The actin-related protein complex (ARP2/3) nucleates filament branching. The complex contains ARP2 and 3, with structures that are similar to monomeric actin, and five other subunits (ARPC1-5). Upon activation, the complex binds to the side of an actin filament and nucleates a filament branch by anchoring the pointed end to the filament and leaving the barbed end open for elongation. The ARP2/3 is activated by a several nucleation-promoting factors, such as Wiskott-Aldrich syndrome protein (WASP) and suppressor of cyclic AMP repressor (SCAR) as well as several other factors.

Severing is another way to enhance filament formation, as more ends will be available upon severing. Cofilin and gelsolin are two proteins with severing activity and therefore can enhance polymerization. Severing creates more ends, barbed ends that can elongate further and pointed ends that will dissociate. Gelsolin has, in addition to the severing activity, a barbed end capping activity that blocks elongation at the barbed end. However, calcium ions are required to activate and stabilize gelsolin in a conformation that can bind to actin filaments.

There are several ABPs, that similar to gelsolin, bind to either end of an actin filament and thereby may block addition or dissociation of actin monomers from that end. The heterodimeric actin capping protein (CP or CapZ, ~64 kDa) is a ubiquitous and abundant barbed end capper. The activity of CP is regulated by proteins that bind to CP and lowers the affinity for the barbed end, like CARMIL (capping protein, ARP2/3 and myosin I linker) but also by formins and other proteins that compete with CP for binding to the barbed end. The capping activity of CP is also inhibited by anionic phospholipids, in particular by phosphatidylinositol 4,5-bisphosphate.

Several actin-binding proteins can cross-link actin filaments into bundles or loose network. Many of these cross-linkers have an actin-binding domain (ABD), consisting of two calponin homology (CH) domains. Fimbrin (~70 kDa, mammalian fimbrin is also known as plastin) contains two ABDs close to each other. Therefore, fimbrin cross-linking results therefore in dense bundles of actin filaments. The cross-linking activity of some fimbrin and plastin isoforms are downregulated by low concentrations of calcium ions.

Tropomodulin (~40 kDa) binds to the pointed end of an actin filament, particularly to filaments coated by tropomyosin. When bound, tropomodulin blocks the spontaneous release of ADP-actin monomers from the pointed end. In the presence of CP and tropomodulin, the length distribution of actin filaments is stabilized, which is of utmost importance in striated muscle.

Due to their ability to dimerize, filamin (~280 kDa) and α-actinin (~102 kDa) also cross-link actin filaments. Monomeric filamin and α-actinin have a single ABD at the N-terminal end, whereas upon dimerization both will have two binding sites for actin and thus able to cross-link filaments (Figure 7.16). In the antiparallel α-actinin dimer, the ABD at each end of the dimer is not as close to each other as in fimbrin. Therefore, bundles cross-linked by α-actinin will not be as dense as those cross-linked by fimbrin. Humans have four α-actinin genes, with distinct tissue expression. The calcium-sensitive α-actinin 1 and 4 are expressed in nonmuscle tissue whereas the calcium-insensitive α-actinin 2 and 3 are expressed in muscle cells. Similar to fimbrin, the actin-binding and thus the cross-linking activity is reduced by calcium ions.

Due to the further distance between the ABDs in filamin, therefore, filamin cross-linking leads to network formation instead of bundles. There are also cross-linking pro-

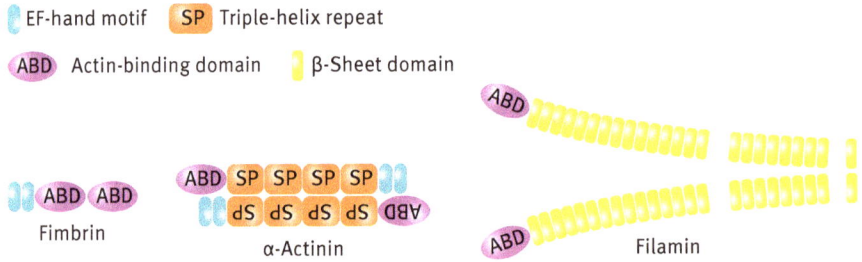

Figure 7.16: Actin filament cross-linking proteins. The domain structures of fimbrin, α-actinin and filamin are illustrated.

teins that do not depend on binding through CH domains. Fascin (~55 kDa) is an example of such a protein. In fascin, two of the four β-trefoil domains bind actin. These two sites are ~5 nm apart, which can be compared with the ~40 nm distance between the ABD in α-actinin. Phosphorylation of a serine residue by protein kinase C in the presumed binding site inhibits the bundling activity of fascin.

In addition to nucleators, cappers and cross-linkers, there are proteins that bind along the side of actin filaments. The coiled-coil protein tropomyosin is one such example. Parallel tropomyosin dimers polymerize end-to-end and bind to the two grooves on actin filaments. This provides structural stability to the filament and protection from severing proteins such as cofilin. In striated muscle, troponin transmits the calcium-signal to tropomyosin that leads to muscle contraction.

Further reading

Andreu, J.M., Ruiz, F.M. and Fernández-Tornero, C. (2023). Conserved GTPase mechanism in bacterial FtsZ and archaeal tubulin filaments. *FEBS J* 290:3527–3532.

Carman, P.J., Barrie, K.R., Rebowski, G. and Dominguez, R. (2023). Structures of the free and capped ends of the actin filament. *Science* 380:1287–1292.

Chothia, C., Levitt, M. and Richardson, D. (1981). Helix to helix packing in proteins. *J Mol Biol* 145:215–250.

Crick, F. (1953). The packing of a-helices – Simple coiled-coils. *Acta Cryst* 6:689–697.

Erickson, H.P. (1995). Ftsz, a prokaryotic homolog of tubulin?. *Cell* 80:367–370.

Etienne-Manneville, S. (2018). Cytoplasmic intermediate filaments in cell biology. *Annu Rev Cell Dev Biol* 34:1–28.

Herrmann, H. and Aebi, U. (2016). Intermediate filaments: Structure and assembly. *Cold Spring Harb Perspect Biol* 8(11), a018242.

Holmes, D.F., Lu, Y., Starborg, T. and Kadler, K.E. (2018). Collagen fibril assembly and function. *Curr Top Dev Biol* 130:107–142.

Kadler, K.E., Baldock, C., Bella, J. and Boot-Handford, R.P. (2007). Collagens at a glance. *J Cell Sci* 120:1955–1958.

Ozsvar, J., Yang, C., Cain, S.A., Baldock, C., Tarakanova, A. and Weiss, A.S. (2021). Tropoelastin and elastin assembly. *Front Bioeng Biotechnol* 9:643110.

Pankov, R. and Yamada, K.M. (2002). Fibronectin at a glance. *J Cell Sci* 115:3861–3863.

Perrin, B.J. and Ervasti, J.M. (2010). The actin gene family: Function follows isoform. *Cytoskeleton (Hoboken)* 67:630–634.

Pollard, T.D. (2016). Actin and actin-binding proteins. *Cold Spring Harb Perspect Biol* 8(8), a018226.

Ricard-Blum, S. (2011). The collagen family. *Cold Spring Harb Perspect Biol* 3:a004978.

8 The midfield

The membrane constitutes a distinct barrier between an inside and an outside. The plasma membrane demarcates the intracellular space from the extracellular space and makes it possible to have a different solute composition inside the cell. For instance, the intracellular concentration of potassium ions is higher than the extracellular concentration, whereas the opposite is true for sodium ions.

8.1 Lipid bilayers

All membranes, irrespective of whether they encompass bacteria, eukaryotic cells or cell organelles, are constructed in the same way from lipids and proteins. The amphiphatic lipids, with polar head groups and hydrophobic tails, form spontaneously bilayers in water, due to the hydrophobic effect, with the polar head groups facing the solvent (Figure 8.1). Phospholipids, glycolipids and sterols are the major classes of membrane lipids. Both phospholipids and glycolipids consist of two fatty acid chains ("the tails") linked to glycerol by an ester bond and either a phosphate or sugar moiety ("the head"). One of the fatty acid chains often contains a double bond in a *cis* conformation, which creates a kink in the chain.

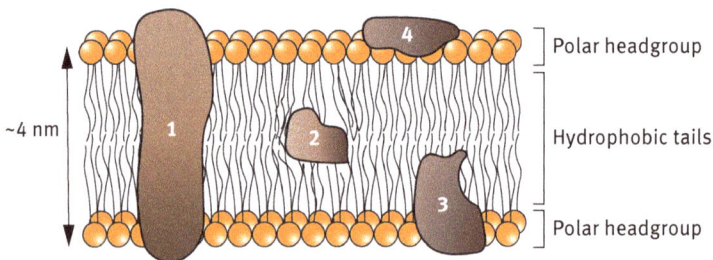

Figure 8.1: A biological membrane. The bilayer is formed spontaneously by the lipids, in such a way that the polar head groups face the surrounding water and the nonpolar hydrocarbon tails of the lipids are placed inside the bilayer, avoiding contact with the polar solvent. Therefore, the surface on both sides of the membrane is hydrophilic, whereas the interior is hydrophobic. Proteins inserted in the membrane can span the entire membrane and protrude on both sides as an integral transmembrane protein does (1). Other types of membrane proteins can be fully inserted (2) or partly (3) inserted in the membrane. Peripheral membrane proteins (4) are attached to the surface of the membrane through interactions with the head groups of the lipids or by fatty acids covalent linked to the protein.

Common phospholipids (or rather glycerophospholipids) are phosphatidylcholine, phosphatidylethanolamine and phosphatidylserin that have a choline, ethanolamine or serine molecule attached to the phosphate group, respectively. Sphingolipids are based on the amino alcohol sphingosine. As sphingosine contains a long-chain hydrophobic

https://doi.org/10.1515/9783111350684-008

tail, only a single fatty acid is required to create a lipid with two tails. An important sphingolipid is sphingomyelin, which contains a phosphocholine group. Glycolipids always have a sugar group in place of the phosphate group present in phospholipids (Figure 8.2).

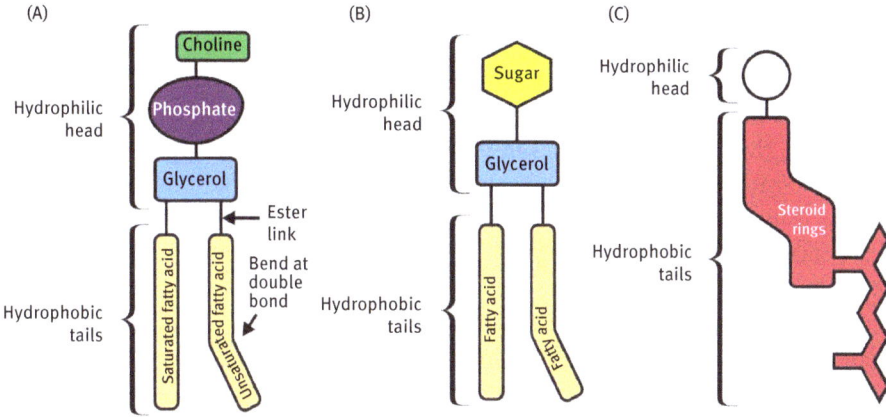

Figure 8.2: Schematic representation of the major types of membrane lipids. (a) The glycerophospholipid phosphatidylcholine. (b) A glyceroglycolipid. (c) A typical sterol. Reprinted with permission from H. Watson (2015) Biological membranes. *Essays Biochem* 59:43–70 (https.//doi.org/10.1042/bse0590043).

Sterols, such as cholesterol and stigmasterol, are important components of animal and plant membranes. Cholesterol and stigmasterol are very rigid and nonpolar compounds. When inserted in membranes, they decrease the fluidity and increase the rigidity of the membrane. The polar "head" of cholesterol is much smaller than those of phospholipids, being only a hydroxyl group.

Bacterial membranes do not contain sterols but instead hopanoids. The structure of the pentacyclic hopanoids is similar to that of cholesterol and stigmasterol. Like cholesterol, hopanoids are rigid and mostly nonpolar and affect the bacterial membrane in the same way as cholesterol.

In addition to cholesterol, the membrane rigidity also depends on the length and saturation of the fatty acid tails. Longer hydrocarbon chains lead to more van der Waals interactions and a stiffer membrane. As a double bond induces a kink in the fatty acid tail, such fatty acids tails cannot be packed as close as fully saturated fatty acid tails. Therefore, increased unsaturation reduces the stiffness and increases the membrane fluidity.

The lipid composition differs between different tissues and organisms as well as between different membranes. Phosphatidylcholine is the major phospholipid in eukaryotes, whereas it is phosphatidylethanolamine and phosphatidylglycerol in most bacteria. The eukaryotic plasma membrane contains very little cardiolipin (diphosphatidylglycerol) in contrast to the mitochondrial inner membrane.

There is also an asymmetric distribution of lipids, particularly in the eukaryotic plasma membrane. For instance, the outer leaflet of red blood cells contains phosphatidylcholine and sphingomyelin, whereas phosphatidylethanolamine, phosphatidylserine and phosphatidylinositol are present in the inner leaflet. The so-called flippases and floppases preserve the phospholipid asymmetry. Flippases transfer phospholipids from the outer leaflet to the inner and floppases transfer phospholipids the other way, albeit at an energy cost.

The charge distribution of the eukaryotic plasma membrane is also asymmetric. As the polar head groups of phosphatidylcholine and sphingomyelin are uncharged in contrast to the polar head groups of phosphatidylserine and phosphatidylinositols that are negatively charged, the lipid asymmetry leads to a charge difference between the two leaflets. The inner leaflet being negative compared to the outer leaflet. At the same time, regions of membrane proteins protruding from the inner leaflet tend to be positive.

The lipid distribution within a membrane is not always uniform. Certain lipids tend to form microdomains or rafts in the membrane. These regions of rafts often have functional implications.

8.2 Membrane proteins

A membrane that contains only lipids can function as an effective insulator, such as myelin sheets that wrap around neurons. By insulating nerve cell axons (in the same way as an electric wire), the myelin sheet increases the speed of the nerve signal. However, most biological membranes contain proteins in addition to the lipids, which give the membrane additional functions. Membrane proteins are either inserted into the membrane or attached to the surface (Figure 8.1). An integral membrane protein is more or less inserted into the membrane. A protein that spans the whole membrane and has protrusions on both sides is an integral transmembrane membrane protein. The protrusion(s) often has a globular structure. Other types of membrane proteins can be fully inserted, without any contacts with the outside, or partly inserted in the membrane, as illustrated in Figure 8.1. Peripheral membrane proteins are attached to the surface of the membrane through interactions with the head groups of the lipids or by fatty acids covalent linked to the protein. A peripheral membrane protein is defined as a protein that can be released from the membrane by high salt concentration, whereas detergent is required to remove an integral membrane protein from the membrane.

Due to the nonpolar hydrocarbon tails of the lipids, the interior or core of the membrane is highly hydrophobic. Therefore, the part of a protein that is inserted into the membrane must be predominantly hydrophobic. In addition, all groups capable of participating in hydrogen bonds should be hydrogen bonded. As the membrane in principle is devoid of water and the hydrocarbon tails cannot participate in hydrogen bonds, hydrogen bonds can only form within a protein region or between different protein regions. An α-helix with nonpolar side chains would be inserted into the membrane without any unfavorable energy cost, as the only polar groups (the amide proton and the carboxyl oxygen) are all hydrogen bonded.

The estimated thickness of a typical biological membrane is ~4 nm, based on the head-to-tail length of the prototypical dipalmitoyl-phosphatidylcholine (~2 nm) and the hydrophobic core of the membrane is ~3 nm. Therefore, an α-helix with around 20 residues would be long enough to span the hydrophobic core of the membrane, as the rise per residue is 0.15 nm or 0.54 nm per turn.

To insert a β-strand is more problematic, as backbone hydrogen bonds form between different β-strands and not within a single β-strand. However, a β-sheet that coils into a β-barrel, where the first strands is hydrogen bonded to the last strand would fulfill the hydrogen bonding requirement. If every second residue in each β-strand is nonpolar, the external surface of the β-barrel would be nonpolar and would fit energetically in the membrane.

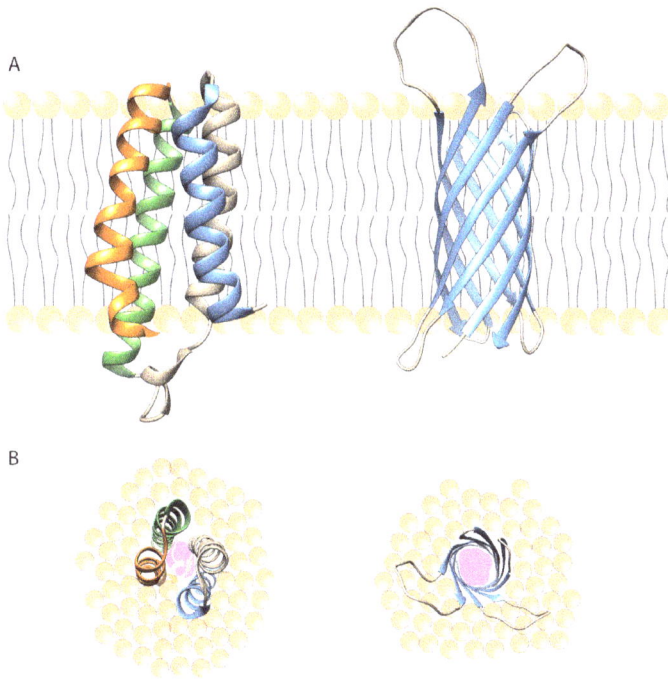

Figure 8.3: Schematic view of a bilayer with transmembrane proteins. Membrane proteins are either all α or all β. An all β protein is always present in the membrane as a β-barrel while an all α protein can have a single α-helix as well as several helices. (A) Side view of a four-helix bundle protein and a β-barrel. (B) Top view of the same protein structures. The pink areas indicate the polar inside of each structure that may function as a channel through the membrane. The four-helix bundle structure is from *Escherichia coli* cytochrome b562 (pdb: 5YM7) and the β-barrel structure is taken from *Escherichia coli* OmpX (pdb: 2M06).

As the distance between the residues in a β-strand is 0.32 (parallel strand) or 0.34 (antiparallel strand), a β-strand with around 10 residues would be long enough to protrude on each side of the nonpolar membrane core.

If the inside of the β-barrel is lined with polar side chains, it would provide a channel for molecules and ions across the membrane. Similarly, four or more amphiphatic α-helices could also form a structure with a nonpolar outside, which allows it to be inserted in the membrane, and a polar inside that could form a channel or pore through the membrane, as illustrated in Figure 8.3.

The region of a membrane protein that is in contact with the hydrophobic core is either all α or all β (using the class definition of SCOP) but never α/β or α + β. In the membrane, β-proteins are always present as β-barrels to fulfill the requirement of hydrogen bonding.

8.3 Glycophorin

The red blood cell protein glycophorin A was one of the first sequenced and characterized membrane proteins. The 131-amino acid long glycophorin A consists of three distinct domains: a 72-amino acid long extracellular N-terminal domain, a transmembrane domain and a 36-amino acid long C-terminal cytoplasmic domain (Figure 8.4). The N-terminal extracellular and the C-terminal cytoplasmic domains both contain mostly polar amino acids, whereas the transmembrane domain contains nearly exclusively nonpolar amino acids. The extracellular domain is heavily glycosylated; 15 serine or threonine side chains are *O*-glycosylated in addition to one asparagine side chain that is *N*-glycosylated. The glycosylation makes the surface of the red blood cell very hydrophilic and also gives it a negative charge. Together this results in less adhering to other cells and vessel walls.

The glycosylation patterns of glycophorins as well as other glycoproteins and glycolipids are determinants of the MN and the better known AB0 blood groupings.

It should be noted that glycosylation occurs only on residues that are located to the extracellular regions of a protein. Although *O*-glycosylation primarily occurs on serine or threonine residues, *O*-glycosylation can modify any side chain with a hydroxyl group, such as tyrosine or hydroxyproline.

The transmembrane α-helical domain consists of 23 residues, long enough to span the hydrophobic core of the membrane. As nearly all amino acid residues in this domain are nonpolar, the surface of the transmembrane domain is also nonpolar.

The C-terminal cytoplasmic domain contains many charged residues. The four basic residues (2 arginines and 2 lysines), most N-terminal of the domain, give this part a positive charge. As the cytoplasmic surface of the membrane is negative (due to the presence of phospholipids), the electrostatic interaction between protein and membrane may function as an anchor and be important for insertion of the protein into the membrane.

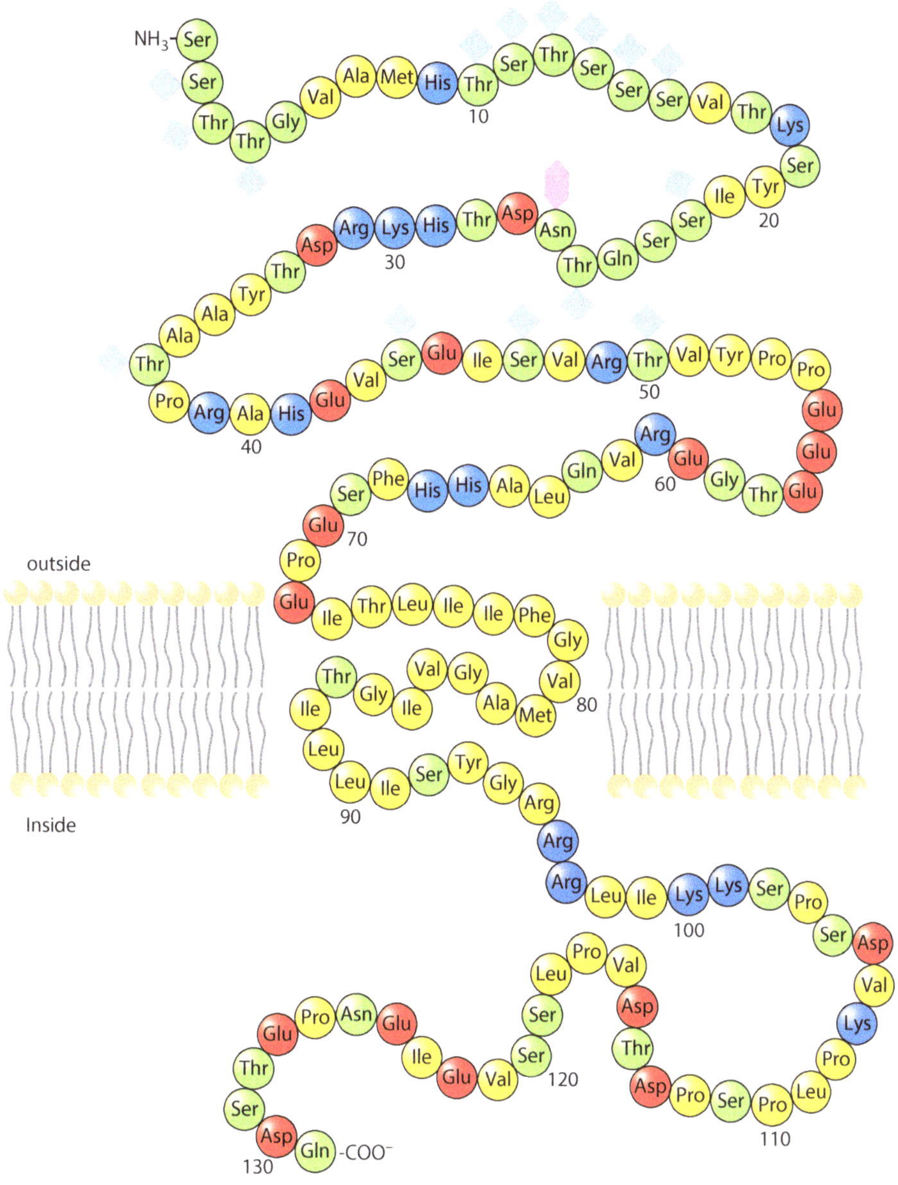

Figure 8.4: Schematic structure of human glycophorin A. The amino acid sequence is taken from the prototypic glycophorin (accession: P02724). The chemical properties of the residues are color-coded: polar and un-charged: green; polar and acidic: red; polar and basic: blue; nonpolar: yellow. Blue squares mark O-glycosylated residues and a pink diamond marks an N-glycosylated residue.

Glycophorin A is synthesized as a 150-amino acid long precursor protein. The extra 19 residues at the N-terminal constitute a localization signal, the so-called signal peptide, that directs the protein to the correct location. Before glycophorin is functional, the signal peptide is proteolytically cleaved. After insertion in the membrane, glycophorin forms a right-handed dimer, which is stabilized by hydrogen bonds and van der Waals interactions, mediated by a seven residue motif LIxxGVxxGVxxT.

8.4 The fluid mosaic membrane model

The present view of biological membranes is based on the fluid mosaic model proposed in 1972 by S. Jonathan Singer and Garth Nicolson. They suggested "that a membrane is an oriented, two-dimensional, viscous solution of amphiphatic proteins and lipids." With time the fluid mosaic model has evolved and become more complex as new knowledge has accumulated.

With the present knowledge it is evident that the membrane is asymmetric, as the lipid and protein composition of the two leaflets differ. That the membrane is fluid implies that molecules in the membrane can move, but are restricted to lateral movement in each leaflet. However, the diffusion rates differ considerably between lipids and proteins as well as between different proteins. When measured in large unilamellar vesicles, lipids move about twice (1 µm/s) as fast as proteins and the diffusion rates decreased with increasing protein density. Lipid diffusion rate in the plasma membrane, as determined in epithelial cells, appears to be around 0.2–0.3 µm/s. In a real membrane, with a protein density around 25,000 proteins per μm^2 (corresponding to an area occupancy ~30% and typical for many membranes), the diffusion rate of proteins is expected to be at least an order of magnitude slower. Protein movement can also be restricted by attachment to the intracellular cytoskeleton or via adhesion to the extracellular matrix. As diffusion is affected by the membrane fluidity (or viscosity), an increased degree of lipid saturation will reduce the diffusion rate.

The red blood cell membrane is probably the best characterized biological membrane. Phospholipids constitute ~60% of the membrane lipids and cholesterol (~30%) and glycolipids (10%) make up the remaining lipids. The phospholipids and glycolipids have an asymmetric distribution, whereas cholesterol is more or less equally distributed between the two leaflets (Table 8.1).

In addition to glycophorins, the red blood cell membrane contains many other integral proteins. Some examples of integral membrane proteins are band 3 (anion exchanger 1 or solute carrier family 4 member 1) that mediates exchange of chloride for bicarbonate, aquaporin that regulate the water content and glucose transporter 1 (GLUT1) that facilitates transport of glucose into cells. A large number of proteins are also attached to the intracellular surface of the red blood cell membrane, forming a membrane skeleton that gives support to the membrane and allows the cell to deform without hemolyzing (rupturing; Figure 8.5).

Table 8.1: Lipid composition of human red blood cells.

Lipid	Percentage of lipids in each leaflet	
	Outer leaflet	**Inner leaflet**
Phosphatidylcholine	~12.9	~3.9
Phosphatidylethanolamine	~3.2	~13.0
Phosphatidylserine	~0	~7.8
Sphingomyelin	~12.9	~2.7
Phosphoinositides[#]	~0.7	~2.9
Glycolipids	~10	~0
Cholesterol	~15	~15

[#]Phosphoinositides include phosphatidic acid, phosphatidylinositol, phosphatidylinositol-4-phosphate, phosphatidylinositol-4,5-bisphosphate. Data from J.M. Boon and B.D. Smith (2002) Chemical control of phospholipid distribution across bilayer membranes. *Med Res Rev* 22:251–281, D.L. Daleke (2003) Regulation of transbilayer plasma membrane phospholipid asymmetry. *J Lipid Res* 44:233–242 and references therein.

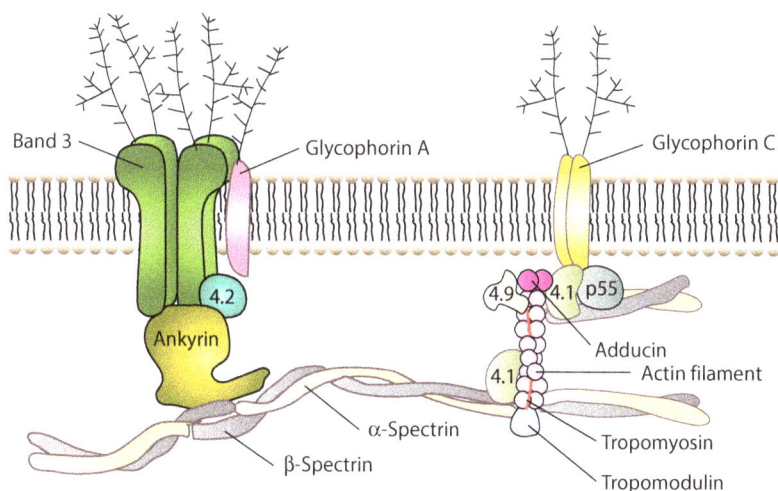

Figure 8.5: Schematic structure of the red blood cell membrane. The intracellular side of the red blood cell membrane is covered by a hexagonal protein network consisting of spectrin tetramers attached to short actin filaments and actin-binding proteins. The membrane skeleton is attached to the membrane by protein complexes interacting with the integral membrane proteins band 3 (anion exchanger) via ankyrin and with glycophorin C. Adapted with permission after R.I. Liem and P.G. Gallager (2005) Molecular mechanisms in the inherited red cell membrane disorders. *Drug Discovery Today: Disease Mechanisms* 2:539–545.

The red blood cell contains at least 2,500 (probably many more) different proteins. Of these close to 500 proteins are believed to be integral to the membrane and around 1000 proteins are associated with the membrane skeleton. Each red blood cell contains more than 1 million band 3 molecules and close to one million actin monomers. The copy numbers of several membrane skeleton proteins are listed in Table 8.2.

Table 8.2: Copy numbers of some membrane skeleton proteins.

Protein	Gene	Copies (x10^{-3})[#]	Copies (x10^{-3})[§]
Band 3 (anion exchanger 1)	SLC4A1	1074	1444
β-Actin	ACTB	908	928
Spectrin dimer	SPTA1 and SPTB	~600	~500
Ankyrin	ANK1	479	305
Protein 4.1	EPB41	350	230
Protein 4.2	EPB42	411	388
Band 4.9 (dematin)	EPB49	153	126
p55	MPP1	181	46
Tropomyosin	TPM3	253	166
Tropomodulin	TMOD1	184	57
Glycophorin A	GYPA	96	288
Glycophorin C	GYPC	~5	31

[#]Data from A.H. Bryk and J.R. Wisniewski (2017) Quantitative analysis of human red blood cell proteome. *J Proteome Res* 16:2752–2761.
[§]Data from E.-F. Gautier et al. (2018) Absolute proteome quantification of highly purified populations of circulating reticulocytes and mature erythrocytes. *Blood Adv* 2: 2646–2657.

The number of some protein molecules associated with the plasma membrane is high, such as band 3 and actin, but most proteins are present in 1,000 or less numbers. This should be compared with the most abundant red blood cell protein hemoglobin. It is estimated that there are more than 200 million copies of hemoglobin molecules per cell. It has been said that if hemoglobin were not packed in red blood cells, blood would be so thick and viscous that it would not be able to flow in our veins.

8.5 Integral membrane proteins

It is much more challenging to determine the structure of an integral membrane protein than of a soluble protein. It must be possible to release the membrane protein from the membrane in an undamaged and stable form. As a membrane protein is partly hydrophobic and partly hydrophilic, it is difficult to keep the protein in solution. These are the first obstacles on the way to a structure. Still, there are two major reasons that motivate the study of these stubborn proteins: they constitute around 30% of the human proteome and around 50% of current drug targets.

It was not until the 1980s proper conditions had been worked out to obtain crystals of membrane proteins. Soon after, the first high-resolution structure of an integral membrane protein was published. Johann Deisenhofer, Robert Huber and Hartmut Michel were able to isolate, crystallize and determine the structure of the photosynthetic reaction center of the purple bacteria *Rhodopseudomonas viridis* by X-ray crystallography.

However, 10 years before a low-resolution model of bacteriorhodopsin, a light-driven hydrogen pump from the purple bacteria *Halobacterium halobium* was presented by Richard Henderson and Nigel Unwin. The structure, determined by electron microscopy, did not display any fine details, such as position of side chains (Figure 8.6).

A   B   C

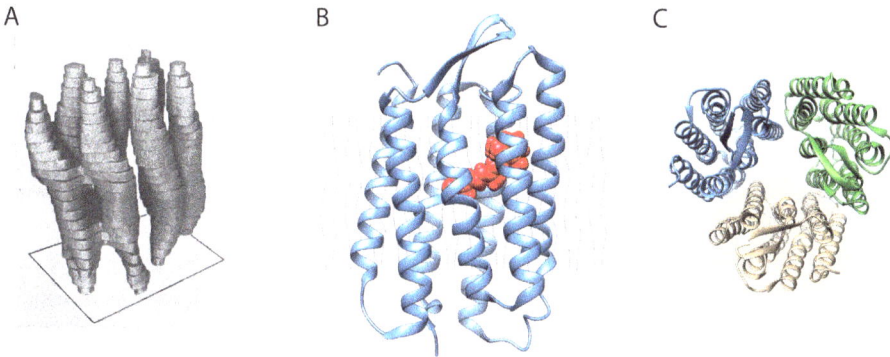

Figure 8.6: Bacteriorhodopsin. (A) The three-dimensional model of rhodopsin from the purple bacteria *Halobacterium halobium*. The model was based on a 7 Å resolution map obtained by electron microscopy. Reprinted with permission from R. Henderson and P.N. Unwin (1975) Three-dimensional model of purple membrane obtained by electron microscopy. *Nature* 257:28–32. (B) Side view of high-resolution structure (1.25 Å) of bacteriorhodopsin from *Halobacterium salinarum* (pdb: 5ZIM). The space-filling model in the structure is retinal. (C) Top view of trimeric bacteriorhodopsin.

The functional form of bacteriorhodopsin consists of three identical subunits. Each polypeptide chain folds into seven hydrophobic membrane spanning α-helices, in a typical antiparallel up-down arrangement. Short loops on each side of the membrane connect the helices to each other. The N-terminal protrudes on the extracellular face of the membrane and the C-terminal on the cytoplasmic side.

A molecule of retinal is bound to a lysine side chain (Lys216) through a Schiff base inside each subunit. It is this retinal molecule that initiates the proton "pumping" from the cytoplasm to the extracellular space. Light induces a photoisomerization that converts the covalent bound all-*trans*-retinal to 13-*cis*-retinal. This causes a conformation change in the structure and a transfer of a proton from the Shiff base to Asp85 (step 1 in Figure 8.7). At the same time a proton extrudes into the extracellular space from the so-called proton release complex that consists mainly of Arg82, Glu194, Glu204 and bound water molecules (step 2). In step 3, a proton is transferred from Asp96 to the Schiff base, followed by uptake of a proton from the cytoplasmic face

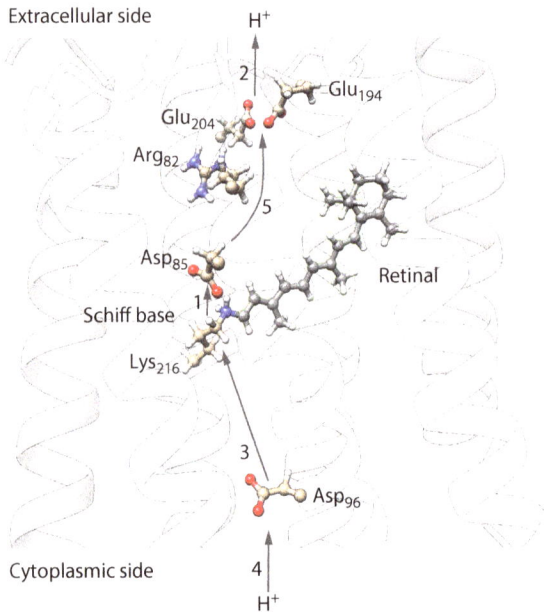

Figure 8.7: Proton pumping by bacteriorhodopsin.

and reprotonation of Asp_{96} (step 4). In the last step (step 5), a proton from Asp85 is transferred to the proton release complex.

The "pumping" of protons out of the cell creates a proton gradient that is used to produce ATP, the universal energy currency. When the protons flow back into the cell, they fuel the ATP synthase and the energy is used to produce ATP from ADP and inorganic phosphorus. In other words, light energy is converted to chemical energy that the cell can use to drive unfavorable reactions.

G-protein coupled receptors (GPCR) have a similar structure as bacteriorhodopsin, with seven slightly tilted transmembrane α-helices, although they have very different functions. One such GPRC is rhodopsin, which is found in the rods of the retina in our eyes. Similar to bacteriorhodopsin, rhodopsin contains a retinal molecule bound to a lysine residue in the center of the structure, which reacts on light. But instead of pumping protons across the membrane, rhodopsin transmits a signal that activates an intracellular associated G-protein (transducin) that initiates a cyclic guanosine monophosphate second messenger cascade. The result of the messenger cascade is that we apprehend a vision.

8.6 The hydropathy index

Great effort has been invested in finding computational methods to predict transmembrane regions as well as secondary structures. Old methods use a hydrophobic scale, based on different theoretical and practical criteria, such as an amino acids tendency to partition between polar and nonpolar environments. Newer methods rely on artificial neural networks that have been trained on membrane proteins with known structure or on hidden Markov models. Although all methods predict transmembrane regions reasonably well, the newer methods appear to be more accurate. Figure 8.8 shows the prediction of transmembrane regions of glycophorin A and bacteriorhodopsin made by the TMHMM program.

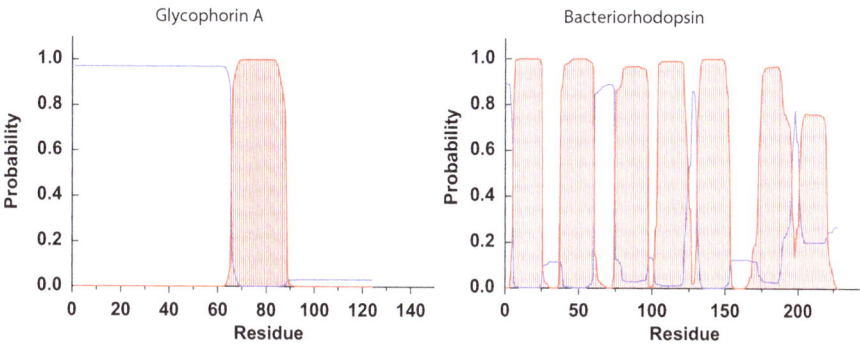

Figure 8.8: Prediction of transmembrane regions. The amino acid sequences of glycophorin A and bacteriorhodopsin were analyzed by the TMHMM program. The red areas indicate the probability of a transmembrane helix and the blue line the probability of a residue to be on the extracellular side of the membrane.

As indicated in Figure 8.8, a single transmembrane spanning helix was predicted in glycophorin A, whereas seven transmembrane helices were predicted in bacteriorhodopsin. In both cases, the predictions agree with known structures. The program also predicted the correct position (cytoplasmic or extracellular) of the termini and loops in both proteins.

8.7 The signal peptide

In eukaryotic cells proteins are produced in the cytoplasm, with the exception of the few proteins that are synthesized in the mitochondria and chloroplast. However, many proteins are not functioning in the cytoplasmic space but in the nucleus or in other cell organelles. Thus, somehow the protein must "know" where to go to fulfill its normal task. This "knowledge" is programmed in the amino acid sequence. Proteins that are supposed to be exported from the cytoplasm to be integrated into a membrane or out of the cell all have an amino acid sequence somewhere in the primary structure, an address tag or topogenic signal, that "tells" the protein where to go (Figure 8.9).

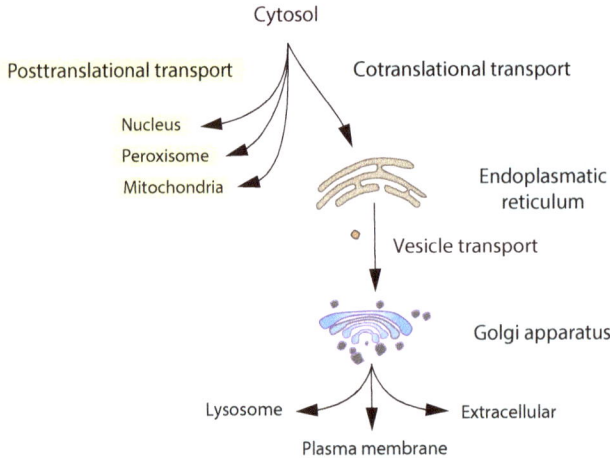

Figure 8.9: Protein transport pathways in eukaryotic cells. If the protein has an N-terminal signal peptide, the protein is exported into the endoplasmatic reticulum during the synthesis. Later, the protein is transported to the Golgi apparatus for further modifications and packaging into secretory vesicles for export to the plasma membrane or the extracellular space or for transport to the lysosome. Proteins without a signal peptide, but with other sorting sequences, take the post-translational pathway to the mitochondria, nucleus, peroxisome or chloroplast. Proteins without an address tag stay in the cytosol.

The address tag hypothesis was worked out by Günter Blobel and colleagues in the 1970s. It was discovered that proteins destined for the plasma membrane contained a signal peptide (also called a leader peptide or leader sequence) at the N-terminal. Later it was found that the "address tag" could be placed nearly anywhere in the amino acid sequence and that the tag itself was enough for proper localization.

The presence of an N-terminal signal peptide usually directs the protein to the co-translational pathway, through the endoplasmatic reticulum and Golgi apparatus for further export to the plasma membrane or extracellular space or into the lysosome.

The signal peptide contains between 13 and to up to 40 amino acid residues. The sequence is divided into three parts. At the very N-terminal there is the N-region that contains positively charged and polar residues. The central H-region is hydrophobic and α-helical, whereas the C-region, at the end of the signal peptide, adopts an extended β-strand conformation in order to facilitate recognition by the cleavage enzyme. Positions −1 and −3 in the C-region very often contain alanine residues (Figure 8.10).

The cotranslational translocation pathway consists of the signal recognition particle complex and the translocon complex. When the nascent polypeptide chain emerges from the ribosome, the signal peptide is recognized by the signal recognition particle,

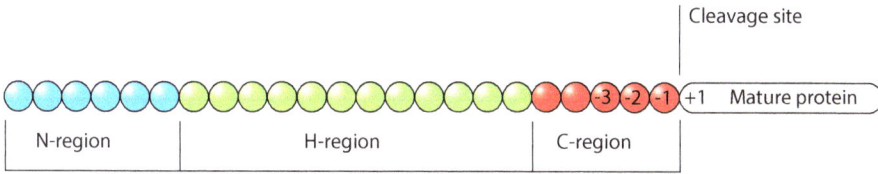

Figure 8.10: The structure of a typical signal peptide. The signal peptide consists of three domains. The N-region consists of positive charged and polar residues, the H-region is hydrophobic and helical and the C-region forms an extended b-strand to facilitate cleavage. Positions −1 and −3 usually contain an alanine residue.

which binds to the ribosome and stalls further translation. The signal recognition particle with the bound ribosome is then recruited by the translocon in the endoplasmatic reticulum membrane. When attached to the translocon, translation resumes and the polypeptide slides into the endoplasmatic reticulum. A special enzyme, a signal peptidase, removes the signal peptide from the mature protein.

After any further post-translational modifications (such hydroxylation or glycosylation), the protein proceeds from the endoplasmatic reticulum to the Golgi apparatus by a transport vesicle. In the Golgi apparatus, the final post-translational modifications take place. When completed the protein, packed in a secretory vesicle, is released from the Golgi apparatus and transported to the final destination. Some proteins contain a C-terminal KDEL (Lys-Asp-Glu-Leu) amino acid sequence that directs them back to the endoplasmatic reticulum.

Proteins using the post-translational pathway have other sorting signals than the signal peptide (Table 8.3). Proteins that are target for the nucleus have 6–20 residues long localization sequences, rich in lysine and arginine. The sequence can be present anywhere in the protein and is not removed after translocation. It has not been possible to find a consensus localization sequence among nuclear-localized proteins, although many imported proteins have an import signal containing several basic residues (Arg and Lys).

Proteins destine for the mitochondria have an N-terminal presequence that forms an amphiphatic α-helix of alternating charged and hydrophobic residues. The amphiphatic helix facilitates the translocation across the mitochondrial membranes and targets the protein for the mitochondrial matrix. Additional signal sequences are required for proteins that should be inserted in either the inner or outer mitochondrial membrane. The presequence of matrix and inner membrane proteins are removed from the mature protein. The presequence of the subunit IV of cytochrome c oxidase directs the protein to the inner mitochondrial membrane, whereas the N-terminal half of the presequence is enough to send a protein to the mitochondrial matrix.

Transit sequences target proteins for the chloroplast. These sequences vary in length (from around 30 to more than 100 residues) depending on where in the chloroplast the protein should end up.

Table 8.3: Examples of topogenic sequence.

Function of sequence	Example of sequence
Import into nucleus	-Pro-Pro-Lys-Lys-Lys-Arg-Lys-Val-[1]
Export from nucleus	-Leu-Lys-Leu-Ala-Gly-Leu-Asp-Ile-[2]
Import into mitochondria	+H3N-Met-Leu-Ser-Leu-Arg-Gln-Ser-Ile-Arg-Phe-Phe-Lys-Pro-Ala-Thr-Arg-Thr-Leu-Cys-Ser-Ser-Arg-Tyr-Leu-Leu-[3]
Import into chloroplast	+H3N-Met-Ala-Phe-Ala-Val-Ser-Val-Gln-Ser-His-Phe-Ala-Ile-Arg-Ala-Leu-Lys-Arg-Asp-His-Phe-Lys-Asn-Pro-Ser-Pro-Thr-Phe-Cys-Ser-[4]
Import into peroxisome	-Ser-Lys-Leu-COO-
Import into ER	+H3N-Met-Met-Ser-Phe-Val-Ser-Leu-Leu-Leu-Val-Gly-Ile-Leu-Phe-His-Ala-Thr-Gln-Ala-[5]
Return to ER	-Lys-Asp-Glu-Leu-COO-

[1]Subunit SPT16 of FACT complex (Q9Y5B9)
[2]cAMP-dependent protein kinase inhibitor (P61925)
[3]Subunit IV of cytochrome c oxidase (P04037)
[4]D-xylose-proton symporter-like 3 (Q0WWW9)
[5]Alpha lactalbumin (P00711)

Further reading

Backes, S. and Herrmann, J.M. (2017). Protein translocation into the intermembrane space and matrix of mitochondria: Mechanisms and driving forces. *Front Mol Biosci* 4:83.

Balashov, S.P. (2000). Protonation reactions and their coupling in bacteriorhodopsin. *Biochim Biophys Acta* 1460:75–94.

Bernardino de la Serna, J., Schutz, G.J., Eggeling, C. and Cebecauer, M. (2016). There is no simple model of the plasma membrane organization. *Front Cell Dev Biol* 4:106.

Blobel, G. (2000). Protein targeting (Nobel lecture). *Chembiochem* 1:86–102.

Bryk, A.H. and Wisniewski, J.R. (2017). Quantitative analysis of human red blood cell proteome. *J Proteome Res* 16:2752–2761.

Deisenhofer, J., Epp, O., Miki, K., Huber, R. and Michel, H. (1985). Structure of the protein subunits in the photosynthetic reaction centre of Rhodopseudomonas viridis at 3a resolution. *Nature* 318:618–624.

Kunze, M. and Berger, J. (2015). The similarity between n-terminal targeting signals for protein import into different organelles and its evolutionary relevance. *Front Physiol* 6:259.

Maraspini, R., Beutel, O. and Honigmann, A. (2018). Circle scanning sted fluorescence correlation spectroscopy to quantify membrane dynamics and compartmentalization. *Methods* 140–141:188–197.

Singer, S.J. and Nicolson, G.L. (1972). The fluid mosaic model of the structure of cell membranes. *Science* 175:720–731.

Spiro, R.G. (2002). Protein glycosylation: Nature, distribution, enzymatic formation, and disease implications of glycopeptide bonds. *Glycobiology* 12:43R–56R.

Tomita, M. and Marchesi, V.T. (1975). Amino-acid sequence and oligosaccharide attachment sites of human erythrocyte glycophorin. *Proc Natl Acad Sci U S A* 72:2964–2968.

Watson, H. (2015). Biological membranes. *Essays Biochem* 59:43–69.

9 The playmaker

Of the many thousands of reactions occurring in any cell at any time, almost all are catalyzed by enzymes. There are two major reasons for this. First of all, an enzyme increases the rate of the reaction. Without a catalyst, the reaction may be so slow that it in practice never occurs. Secondly, the enzyme-catalyzed reaction gives only one product; no unwanted side reactions occur giving the wrong products. This also makes it possible to control and regulate the flow of substances through the various reaction pathways occurring in a living cell.

9.1 A little bit of thermodynamics

The reversible conversion of a substrate S to a product P can be described by

$$S \rightleftharpoons P \tag{9.1}$$

For this reaction to proceed from S to P, it is necessary that Gibbs free energy of the product (G_P) is lower than the free energy of the substrate (G_S)

$$\Delta G_{S \to P} = G_P - G_S < 0 \tag{9.2}$$

Under the same conditions, the change in free energy for the reverse reaction (P → S) is

$$\Delta G_{P \to S} = G_S - G_P > 0 \tag{9.3}$$

As the change in free energy is positive, P would not be converted to S spontaneously.

ΔG depends on the standard free energy change and the factual concentrations of substrate and product

$$\Delta G = \Delta G^\circ + RT \cdot \ln K \tag{9.4}$$

where R is the gas constant, T the temperature in Kelvin and K the equilibrium constant. ΔG° is the standard free energy change, defined at the standard state which in biological systems is at pH 7 and all substrates and products at one molar concentration.

At equilibrium, when the forward reaction rate equals the reverse reaction rate, $\Delta G = 0$ and therefore

$$\Delta G^\circ = -RT \cdot \ln K \tag{9.5}$$

The equilibrium constant is defined as the concentration of the product divided by the concentration of the substrate:

$$K = \frac{[P]}{[S]} \tag{9.6}$$

https://doi.org/10.1515/9783111350684-009

Therefore ΔG can be expressed as

$$\Delta G = \Delta G^{\circ} + RT \cdot \ln \frac{[P]}{[S]} \tag{9.7}$$

However, even if the change in free energy is negative and large, it does not necessarily mean that the reaction will occur. There is a "barrier" that S must "climb" to transform into P. The barrier represents the activation energy ($\Delta G^{\#}$) and is the energy S must gain to transform into P (Figure 9.1).

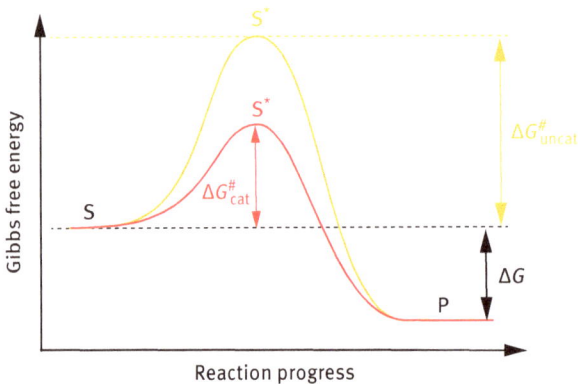

Figure 9.1: Transition state diagram of a spontaneous reaction $S \rightleftharpoons P$. ΔG is the change in Gibbs free energy, $\Delta G^{\#}_{cat}$ and $\Delta G^{\#}_{uncat}$ are the activation energy of the catalyzed and uncatalyzed reactions, respectively, and S^{*} is the transition state of the substrate. In the presence of a catalyst (red), the reaction proceeds faster, as the required activation energy is lower. However, the reaction equilibrium is not changed as the catalyst does not perturb the change in Gibbs free energy.

When S reaches the top of the activation energy barrier, something has happened to S. It is not any longer identical to S but rather in a state partly similar to the substrate and partly similar to the product. S has reached the transition state S^{*}. From the transition state, the reaction can proceed to give P or S. The probability that S^{*} converts to P or returns to S depends on the value of ΔG. A large negative value indicates a high probability of conversion to P. Transition state theory assumes that S^{*} is metastable and in rapid equilibrium with the reactants.

It is said the diamonds last forever but the Gibbs free energy change for the transformation of diamond to graphite is in fact negative and implies that a diamond would turn into dust spontaneously but we all know that that is not the case.

The reaction rate is only dependent on $\Delta G^{\#}$, not on ΔG, the higher the barrier the slower is the reaction rate. The activation energy for the reaction $C_{diamond} \rightarrow C_{graphite}$ is very large and normal thermal atomic motions are not energetic enough for the reaction to proceed. That is the reason why diamonds do not disappear.

A catalyst lowers $\Delta G^{\#}$ which means that less energy is required to climb the barrier and the reaction will proceed faster. It is important to realize that even though the reaction rate may be increased millions of times, the reaction equilibrium is not changed, as the Gibbs free energies of both substrate and product are unperturbed by the catalyst.

Another way to reduce the activation energy is to increase the free energy of the substrate by increasing the temperature. This is nearly always the means to increase the rate of a chemical reaction in a test tube. However, it is not particularly useful to increase the temperature to make a reaction to proceed faster in a human being.

If the catalyst reduces the activation energy for the reaction S → P, the same catalyst also reduces the activation energy for the reaction P → S.

9.2 The biological workhorse

For the reaction to occur, the enzyme must "find" the substrate S that will be converted to a product P and bind it to a special site, the so-called active site on the enzyme. The substrate as well as the product usually binds to the active site by non-covalent bonds (hydrogen bonds, ion-ion bonds and van der Waals interaction), although there are substrates that bind covalently. The binding of the substrate to the active site facilitates the breaking of old bonds and forming of new ones by stabilizing the transition states.

Thus, when the substrate binds to the enzyme, an enzyme-substrate complex ES forms, which proceeds to the transition state ES*. At this stage, the substrate is no longer identical with the free substrate (nor with the free product); it is in transition between substrate and product. When the reaction has occurred, the product is still bound to the active site of the enzyme (EP) and must be freed from the active site before the reaction is completed. The whole reaction scheme is described by

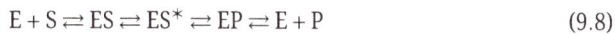

$$E + S \rightleftharpoons ES \rightleftharpoons ES^* \rightleftharpoons EP \rightleftharpoons E + P \tag{9.8}$$

The free energy of the different steps is illustrated in Figure 9.2. The binding of the substrate to the active site of the enzyme, lowers the free energy of ES compared to E + S. The more the substrate-binding lowers the free energy, the more ES is formed, but it is also contraproductive as a lowered free energy for the binding increases the activation energy for the transition state ($\Delta G_{cat}^{\#}$).

Again, if an enzyme catalyzes a reaction S → P, the very same enzyme also catalyzes the reverse reaction P → S. However, as can be seen in Figure 9.2, the activation energy for the reverse reaction is much larger ($\Delta G_{reverse}^{\#} = G_{EP} - G_{ES^*}$), implicating that the reverse reaction is much slower.

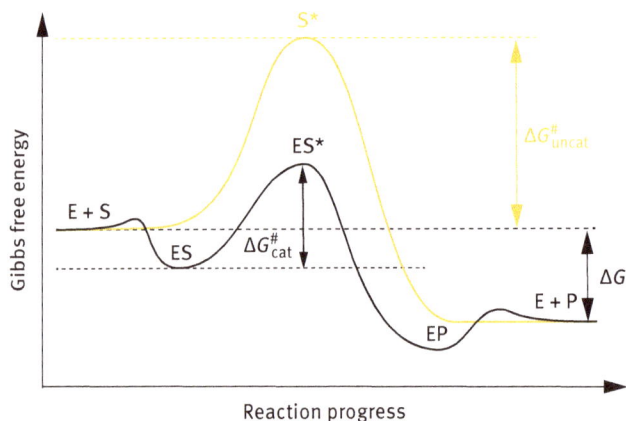

Figure 9.2: Transition state diagram of an enzyme-catalyzed reaction. ΔG is the change in Gibbs free energy for the reaction. $\Delta G^{\#}_{cat}$ and $\Delta G^{\#}_{uncat}$ are the activation energy of the enzyme catalyzed and uncatalyzed reactions, respectively. ES is the enzyme-substrate complex, ES* is the transition state of ES and EP is the enzyme-product complex.

9.3 Reaction rate

Before a reaction can take place, the substrate must not only encounter the enzyme but also find the active site. Therefore, the concentration of both enzyme and substrate will influence the reaction rate. The higher the concentration of substrate, the better is the chance a substrate molecule will find the enzyme in proper orientation and bind to the active site.

The reaction rate or velocity (v) of the irreversible conversion of S to P

$$S \xrightarrow{k_1} P \tag{9.9}$$

can be described by

$$v = \frac{d[P]}{dt} = k_1[S] \tag{9.10}$$

where k_1 is the rate constant. By convention, forward rate constants have a positive subscript, whereas reverse rate constants have a negative subscript. A rate constant is a proportionality constant and can be used to link the rate of a reaction with the concentrations of the substrates.

If the reaction were reversible

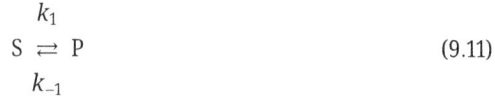

$$S \underset{k_{-1}}{\overset{k_1}{\rightleftharpoons}} P \tag{9.11}$$

and the velocity would be

$$v = \frac{d[P]}{dt} = k_1[ES] - k_{-1}[E][P] \tag{9.12}$$

Since the reverse reaction (P → S) will reduce the velocity also the reverse reaction must be considered.

Going back to the enzyme catalyzed reversible reaction S ⇌ P and including all rate constants, eq. (9.8) turns into

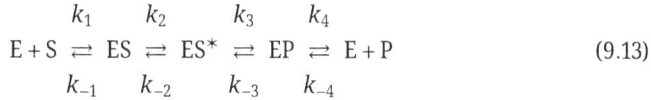

$$E + S \underset{k_{-1}}{\overset{k_1}{\rightleftharpoons}} ES \underset{k_{-2}}{\overset{k_2}{\rightleftharpoons}} ES^* \underset{k_{-3}}{\overset{k_3}{\rightleftharpoons}} EP \underset{k_{-4}}{\overset{k_4}{\rightleftharpoons}} E + P \tag{9.13}$$

To formally analyze the reaction, the concentration of all species and all rate constants should be included in the analysis. However, it is very unlikely that the concentration ES, ES* and EP can be determined. Therefore, the reaction scheme is simplified and reduced to

$$E + S \underset{k_{-1}}{\overset{k_1}{\rightleftharpoons}} ES \overset{k_{cat}}{\rightarrow} E + P \tag{9.14}$$

where k_{cat} is the rate constant for the final reaction step, the step where the product forms. k_{cat} and k_{-1} are first-order rate constants whereas k_1 is a second-order rate constant. The unit of a first-order rate constant is s^{-1} and that of a second-order rate constant is $s^{-1} \cdot mol^{-1} \cdot L$ (or $s^{-1} \cdot M^{-1}$).

Now it is possible to formulate an expression for the velocity of the enzyme-catalyzed reaction

$$v = \frac{d[P]}{dt} = k_{cat}[ES] \tag{9.15}$$

The unit of the velocity v is therefore in $mol \cdot L^{-1} \cdot s^{-1}$ ($M \cdot s^{-1}$).

In any multistep reaction, one of the steps must be rate-limiting and therefore must determine the over-all rate of the reaction. If $k_1 > k_{cat}$ in eq. (9.14), it implies that ES forms faster than its dissociation and that the product formation is the rate-limiting step.

9.4 The steady-state assumption

When a substrate, in great excess, is mixed with an enzyme, there is an initial tran-sient phase during the build-up of the enzyme-substrate complex. After this usually very short phase (milliseconds), the concentrations of the enzyme-substrate complex and free enzyme are nearly constant as long as the substrate is in excess as illustrated in Figure 9.3. This implies that the enzyme-substrate complex is formed as fast as it is dissociated and the change in concentration is very close to zero:

$$\frac{d[ES]}{dt} \approx 0 \tag{9.16}$$

This is the steady-state assumption, first proposed 1925 by George E. Briggs and John B. S. Haldane.

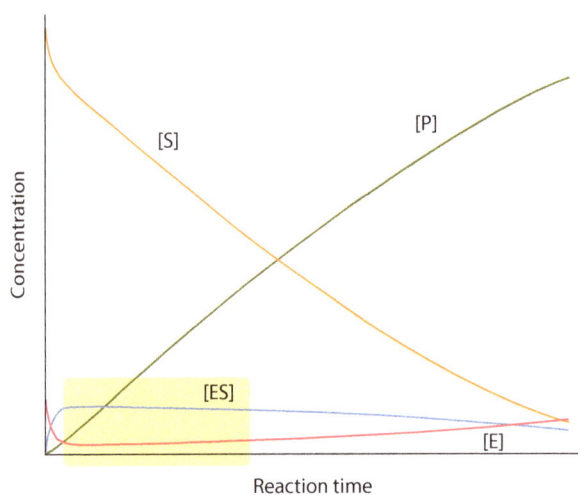

Figure 9.3: Progress curves for the components of an enzyme-catalyzed reaction S $\rightleftharpoons$ P. After the initial transient phase of the reaction (before the shade block), the concentrations of ES and E are nearly constant as long as the substrate is in great excess over the enzyme.

When the substrate is in great excess over enzyme, it can be assumed that the active site of all enzyme molecules contains a substrate molecule after the initial transient build-up of the ES-complex. When the reaction proceeds and the product is released from the enzyme, another substrate molecule binds immediately to the active site and the enzyme is ready to catalyze the reaction once more. The reaction is as fast as it can be as all enzyme molecules are engaged and the enzyme works at its the maximal reaction velocity, V_{max}. A larger excess of substrate over enzyme would not increase the rate, as all enzyme molecules are already saturated with substrate. V_{max} can be expressed as

$$V_{max} = k_{cat}[E]_T \tag{9.17}$$

since

$$[E]_T = [E] + [ES] \tag{9.18}$$

where $[E]_T$ and $[E]$ are the total and free enzyme concentrations, respectively.

The assumption that the reverse reaction can be ignored, as stated in eq. (9.14), is only valid as long as no product is present. As soon as the product is formed, the assumption fails. However, initially (at time $t = 0$) the rate of formation of ES is

$$\frac{d[ES]}{dt} = k_1[E][S] - k_{cat}[ES] - k_{-1}[ES] \tag{9.19}$$

Combining this expression with the steady-state assumption (eq. 9.16) and the conservation condition (eq. 9.18) yields

$$0 = k_1([E]_T - [ES])[S] - k_{-1}[ES] - k_{cat}[ES] \tag{9.20}$$

that upon rearrangements gives

$$[ES](k_1[S] + k_{-1} + k_{cat}) = k_1[E]_T[S] \tag{9.21}$$

Dividing both sides with k_1 and solving for [ES]

$$[ES] = \frac{[E]_T[S]}{K_M + [S]} \tag{9.22}$$

where K_M is the Michaelis–Menten constant

$$K_M = \frac{k_{-1} + k_{cat}}{k_1} \tag{9.23}$$

The unit of K_M is $mol \cdot L^{-1}$ (M).

The velocity of product formation can now be expressed as

$$v_0 = \frac{k_{cat}[E]_T[S]}{K_M + [S]} \tag{9.24}$$

where v_0 is the initial velocity, before any product has formed.

As $k_{cat}[E]_T$ equals V_{max}, the expression can be rewritten as

$$v_0 = \frac{V_{max}[S]}{K_M + [S]} \tag{9.25}$$

This expression, the Michaelis–Menten equation, is strictly valid only when no product is present ($[P] = 0$) and the substrate concentration is much larger than the enzyme concentration ($[S] \gg [E]$) in order to attain steady-state as fast as possible. As soon as the product begins to form, the reverse reaction (with the rate constant k_{-2}) must be considered. Therefore, it is essential to measure the initial velocity whenever the activity of an enzyme is to be determined.

9.5 The Michaelis–Menten constant

The Michaelis–Menten equation, named after Leonor Michaelis and Maud L. Menten, describes how the activity of an enzyme depends on the substrate concentration, as illustrated in Figure 9.4.

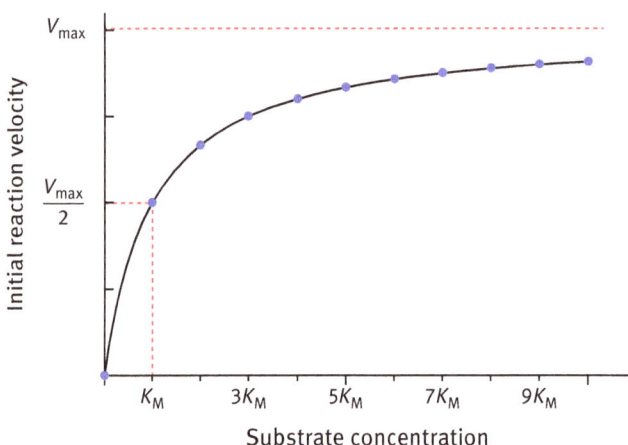

Figure 9.4: Graph of the initial velocity versus the substrate concentration of a Michaelis–Menten enzyme.

It can easily be shown that when the substrate concentration equals the Michaelis–Menten constant K_M, the reaction velocity is $V_{max}/2$. The initial reaction velocity approaches asymptotically V_{max} when the substrate concentration increases. However, it should be noted that even at a very large excess of substrate over enzyme V_{max} is not attained. At a substrate concentration corresponding to $10 \cdot K_M$, the velocity is still only 90% of V_{max}.

The expression for K_M (eq. 9.23) can be rearranged

$$K_M = \frac{k_{-1}}{k_1} + \frac{k_{cat}}{k_1} \tag{9.26}$$

The first term (k_{-1}/k_1) is the dissociation constant K_d of the first step of the enzyme reaction $E + S \rightleftharpoons ES$

$$K_d = \frac{k_{-1}}{k_1} = \frac{[ES]}{[E] + [S]} \tag{9.27}$$

Therefore, whenever k_{-1} is much larger than k_{cat} ($k_{-1} \gg k_{cat}$), the value of K_M is a measure of the affinity of the substrate for the enzyme. Under these conditions, a small K_M-value indicates strong binding. However, it is important to remember that K_M is not a true dissociation constant.

k_{cat} is a direct measure of the catalytic efficiency of the enzyme and a large k_{cat} indicates an efficient enzyme with a large turnover number. The turnover number varies widely from enzyme to enzyme (Table 9.1), from a few molecules per second to more than a million. K_M is usually in the milli- to nanomolar range.

Table 9.1: Turnover numbers for some uncatalyzed and enzyme catalyzed reactions.

Enzyme	Nonenzymatic half-life	k_{uncat} s^{-1}	k_{cat} s^{-1}	Rate enhancement k_{cat}/k_{uncat}
OMP decarboxylase[a]	78,000,000 years	$2.8 \cdot 10^{-16}$	39	$1.4 \cdot 10^{17}$
Staphylococcal nuclease	130,000 years	$1.7 \cdot 10^{-13}$	95	$5.6 \cdot 10^{14}$
Adenosine deaminase	120 years	$1.8 \cdot 10^{-10}$	370	$2.1 \cdot 10^{12}$
Carboxypeptidase A	7.3 years	$3.0 \cdot 10^{-9}$	578	$1.9 \cdot 10^{11}$
Ketosteroid isomerase	7 weeks	$1.7 \cdot 10^{-7}$	66,000	$3.9 \cdot 10^{11}$
Triosphosphate isomerase	1.9 days	$4.3 \cdot 10^{-6}$	4,300	$1.0 \cdot 10^{9}$
Chorismate mutase	7.4 h	$2.6 \cdot 10^{-5}$	50	$1.9 \cdot 10^{6}$
Carbonic anhydrase	5 s	$1.3 \cdot 10^{-1}$	1,000,000	$7.7 \cdot 10^{6}$

[a]OMP decarboxylase: orotidine monophosphate decarboxylase.

When $[S] \ll K_M$, which probably is the case for many enzymes under cellular conditions, the concentration of free enzyme is close to the total enzyme concentration ($[E]_T$) as very little [ES] is formed. Therefore, the Michaelis–Menten equation (eq. 9.24) reduces to

$$v_0 \approx \frac{k_{cat}}{K_M} [E]_T [S] \approx \frac{k_{cat}}{K_M} [E][S] \qquad (9.28)$$

where k_{cat}/K_M is a second-order rate constant. Under these conditions, v_0 is directly dependent on how often the substrate collides with the enzyme correctly to bind to the active site. As k_{cat}/K_M is both related to the turnover number and the substrate affinity, it is a useful measure of the efficiency of an enzyme. In addition, it can also be used to compare the efficiency of different enzymes.

By inserting eq. (9.23), k_{cat}/K_M can be rewritten as

$$\frac{k_{cat}}{K_M} = \frac{k_{cat} \cdot k_1}{k_{-1} + k_{cat}} < k_1 \qquad (9.29)$$

This indicates that there is an upper limit of an enzyme's efficiency; k_{cat}/K_M is maximal when $k_{-1} \ll k_{cat}$ but it can never exceed k_1. In other words, the limiting factor of a highly efficient enzyme is how fast ES can form. The enzyme and substrate must encounter each other to form the ES complex, as molecules move by diffusion in solution, the encounter rate is limited by diffusion. In solution the diffusion of small molecules is limited in the range of 10^8 to 10^9 $M^{-1} \cdot s^{-1}$. An enzyme with a k_{cat}/K_M values in this order would therefore be considered as a "perfect enzyme".

Triosphosphate isomerase, which catalyzes the reversible conversion between dihydroxyacetone phosphate and glyceraldehyde 3-phosphate, is an example of a "perfect" enzyme with a k_{cat}/K_M value of $2.4 \cdot 10^8$ $M^{-1} \cdot s^{-1}$. For a perfect enzyme, nearly every time the substrate encounters the enzyme, it must collide in the correct way to bind to the active site and form a productive enzyme-substrate complex.

The parameters K_M and V_{max} (or k_{cat}) can be obtained from experiments where the initial velocity is determined at varying substrate concentrations. Nonlinear regression analysis of a graph as the one shown in Figure 9.4, would give K_M and V_{max}. If the enzyme concentration is known, k_{cat} is obtained from $V_{max} = K_{cat}[E]_T$ (eq. 9.17).

However, traditionally the Michaelis–Menten parameters are obtained from linear graphs. The hyperbolic Michaelis-Menten equation can be transformed into several linear equations. Lineweaver–Burk, Eadie–Hofstee and Hanes–Wolf are the most popular used transformations (Figure 9.5).

Lineweaver–Burk plot is a double reciprocal plot where

$$\frac{1}{v_0} = \frac{1}{V_{max}} + \frac{K_M}{V_{max}} \cdot \frac{1}{[S]} \tag{9.30}$$

The y- and x-intercepts are $1/V_{max}$ and $-1/K_M$, respectively.

Eadie–Hofstee is a semi reciprocal plot where

$$v_0 = V_{max} - K_M \cdot \frac{v_0}{[S]} \tag{9.31}$$

The y-intercept is V_{max} and the slope gives $-K_M$.

In a Hanes–Wolf plot $[S]/v_0$ is plotted against [S]. The equation is given by

$$\frac{[S]}{v_0} = \frac{K_M}{V_{max}} + \frac{1}{V_{max}} \cdot [S] \tag{9.32}$$

where the y-intercept is K_M/V_{max} and the slope is $1/V_{max}$.

While determining the kinetic parameters, it is important to measure v_0 at constant buffer and salt concentrations as well as constant temperature and pH. Like all chemical reactions, also enzyme-catalyzed reaction rates also increase with temperature, at least to a certain point. A too high temperature may denature the enzyme and thus "kill" the enzyme. As acidic or basic groups may be involved in the catalytic process, an altered pH may protonate a group that should be ionized or vice versa. It is also essential to make sure that the determined v_0 is the actual initial velocity.

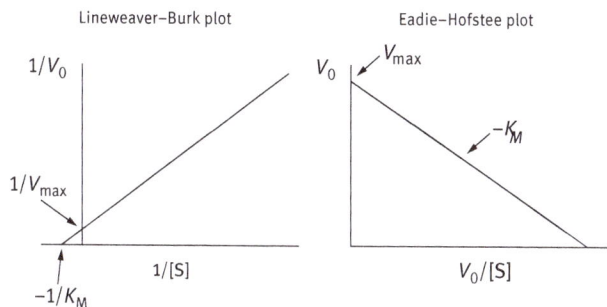

Figure 9.5: Lineweaver–Burk and Eadie–Hofstee plots.

9.6 The active site

The active site constitutes in general only a minor fraction of the total surface of the enzyme. The active site is usually a cavity, a cleft or a gorge in the enzyme lined by amino acid side chains. The amino acid residues in the active site need not to be close to each other in sequence but the folding process brings them in proximity. It is possible to distinguish two groups of side chains in the active site; those with a structural function and those with a catalytic function. The structural side chains bind and place the substrate or substrates correctly and in such a way that the catalytic side chains can catalyze the reaction. When the substrate binds, side chains in the active site induce a delocalization of electrons in one or more bonds in the substrate, converting the substrate to its transition state analogue.

In 1894 Emil Fischer proposed a model for the active site and named it "Schloss und Schlüssel" (lock-and-key). In this model it is presumed that the substrate fits perfect in the active site, as illustrated in Figure 9.6. Although the lock-and-key model is attractive, it is evident that it cannot describe how the substrate transforms to the transition state, as this must involve conformational changes in either substrate or enzyme.

More than 60 years later, Daniel E. Koshland Jr. suggested that binding of the substrate to the active site induces a conformational change of the active site, which would also explain the transition of the substrate to the transition state.

Later, a variation of the induced fit model has been proposed. In this model, called conformational selection, it is assumed that the active site undergoes continues conformational changes and that the binding per se does not induce the conformational change.

The first close look at an active site was presented in the mid-1960s with the structure determination of hen egg white lysozyme by David Phillips and coworkers. The structure gave an insight in how lysozyme is able to break the bacterial cell wall by hydrolysis of the saccharide part of the peptidoglycan.

Lock-and-key

E + S	ES*	E + P

Induced fit

E + S	ES	ES*	E + P

Conformation selection

E + S	E + S	ES*	E + P

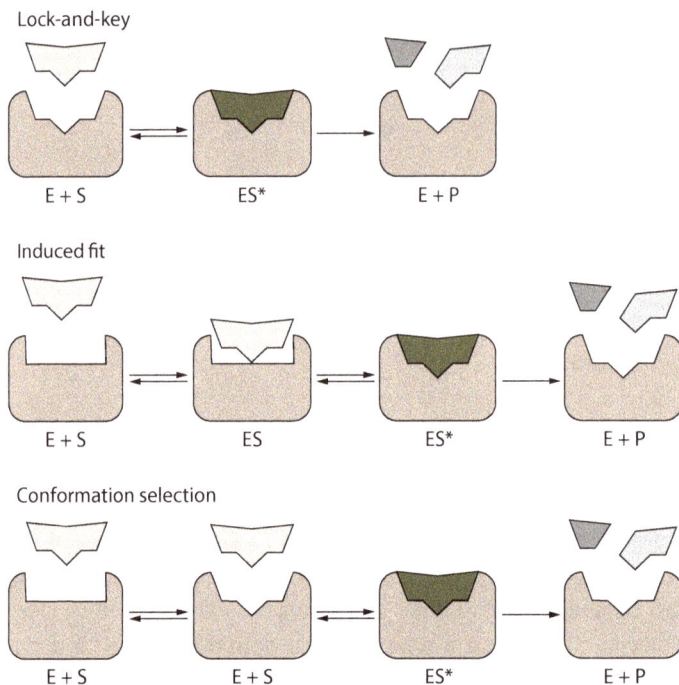

Figure 9.6: Models of the actives site. In the lock-and-key model, the substrate fits perfectly in the active site and binding leads to the transition state complex ES*. In the induced fit model, binding to the active site (ES in the figure) leads to a conformational change, which brings the substrate to the transition state. Conformational selection implies that the active site undergoes conformational changes continuously and that the substrate binds when the conformation is right. In the last step, the product is dissociated from the active site and the enzyme is ready for another catalytic cycle.

9.7 Lysozyme

The natural substrate of lysozyme is the peptidoglycan of Gram-positive bacteria. The peptidoglycan is composed of cross-linked and alternating N-acetylmuramic acid (NAM) and *N*-acetylglucose amine (NAG). Lysozyme catalyzes the hydrolytic cleavage of a glycoside bond in the polysaccharide. The active site is a deep groove that runs across the surface of lysozyme and can accommodate a saccharide with six sugar rings (Figure 9.7). There are six subsites in the groove (termed −4 to +2) that each bind a sugar ring of the substrate. Due to NAM's lactyl group, subsites −4, −2- and +1 cannot accommodate NAM. Since analysis has indicated that the enzyme catalyzes hydrolysis of the β(1→4) glycoside bond between the sugars placed in subsite −1 and +1, the hydrolysis must occur between NAM and NAG.

When the substrate binds, the conformation of the sugar ring in subsite −1 distorts, from a chair to a half-chair conformation, due to hydrogen bonds with Asn59 and

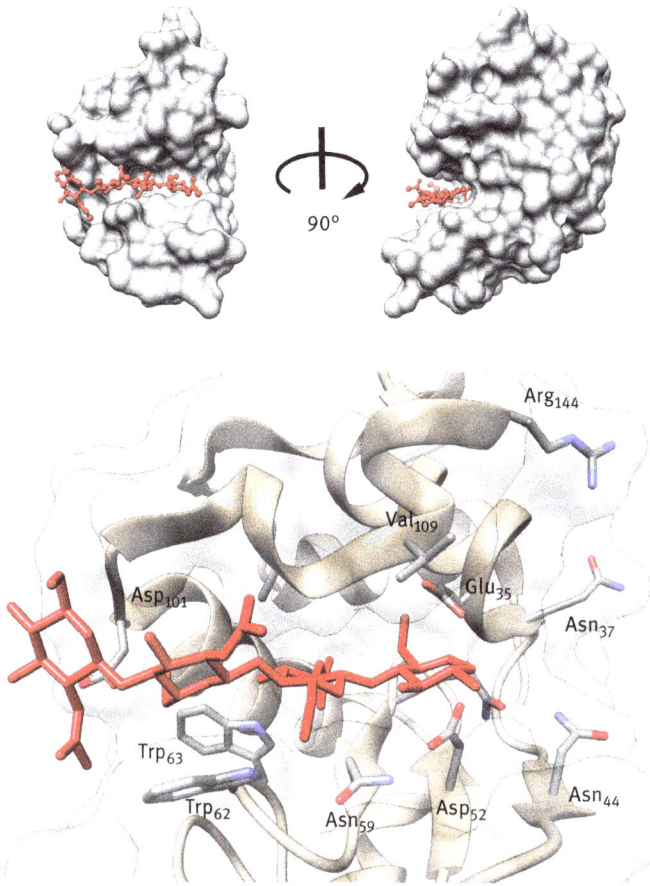

Figure 9.7: The structure and active site of hen egg white lysozyme. (Top) Surface representation of lysozyme with a substrate analogue (4-*O*-β-tri-*N*-acetylchitotriosylmoranoline) in the active site (pdb: 4HP0). (Bottom) Close-up of the active site. The catalytic residues Glu35 and Asp52 as well as residues involved in binding the substrate are indicated (pdb: 4HP0).

Val109, and places the bond to be hydrolyzed in close vicinity of the catalytic residues Glu35 and Asp52. Other residues in the active site form several hydrogen bonds and van der Waals interactions with the substrate.

The reaction catalyzed by lysozyme is in principle the same as the nonenzymatic reaction: an acid catalyzed hydrolysis of an acetal to a hemiacetal (Figure 9.8). Due to the nonpolar surroundings that increases the pK_a, Glu35 retains the carboxyl group in a protonated state and can therefore acts an acid catalyst. The other catalytic residue, Asp52, is surrounded by polar groups and therefore deprotonated and negatively charged at pHs above 3.

The breakdown of the bacterial cell wall begins when lysozyme binds to a hexasaccharide of the peptidoglycan. Binding of the substrate distorts the sugar ring

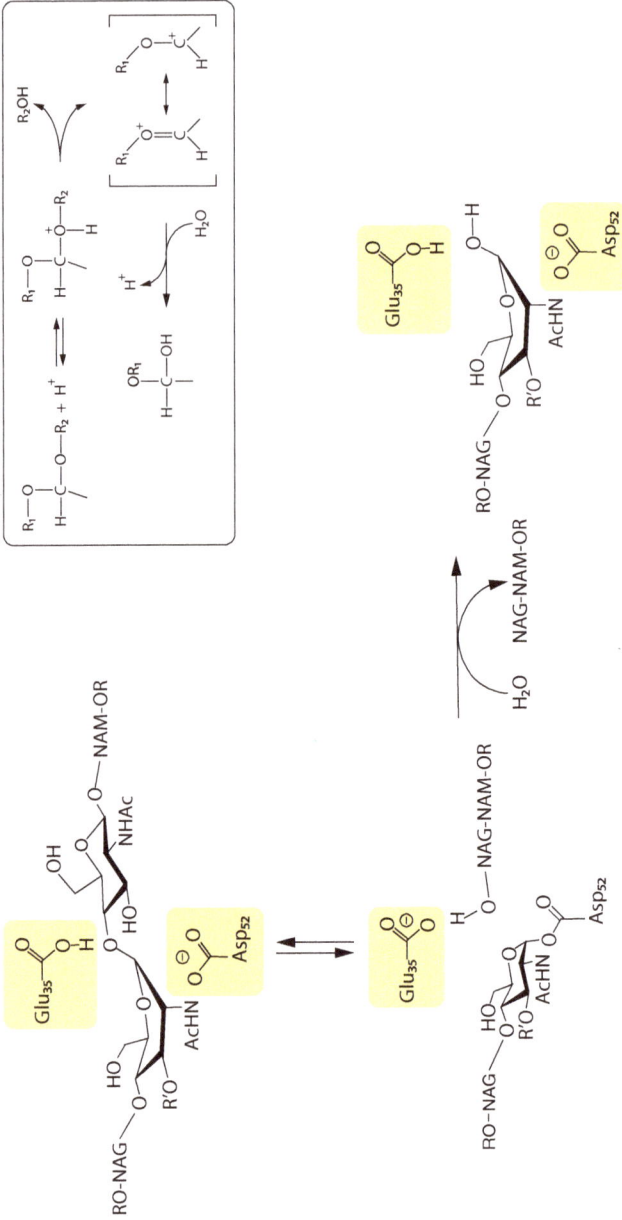

Figure 9.8: The catalytic mechanism of hen egg white lysozyme. Insert: The nonenzymatic catalysis of an acetal to a hemiacetal. Adapted with permission from D.J. Vocadlo, G.J. Davies, R. Laine and S.G. Withers (2001) Catalysis by hen egg-white lysozyme proceeds via a covalent intermediate. *Nature* 412:835–838.

(NAM) bound to subsite −1. Glu35 is close enough and acts as a general acid catalyst by transferring the carboxyl proton to the oxygen in the glycoside bond between NAM in subsite −1 and NAG in subsite +1. This leads to hydrolysis of the bond between NAM and NAG and formation of a covalent glycosyl-enzyme intermediate between NAM in subsite −1 and Asp52. Addition of water from the solvent then reprotonates Glu_{35} and releases the product.

9.8 Serine proteases

All serine proteases have an extraordinarily active serine residue in the active site, hence the name of the protein family. The protein family includes more than 200 different enzymes, all catalyzing the hydrolytic cleavage of covalent bonds, such as C-N bonds in peptides and C-O, C-S or C-C bonds in structurally diverse metabolites. The active sites of these enzymes contain three common residues: in addition to the serine residue, a histidine and an aspartate. Together, these three residues comprise a catalytic triad and cooperate in the catalytic events.

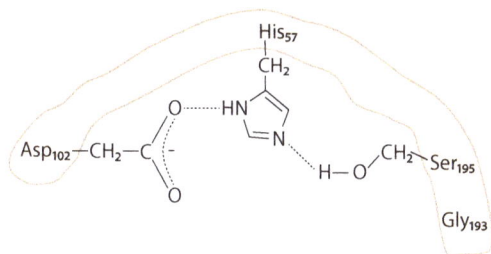

Figure 9.9: The catalytic triad of a serine protease. Residue numbers from the sequence of chymotrypsin.

The residues of the catalytic triad are not close to each other in the amino acid sequence. In chymotrypsin, the triad comprises residues Ser195, His57 and Asp102, whereas in human pancreatic lipase it is residues Ser153, His244 and Asp177. Although these residues are far apart in the polypeptide chain, folding brings them into close proximity in the active site together with residues that determine the specificity of the enzymes (Figure 9.9).

A serine residue at position 214 (numbering based on the chymotrypsin sequence) is conserved and is present in nearly all serine proteases. It is hydrogen bonded to both Asp102 and the main chain nitrogen of the bond to be hydrolyzed. Further, it is believed that Ser214 contribute to stabilize the negative charge of the buried Asp102. The main chain amide N-H groups of Gly193 and Ser195, in the so-called oxyanion hole, form hydrogen bonds with the carbonyl oxygen in the scissile bond.

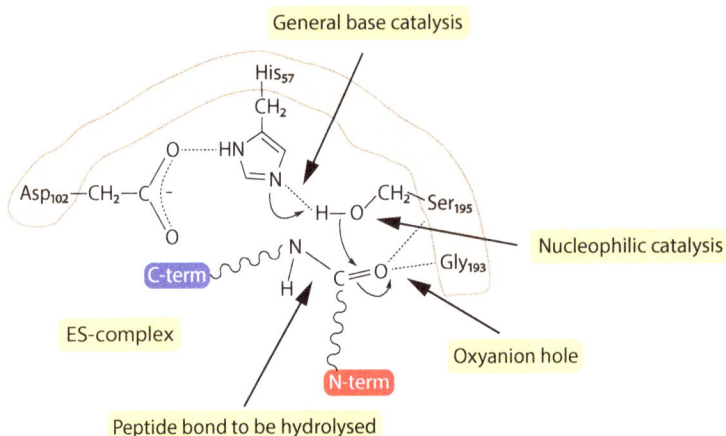

Figure 9.10: The active site with a peptide substrate. Adapted after R. Wieczorek, K. Adamala, T. Gasperi, F. Polticelli and P. Stano (2017) Small and random peptides: An unexplored reservoir of potentially functional primitive organocatalysts. The case of seryl-histidine. *Life* 7(2), 19. Licensed by CC BY (creativecommons.org/licenses).

The residue before the scissile peptide bond determines whether the substrate binds to the active site or not. For instance, in chymotrypsin there is a binding pocket (called S1) that fits large hydrophobic side chains like those of phenylalanine, tryptophan and tyrosine, whereas the binding pocket in trypsin favors basic side chains like arginine and lysine. Upon binding the peptide bond to be hydrolyzed is placed in close proximity to the active serine (Ser195 in chymotrypsin) and the carbonyl oxygen of the substrate is placed in the oxyanion hole by hydrogen bonds to main chain N-H groups (Figure 9.10).

The catalytic steps in peptide bond hydrolysis catalyzed by serine proteases and in particular by chymotrypsin have been worked out in detail. The reaction mechanism of chymotrypsin follows below and can be applied on all serine proteases.

After the peptide substrate binds, His57 abstracts a proton from Ser195 (step 1 in Figure 9.11). This generates a strong alkoxide ion on Ser195 that makes a nucleophile attack on the carbonyl group of the substrate and a tetrahedral intermediate forms (step 2). The negative charge on the carbonyl oxygen (the oxyanion) is stabilized by hydrogen bonds with the main chain amide N-H groups of Ser195 and Gly193. It is believed that the negatively charged Asp102 helps to stabilize the positive charge on His57.

The peptide bond is cleaved as His57 donates a proton to the amide group of the peptide substrate. The other part of the peptide substrate forms an acyl-enzyme intermediate (step 3). Here His57 functions as an acid by donating a proton to the substrate, whereas the first step is an example of general base catalysis.

The released C-terminal part of the substrate peptide is expelled from the active site and a water molecule enters the site instead (step 4). Deacylation and release of the

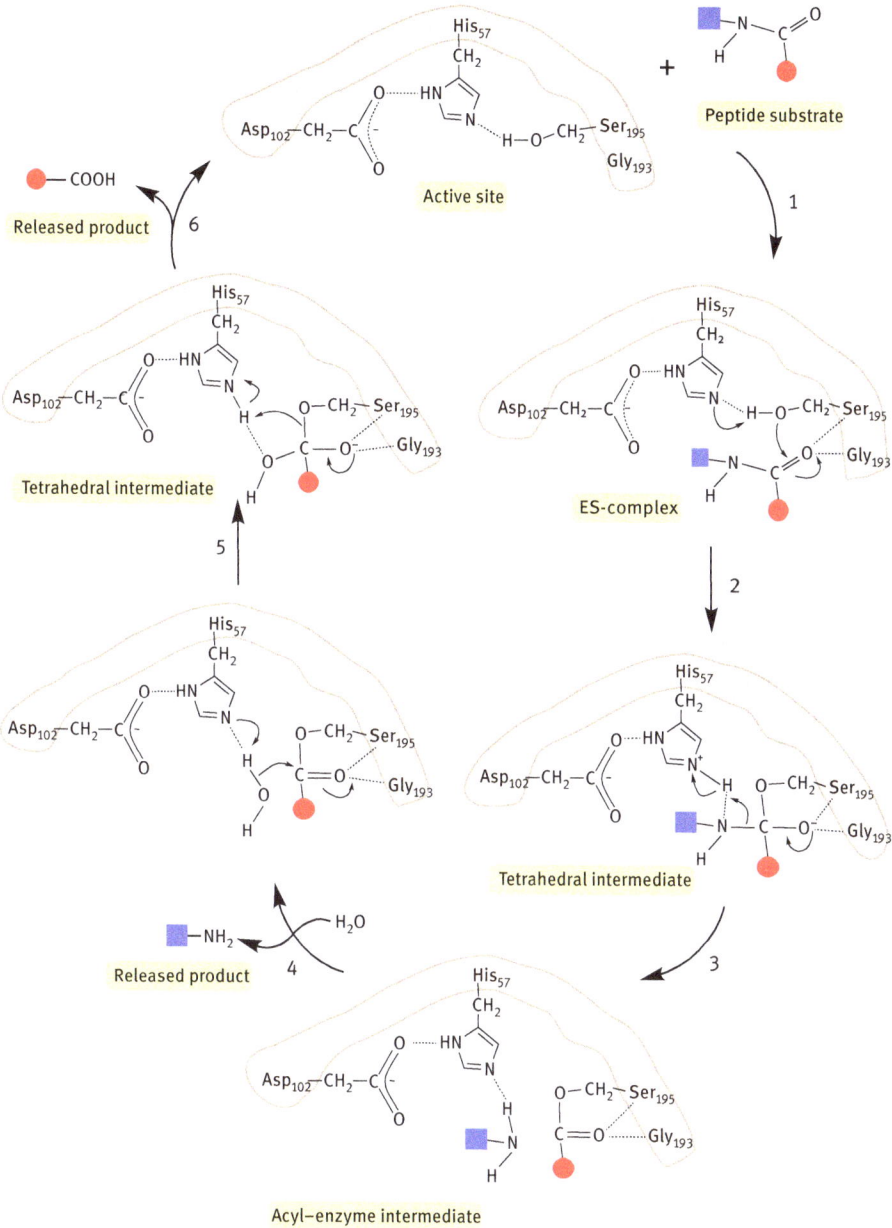

Figure 9.11: Reaction mechanism of chymotrypsin. See text for details. Adapted after R. Wieczorek, K. Adamala, T. Gasperi, F. Polticelli and P. Stano (2017) Small and random peptides: An unexplored reservoir of potentially functional primitive organocatalysts. The case of seryl-histidine. *Life* 7(2), 19. Licensed by CC BY (creativecommons.org/licenses).

remaining substrate occurs by a reversal of the previous steps. Again, His57 abstracts a proton but this time from water that becomes a strong nucleophile that attacks the ester bond between substrate and Ser195 and a second tetrahedral intermediate forms. In the final step (step 5), His57 donates a proton to Ser195, the ester bond breaks and the remaining substrate is released and expelled from the active site.

Although the peptide bond is very stable, polypeptides hydrolyzes (though very slowly) in aqueous solutions even in the absence of a catalyst. This implies that the catalyzed reaction follows an alternative pathway as the uncatalyzed reaction mechanism does not include the formation of a covalent intermediate.

Mutation analysis has been used to demonstrate the importance of the residues in the catalytic triad as well as those residues involved in the hydrogen bonding network. Any changes to the residues implicated in the reaction mechanism reduce the activity. Interestingly, even after the Asp-His-Ser triad is removed, the activity is still higher than that of the uncatalyzed reaction- This has been taken as an indication that the substrate is bound in a form that resembles the transition state, probably due to the positioning of the substrate in the oxyanion hole and the stabilization of the charged transition state.

Kinetic experiments, using p-nitrophenylacetate as substrate, have shown that chymotrypsin initially catalyzes a fast so-called "burst phase", followed by a slower steady-state phase (Figure 9.12). During the burst phase, the covalent acyl-enzyme intermediate builds up quickly and the first product p-nitrophenol is released. When all enzyme molecules have been acylated, the reaction slows down. At this stage, the reaction follows steady-state kinetics ($d[ES]/dt \approx 0$) as the acylated enzyme must be deacylated in order to catalyze another reaction. This indicates that deacylation is the slowest step and the step that determines the overall reaction rate. By extrapolating back to zero time, it was found that here is a nearly one-to-one ratio between p-nitrophenol and enzyme. In other words, it is possible to obtain the enzyme concentration from activity measurements.

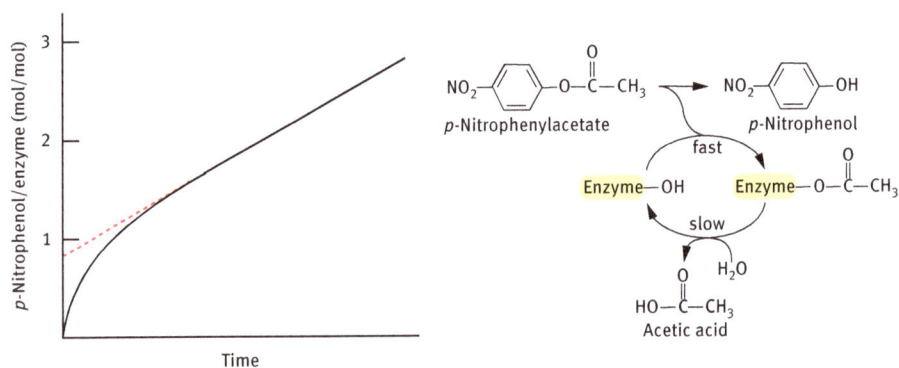

Figure 9.12: Hydrolysis of p-nitrophenylacetate catalyzed by chymotrypsin. The initial burst phase indicates the presence of an acylated enzyme intermediate in the reaction.

The digestive proteases chymotrypsin, trypsin and elastase all catalyze hydrolysis of peptide bonds but with different specificity. The selectivity depends mainly on the properties of binding pocket S1 as illustrated in Figure 9.13. In chymotrypsin S1 can accommodate large bulky hydrophobic residues (phenylalanine, tyrosine and tryptophan), whereas S1 of trypsin hydrolyzes the bond after basic residues (lysine or arginine) and elastase after small uncharged residues. In trypsin, a negatively charged aspartate can bind to the positively charged lysine or arginine in the substrate. S1 of elastase is lined with two valines that allow only small uncharged residues to enter the binding pocket.

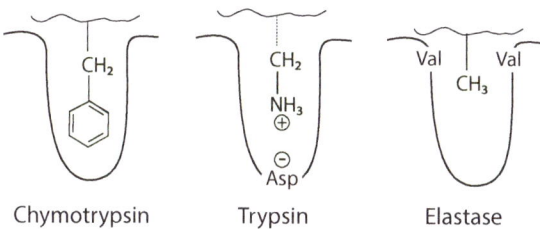

Figure 9.13: Binding pockets in serine proteases. The binding pocket S1 in chymotrypsin is large and binds hydrophobic residues, whereas in trypsin there is a negatively charged aspartate at the bottom that attracts basic residues. The binding pocket of elastase is narrow as it is lined with two valines allowing only small uncharged residues to bind.

9.9 Enzyme control

Many metabolites in our cells have different fates. An example is glucose that is a primary source of energy, but it is also stored for later use as glycogen. The use of glucose is determined by the energy status of the cell (or tissue). In good times, glucose is used to synthesize glycogen, whereas in bad times, it is used to produce ATP that is a much more useful form of energy. Therefore, it is necessary to control these two pathways in such a way that both are not active at the same time. This is done by controlling the activity of an enzyme in each of the two pathways. By reducing the activity of an enzyme in the pathway, the flow through the pathway is also reduced. Often one of the first enzymes in a pathway is the one that controls the flow and therefore also determines how fast the final product is formed.

The activity of an enzyme can be controlled in several ways. Most control mechanisms are reversible and make it possible to turn the activity on or off as required. Irreversible control mechanisms are more abiding as there is no way to reverse the inactivation. The conversion of a zymogen (inactive proenzyme) to an active protease by proteolysis cannot be undone and the binding of the trypsin inhibitor to trypsin is in principle irreversible.

A single enzyme molecule can only work at full speed or not at all. In an ensemble of a certain enzyme, the maximal activity will occur when all enzyme molecules are active. If some of the enzyme molecules are turned off, the activity of the ensemble would be reduced. Small molecules and metabolites can turn off the activity of an enzyme by two general ways: either by binding to the active site and inhibit the binding of the substrate or by binding somewhere else on the enzyme and somehow inhibit the catalytic reaction.

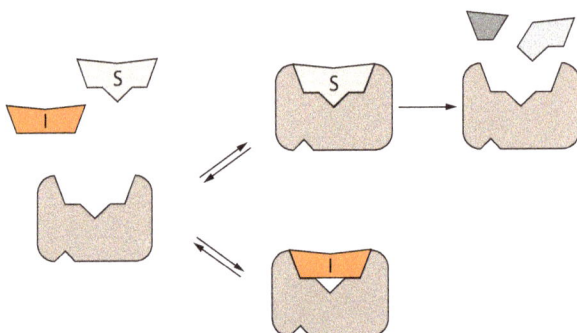

Figure 9.14: Schematic representation of competitive inhibition. The substrate (S) and inhibitor (I) both can bind to the active site and therefore compete for binding. However, only the substrate can be converted to products.

A competitive inhibitor, usually has a structure that partly resembles the substrate, and can therefore bind to the active site and hinders the substrate to bind and therefore reduces the activity of the enzyme ensemble (Figure 9.14). As inhibitor and substrate both can bind to the active site, they compete for binding. The activity inhibition depends on the ratio between substrate and inhibitor. Increasing substrate concentration at constant inhibitor concentration leads to reduced inhibition. At a very large excess substrate, no inhibitor molecule will bind and the enzyme works at full speed. The reaction scheme can be described as

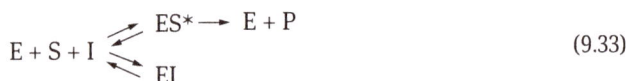

$$E + S + I \; \underset{\longleftarrow}{\overset{\longrightarrow}{\rightleftarrows}} \; \begin{array}{l} ES^* \longrightarrow E + P \\ EI \end{array} \tag{9.33}$$

As before, ES^* is the transition state that dissociate to give either the product or substrate and free enzyme. EI is the inhibitor-enzyme complex with inhibitor bound to the active site and this complex is unproductive.

In contrast to a competitive inhibitor, a noncompetitive inhibitor binds somewhere else on the enzyme but the active site. The binding of a noncompetitive inhibitor affects the activity somehow and turns off the enzyme (Figure 9.15).

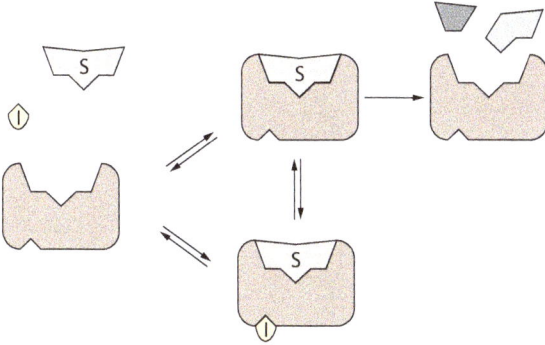

Figure 9.15: Schematic representation of noncompetitive inhibition. The binding of the substrate (S) to the active site is not affected by the inhibitor (I).

A noncompetitive inhibitor binds somewhere on the enzyme and inhibits the catalytic activity, often by inducing a conformational change in the enzyme. As the inhibitor does not bind to the active site as a competitive inhibitor does, the inhibition cannot be reduced by a higher substrate concentration. The reaction scheme can be described as

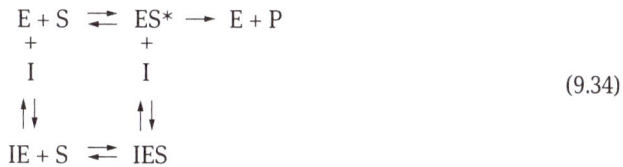

$$
\begin{array}{ccc}
\text{E} + \text{S} & \rightleftharpoons \ \text{ES}^* \ \longrightarrow \ \text{E} + \text{P} \\
+ & + \\
\text{I} & \text{I} \\
\updownarrow & \updownarrow \\
\text{IE} + \text{S} & \rightleftharpoons \ \text{IES}
\end{array}
\tag{9.34}
$$

The substrate can bind to the active site independent of whether the inhibitor is bound or not, but the product can only be formed from the transition state complex ES*. The inhibitor-substrate-enzyme complex is unproductive.

There is also a third type of inhibition called uncompetitive inhibition. In this case, the inhibitor only binds to the complex formed between substrate and enzyme, the ES-complex.

The type of inhibition can be distinguished experimentally by determining the reaction velocity at varying substrate concentrations in the absence and presence of the inhibitor (Figures 9.16 and 9.17).

Since a competitive inhibitor competes with the substrate for binding to the active site, the enzyme is not inhibited when the substrate is in great excess; V_{max} is the same. A noncompetitive inhibitor binds somewhere else on the enzyme; therefore, the enzyme is inhibited even at a large excess of substrate. The apparent V_{max}^{app} is always less whereas K_M is unaffected.

By determining the maximal reaction velocity and the Michaelis–Menten constant for the uninhibited and inhibited enzyme, the dissociation constant of the inhibitor can be calculated from

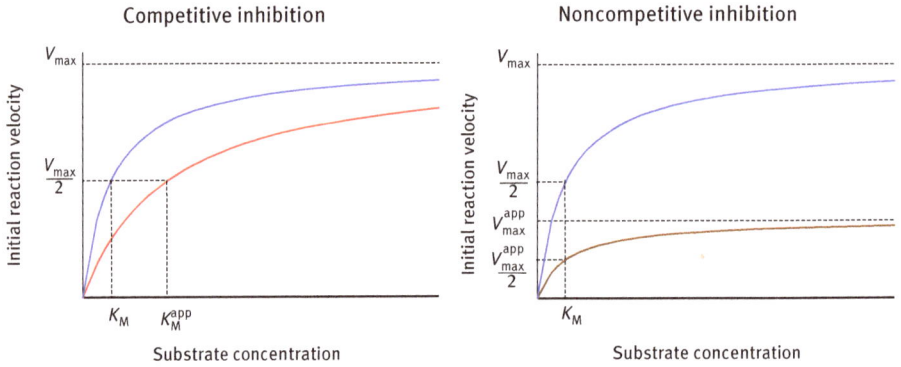

Figure 9.16: Graphs of the initial velocity versus the substrate concentration of a Michaelis–Menten enzyme inhibited by a competitive (red) and noncompetitive (brown) inhibitor.

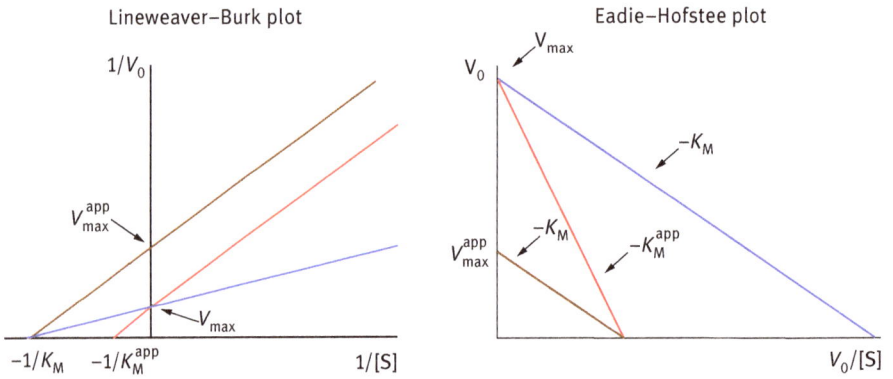

Figure 9.17: Lineweaver–Burk and Eadie–Hofstee plots of a Michaelis–Menten enzyme inhibited by a competitive (red) and noncompetitive (brown) inhibitor.

Competitive inhibition:

$$K_M^{app} = K_M \cdot \left(1 + \frac{[I]}{K_I}\right) \tag{9.35}$$

Noncompetitive inhibition:

$$V_{max} = V_{max}^{app} \cdot \left(1 + \frac{[I]}{K_I}\right) \tag{9.36}$$

where $[I]$ and K_I are the inhibitor concentration and the dissociation constant, respectively.

9.9.1 Feedback inhibition

All pathways in our metabolism are controlled to make sure that they are active only when necessary, thereby not wasting energy or metabolites unnecessarily. Often the end product of the pathway inhibits one of the enzymes in the beginning of the pathway. For instance, the main purpose of our energy metabolism is to convert food stuff to chemical energy in the form of ATP. As soon as the ATP level increases this is sensed by enzymes that catalyze one of the initial steps of the metabolic pathways, and the flow of metabolites through the pathways are inhibited as well as the production of more ATP. This type of control loop is called feedback inhibition (Figure 9.18).

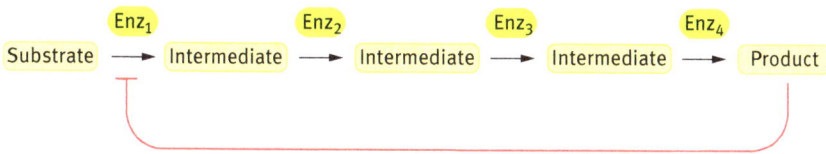

Figure 9.18: Feedback inhibition. The end product inhibits the first enzyme of the pathway, thereby turning off the flow of metabolites and the product formation.

9.9.2 Alloster control

Many, if not all, metabolic pathways are under alloster control. The catalytic ability of one enzyme, usually one of the first enzymes of the pathway that catalyzes the committed step, depends on the substrate concentration, such that at high concentrations the enzyme becomes more efficient and is a better catalyst. Graphically it is seen as a

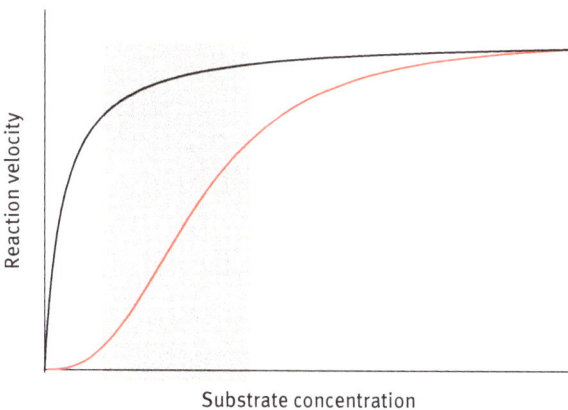

Figure 9.19: Alloster control of an enzyme reaction. The activity of an allosteric (red line) and Michaelis–Menten (black line) plotted versus substrate concentration.

sigmoid reaction velocity curve and not a hyperbolic curve as enzymes following Michaelis–Menten kinetics would display.

If the substrate concentration varies in the range indicated by the colored region in Figure 9.19, the reaction velocity of a Michaelis–Menten enzyme would vary from 80 to nearly 100% of the maximal velocity. In the same substrate concentration range, the reaction velocity of an allosteric enzyme would vary from around 10 to 70% of the maximal velocity. Thus at low concentrations, the allosteric enzyme would catalyze the reaction very slowly whereas with increasing concentrations, the enzyme would catalyze the reaction faster and faster.

The common definition of an allosteric enzyme states that the enzyme changes conformation upon binding of a regulatory molecule (effector) and this results in a change in binding affinity of the substrate. If the effect is homotropic, the effector is the normal substrate and the enzyme must consist of two or more subunits. If the regulatory molecule binds to a site different from the active site, the allostery is heterotropic and the enzyme may comprise a single polypeptide. The allosteric effector can have a positive effect on the substrate binding affinity and can increase the reaction velocity or a negative effect and reduce the velocity. In general, allosteric effects results from subunit interactions in oligomeric enzymes.

Aspartate transcarbamoylase (ATCase) catalyzes the reaction of carbamoyl phosphate with *L*-aspartate to form *N*-carbamoyl-*L*-aspartate and inorganic phosphate. The reaction is the committed step in the biosynthesis of pyrimidines (cytidine 5-triphosphate, CTP and uridine 5-triphosphate, UTP), which are required for synthesis of DNA and RNA. ATCase is a large enzyme, consisting of six catalytic and six

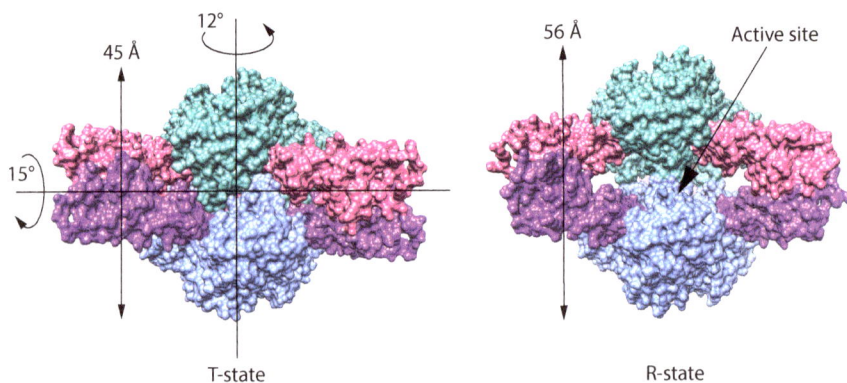

Figure 9.20: Side view of aspartate transcarbamoylase (ATCase). ATCase consists of two catalytic trimers (blue and green) facing each other and the three regulatory dimers (pink and magenta). Upon binding of a substrate analog (*N*-phosphonoacetyl-*L*-aspartate, PALA) a conformational change is induced, from the inactive T-state to the active R-state. The catalytic trimers move away from each other by 11 Å and there is a rotation along the threefold axis by 12° and along the pseudo twofold axis by 15°. The position of the active site in one catalytic subunit is indicated by the arrow. The six regulatory subunits contain the effector sites. The structures are from pdb: 1Q95 and 1R0C.

regulatory subunits, where the catalytic subunits form two trimers and the regulatory subunits three dimers (Figure 9.20).

Binding of substrates to one of the active sites of ATCase induces a large conformational change. The catalytic trimers move 11 Å from each other and one trimer rotates 12° in relation to the other trimer. There is also a 15° rotation along the pseudo twofold axis. The effector sites on the regulatory subunits bind CTP or ATP. CTP, being the end product of the pyrimidine pathway, binds to the effector sites, stabilizes the inactive T-state, and turn off its own synthesis. ATP, which signals a high level of energy and need to synthesize CTP, also binds to the effector sites, stabilizes the active R-state, and activates ACTase instead of inhibiting it. Thus, CTP is an allosteric feedback inhibitor whereas ATP is an allosteric activator. The distance between the active sites is around 22 Å, and the distance between active site and effector sites is around 60 Å. Effector binding, causes a conformational change which is "sensed" a long way.

9.9.3 Covalent modification

The most common covalent modification of enzymes is phosphorylation. A protein kinase catalyzes the transfer of a phosphate group from ATP to a hydroxyl group on a serine, threonine or tyrosine residue in the enzyme. Thereby the enzyme is turned "on" (or "off") and active. A protein phosphatase catalyzes the removal of the phosphate group from the enzyme, turning the enzyme "off" (or "on"), as illustrated in Figure 9.21.

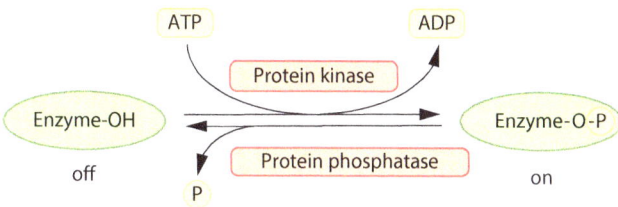

Figure 9.21: Activation of an enzyme by phosphorylation.

Pyruvate kinase (PK) catalyzes the last step in the glycolysis, the formation of ATP from phosphoenolpyruvate (Figure 9.22). Liver PK is under several layers of control. Phosphorylation by protein kinase A makes liver PK less active than in the dephosphorylated state. Dephosphorylated liver PK is inhibited by its own product, ATP. In other words, ATP inhibits its own synthesis. Fructose 1,6-bisphosphate, an important intermediate in glycolysis, activates liver PK. The liver isoform of PK is also indirectly regulated by the blood glucose level, as blood glucose controls the secretion of insulin and glucagon. A low level of blood glucose increases secretion of glucagon and sends a signal to the liver (and kidney) to produce more glucose by a process called gluconeogenesis. Therefore, further synthesis of ATP by the liver is inhibited by phosphorylation of liver PK. Instead the glycolytic metabolites are used to produce glucose for export from the liver to other tissues, such as brain and muscle tissues.

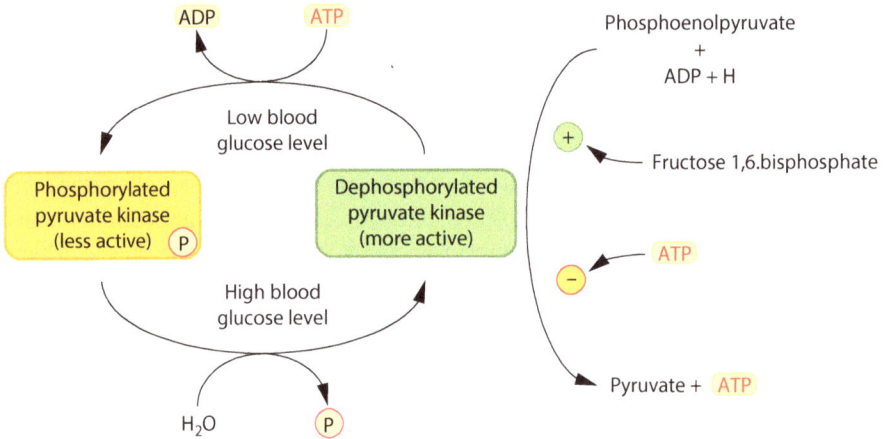

Figure 9.22: Control of liver pyruvate kinase.

Phosphorylation is by no means the only covalent modification of enzymes that affects the activity. Many enzymes require a covalent attached prosthetic group to be functional, such as biotin and lipoate that are bound to the amino group of a lysine's side chain in carboxylases and acetyltransferases, respectively. The most frequent modifications are summarized in Table 9.2.

Table 9.2: Common covalent posttranslational modifications of enzymes.

Type of modification	Localization for modification	Target amino acid
Biotin or lipoate binding	Mitochondria	Lys
N-glycosylation	ER, Golgi	Asn
O-glycosylation	Golgi	Ser, Thr
Ubiquitylation	Cytoplasm	Lys
Methylation	Nucleus	Lys, Arg
Prenylation	Cytoplasm	Cys
Acetylation	Nucleus	Lys
Myristoylation	Cytoplasm	Gly (N-term)
Palmitoylation	Cytoplasm	Cys
ADP ribosylation	Nucleus	Arg, Ans, Cys
Hydroxylation	Nucleus	Pro, Lys, Asn
Glu carboxylation	Extracellular	Glu

Data from H. Ryšlavá, V. Doubnerová, D. Kavan and O. Vaněk (2013) Effect of posttranslational modifications on enzyme function and assembly. *J Proteomics* 92:80–109.

9.9.4 Proteolytic control

The activity of a protease must to be under strict control; it should only be active when needed. Therefore, proteases are synthesized as inactive precursors, so-called zymogenes. In contrast to phosphorylation and other covalent modifications, proteolytic activation is irreversible. Once the zymogen has been activated, there is no way turning back.

Chymotrypsin is secreted from the pancreas as an inactive precursor, chymotrypsinogen, like many other proteases. After transport to the intestinal tract, chymotrypsin is activated by proteolysis. Trypsin hydrolyzes the peptide bond between Arg15 and Ile16, to give Π-chymotrypsin that in turn cleaves off Ser14-Arg15 and Thr147-Asn148 of other Π-chymotrypsin molecules to give active α-chymotrypsin (Figure 9.23). The cleavages generate new N-terminals and organize the architecture of the active site, especially the oxyanion hole.

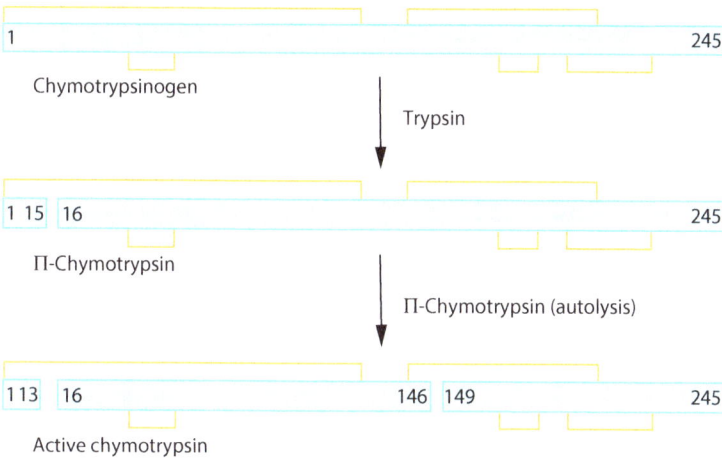

Figure 9.23: Activation of chymotrypsinogen. The peptide bond between Arg_{15} and Ile_{16} is hydrolyzed by trypsin, to give Π-chymotrypsin that hydrolyzes two more peptide bonds to produce active α-chymotrypsin. Yellow lines indicate disulfide bridges.

Since a single molecule of a serine protease, in particular trypsin, can activate many more protease molecules, which in turn activate even more, it is important to avoid premature activation of this group of enzymes. In the pancreas, a specific trypsin inhibitor, pancreatic secretory trypsin inhibitor (PSTI), is synthesized. PSTI belongs to the widely distributed family of Kazal-type serine protease inhibitors. The human PSTI gene encodes a 79 amino acid long protein. PSTI has a 23 amino acid signal peptide that directs it to the endoplasmatic reticulum where the maturation process produces a 56 residue long peptide. Mature PSTI binds very strongly to the active site of trypsin; it is one of the strongest noncovalent bindings known in biological systems, and inactivates trypsin very fast and efficiently. Only a small amount of PSTI is produced, much less than to inactivate all trypsin molecules but enough to inactivate premature activated trypsin in the pancreas under normal conditions. However, if PSTI cannot inhibit trypsin completely, the protease may start to degrade the pancreatic tissue, causing acute pancreatitis, a painful and sometimes fatal condition.

Clotting of blood is another example on a remarkable teamwork among serine proteases (Figure 9.24). The final step in blood coagulation is the conversion of fibrinogen to fibrin that forms a clot and stops the bleeding. An injury to a blood vessel damages the endothelium lining of the vessel and exposes the subendothelial. This activates blood platelets and allows a tissue factor to interact with factor VII, a serine protease, that in turn

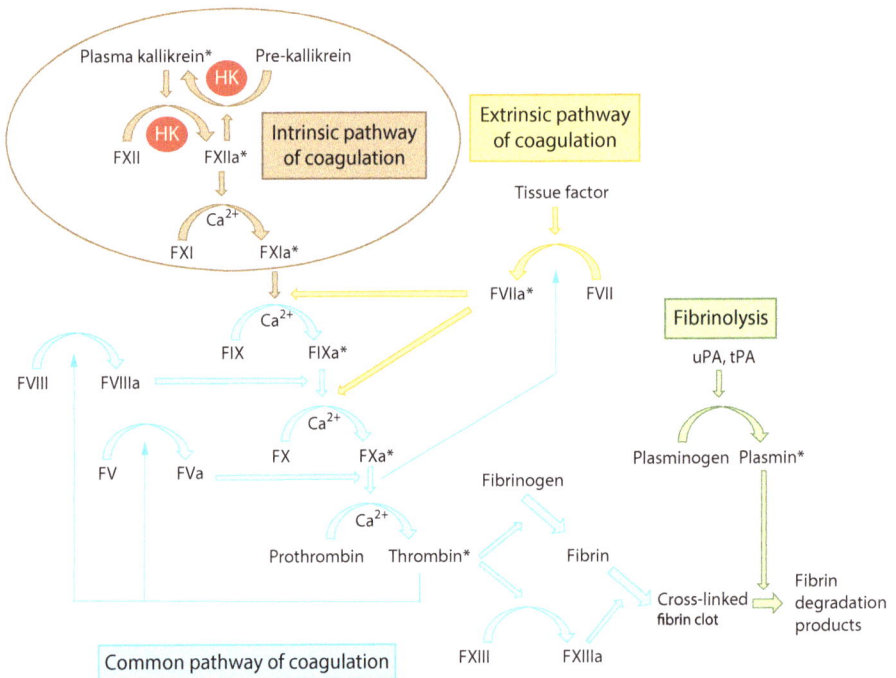

Figure 9.24: Schematic illustration of the blood coagulation cascade. When necessary, the coagulation cascade is activated by the extrinsic pathway (yellow arrows) and by the intrinsic pathway (brown arrows). The cascade converges on the common pathway of coagulation (blue arrows) that activates factor X and subsequently thrombin, which in turn catalyzes formation of fibrin from fibrinogen and activates factor XIII. Factor XIII cross-links fibrin and stabilizes the blood clot. At the same time the coagulation cascade initiates, the fibrinolysis is activated that eventually leads to the removal of the blood clot. Coagulation factors are indicated with "F" followed by a roman numeral, an additional "a" denotes the active form. HK; high molecular weight kininogen; uPA, urokinase plasminogen activator; tPA, tissue plasminogen activator, * denotes a serine protease. Adapted after T.G. Loof, C. Deicke and E. Medina (2014) The role of coagulation/fibrinolysis during *Streptococcus pyogenes* infection. *Front Cell Infect Microbiol* 4:128. Licensed by CC BY (creativecommons.org/licenses).

starts a cascade reaction of activating a series of other serine proteases and cofactors that normally circulate in the blood stream as inactive zymogenes.

The extrinsic pathway or tissue factor pathway is probably the most important for blood clotting. The binding of tissue factor to factor VII, converts it to the proteolytic active form factor VIIa, which converts the inactive factors IX and X to the active forms. Factor Xa associates with factor Va to form the prothrombinase complex and catalyzes the proteolytic conversion of prothrombin to thrombin. Thrombin then catalyzes the proteolytic activation of fibrinogen to fibrin. Thrombin also activates factor XIII and several other factors. Factor XIII is a transglutaminase that cross-links and stabilizes fibrin. Degradation of the fibrin network, fibrinolysis, activates at the same time as the coagulation system but is slower and removes the clot with time.

Most of the coagulation factors are synthesized in the liver. After their synthesis several are modified both by glycosylation of serine, threonine and asparagine residues and by carboxylation of several glutamic residues creating γ-carboxyl-glutamate. The carboxylation is essential for their active calcium-dependent conformation and ability to bind to membranes.

The inherited genetic disorder hemophilia leads to inability to stop bleeding. The major type, hemophilia A, occurs due to a lack of coagulation factors VIII. The disorder is transmitted by a nonfunctional gene on the X chromosome and only affects men.

Once the blood clot has formed and the bleeding stopped, thrombin must be deactivated. There are several inhibitors present in circulating blood that can associate with and inhibit the factors that activate thrombin. In addition, thrombin forms a large complex with receptors and other proteins that leads to formation of another membrane-bound complex. This complex catalyzes proteolytic inactivation of factors Va and VIIIa that stop further activation of prothrombin.

9.9.5 Regulatory proteins

Extracellular signals, such as hormones, neurotransmitters or cytokines, are transmitted across the plasma membrane by various receptors. The binding of the signal molecule induces a conformational change in the receptor that induces an intracellular signal, a secondary messenger, often cyclic AMP or calcium ions. A prominent target for the calcium ions is calmodulin. The dumbbell-shaped protein consists of two globular lobes connected by a flexible linker. Each lobe contains a pair of EF-hands motifs. Calcium-binding induces a large conformational change in the linker that exposes a hydrophobic surface allowing calmodulin to interact with the target protein.

Activation of smooth muscle contraction requires that the myosin light chain of the contractile machinery is phosphorylated. The phosphorylation is catalyzed by myosin light chain kinase, that is activated by calcium-bound calmodulin. Figure 9.25 shows the conformational change when calmodulin binds to a peptide of myosin light chain kinase. The conformational change due to calcium-binding moves the two glob-

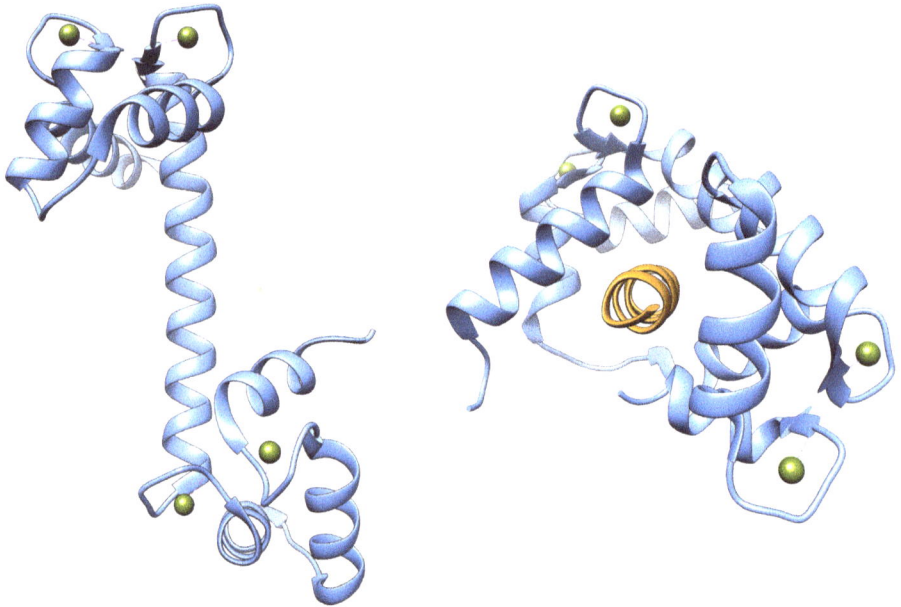

Figure 9.25: Conformational change of calmodulin. Left: The calcium-bound structure of chicken calmodulin (pdb: 1UP5). Right: A peptide of myosin light chain kinase (golden) encompassed by calmodulin (blue) (pdb: 2O5G). Green balls: calcium ions.

ular lobes away from each other, exposing the central part of calmodulin. Upon binding, calmodulin wraps around the target site that leads to activation of the kinase and the kinase catalyzes the necessary transfer of a phosphate group from ATP to the myosin light chain and the muscle contracts.

Myosin light chain kinase is by no means the only target regulated by calmodulin. There are numerous enzymes as well as proteins that rely on calmodulin for proper function, such as phosphodiesterase, phospholipase, cyclases, kinases and many more.

There are several other proteins that regulate cellular processes. Both monomeric and trimeric G-proteins respond to extracellular signals and turn on or off cellular processes. Another example is adenylate cyclase that catalyzes ATP → cyclic AMP (cAMP). cAMP was the original secondary messenger to be discovered. cAMP is considered as a "hunger" signal as it turns on glycogen degradation and turns off glycogen synthesis, which increases the blood glucose level. It should be noted that cAMP has many other cellular targets.

9.9.6 Synthesis control

The number of a certain enzyme will have a strong influence on how fast it can convert substrate to product; the more copies the faster reaction rate (if not the substrate is limiting). Thus, by adjusting the amount of enzyme, the reaction rate can be regulated without an inhibitor. This can be done by regulating the synthesis or the life length of the enzyme. Protein synthesis depends on both transcription and translation. If a gene is transcribed often, it will generate more copies of the enzyme compared to a gene that is transcribed seldom. Some genes are only transcribed under specific conditions, such as those required for metabolism of lactose by bacteria. In the absence of lactose, the enzymes needed for metabolism of lactose are not transcribed or translated but as soon as lactose is present the bacteria start to produce the necessary enzyme for lactose metabolism.

The degradation rate of proteins varies from seconds to many years. For some proteins, the life length depends on the N-terminal amino acid. If the N-terminal residue is Phe, Leu, Asp, Lys or Arg, the life length is 2–3 min, whereas if the first residue is Met, Ser, Ala, Thr, Val or Gly, the expected life length is >20 h. This is called the N-terminal rule.

A PEST sequence, a sequence region rich in Pro (P), Glu (E), Ser (S) and Thr (T), has been noticed in some short-lived proteins. However, the PEST sequence does not automatically lead to degradation. Some proteins need to be phosphorylated in the PEST sequence before degraded by the ubiquitin system.

Ubiquitin is a 76 amino acid peptide that is used to mark a protein for degradation by the proteasome (Figure 9.26). A ubiquitin-activating enzyme, ubiquitin-conjugating

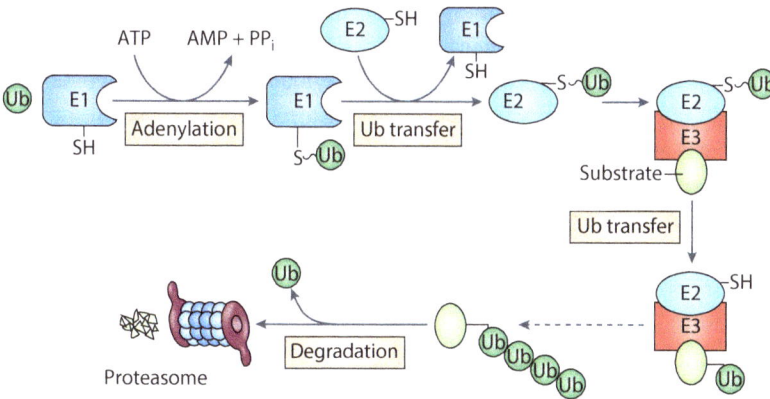

Figure 9.26: Degradation by the proteasome. By the concerted action of a ubiquitin-activating enzyme (E1), a ubiquitin-conjugating enzyme (E2) and a ubiquitin ligase (E3) adds ubiquitin (Ub) to the protein for degradation by the 26S proteasome. E1 adenylates the C-terminal's carboxyl group of Ub, forming an Ub-AMP intermediate and then an E1-Ub intermediate. Ub is then transferred to the target protein via E2 and by assistance from E3. In the last step of ubiquitylation, Ub is transferred to the ε-amino group of a lysine side chain of the target protein or to a growing ubiquitin chain. In the proteasome, deubiquitylating enzymes release and recycle Ub and the target protein is degraded. Reprinted with permission from J. Maupin-Furlow (2011) Proteasomes and protein conjugation across domains of life. *Nature Rev Microbiol* 10:100–111.

enzyme and a ubiquitin ligase cooperate to add a series of ubiquitin molecules to lysine residues in the protein to be degraded.

9.10 Enzyme evolution

Enzymes are used in a wide range of industrial applications; from production of food and beverage to production of pharmaceuticals and fine chemicals. One of the best known applications is the use of yeast in wine and beer making, where the yeast enzymes catalyze the fermentation of sugar into ethanol.

Due to the high specificity and the mild reaction conditions, enzymes have certain advantages compared to nonbiocatalysts. However, it is not always that the natural enzyme is "good enough"; it may be unstable and degrades too rapidly or catalyzes the reaction too slow or not even catalyzing the required reaction. There are several ways to overcome these problems.

Rational design and directed evolution of enzymes can broadly be divided in two groups; to increase stability and activity of the target enzyme or to create an enzyme with new catalytic capabilities. Directed evolution involves the use of random or focused mutations of the target enzyme while rational design exploits structural and mechanistic knowledge of the enzyme combined with computational methods to introduce mutations in target gene. In both cases, usually a high-throughput assay is used to select for improved versions of the enzyme. Then, these improved versions of the enzyme are used for the next round of mutations (Figure 9.27).

In a series of groundbreaking experiments in the 1990s, Frances H. Arnold and coworkers showed the usefulness and applicability of directed evolution to improve the stability and activity of subtilisin E, an important ingredient in washing powder. By combining random mutagenesis (by error-prone PCR) and recombination (by a staggered extension process), they were able to, among other things, to increase the half-life of subtilisin E at 65 °C ~200 times and the catalytic efficiency ~5 times compared to the wild-type enzyme.

Organic synthesis is used to produce a wide variety of important substances, including pharmaceuticals, plastics and many more. However, the reactions often require extreme temperatures, harsh organic solvents and are generally not environmental friendly. In addition, the reactions usually give by-products that must be removed, which can be difficult and time consuming. By using a biocatalyst, would not only be sustainable but also catalytic advantageous as the enzyme catalyzed reaction is specific and often stereoselective, and does not produce by-products that must be removed.

To tame an enzyme, a general starting point is to find an enzyme that catalyzes a reaction that is similar to the wanted reaction. However, the nonnatural substrate probably does not fit the active site. Therefore, residues in the active site, both those that bind and place the substrate in the active site and those involved in the catalytic steps, probably must be modified. This may involve changing residues that hinder the

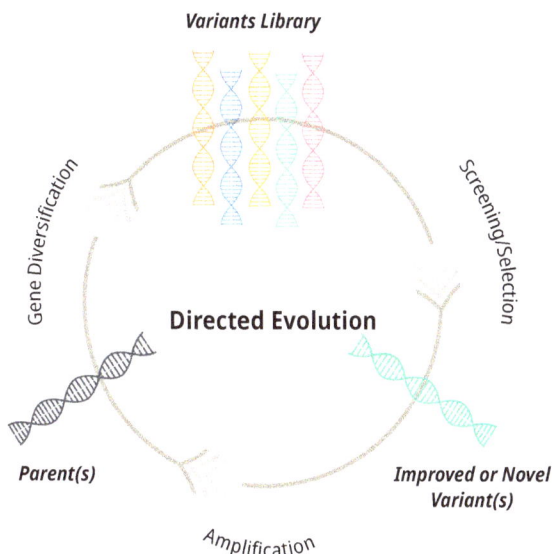

Figure 9.27: Principle of directed evolution. The parent gene of the target enzyme is mutated (by error prone PCR, DNA shuffle, staggered recombination or any mutation method) to diversify the gene pool. The gene pool is then expressed and screened for a "better" enzyme. The best variant(s) is selected, amplified and becomes the "new" parent. The process is repeated until an enzyme with the required catalytic and stability properties is obtained. Adapted with permission from Y. Wang, P. Xue, M. Cao, T. Yu, S.T. Lane and H. Zhao (2021) Directed Evolution: Methodologies and Applications. *Chem Rev* 121:12384–12444. Copyright 2021 American Chemical Society.

substrate to enter the active site or to enlarge the space of the active site to accommodate the substrate. In addition, catalytic groups need to be correctly placed to catalyze the reaction. Such changes may require mutations not only of residues in the active site but also residue in other parts of the enzyme. When the enzyme has been tamed and catalyzes the wanted reaction, it is probably necessary to improve the enzyme further to increase the reaction rate and/or stability.

Alcohol dehydrogenases (ADH) catalyze the reduction of an aldehyde, such as acetaldehyde, or oxidation of an alcohol, like ethanol. ADHs have proven to be suitable to repurpose to reduce ketons to secondary alcohols.

Rational design was used to change the specificity of *Thermoanaerobacter brockii* ADH to catalyze the reduction of "difficult-to-reduce bulky" ketones. Structure analysis of *T. brockii* ADH indicated that there are two hydrophobic pockets that bind the substrate. By introducing single mutations (A85G or I86L) in the active site, the enzyme catalyzed the reduction but at a very low rate, while mutating both residues increased the rate nearly 10 times. A third mutation (Q101A or C295A), also of residues in the active site, gave a further increase of the rate. In addition, the reaction was nearly 100% stereospecific (Figure 9.28).

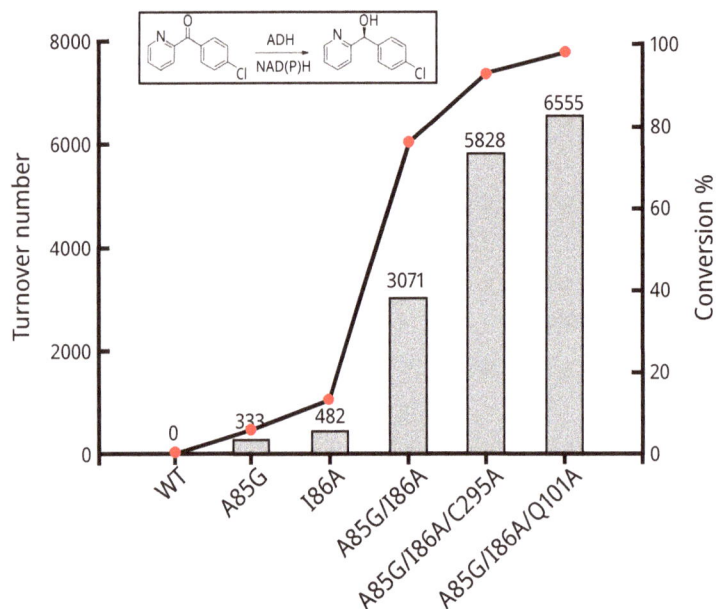

Figure 9.28: Rational design was used to change the specificity of *Thermoanaerobacter brockii* alcohol dehydrogenase to catalyze the reduction of a "difficult-to-reduce bulky" ketone. By introducing mutations in the substrate binding pockets, the active site could accommodated the substrate (the "difficult-to-reduce bulky" ketone) and convert the ketone to a secondary alcohol with a nearly 100% enantioselectivity. Adapted after G. Qu, B. Liu, Y. Jiang, Y. Nie, H. Yu and Z. Sun (2019) Laboratory evolution of an alcohol dehydrogenase towards enantioselective reduction of difficult-to-reduce ketones. *Bioresour Bioprocess* 6:18. Licensed by CC BY (creativecommons.org/licenses).

9.11 Enzyme classes

The unambiguous identification of any enzyme relies on a numerical system (EC number) devised by The International Union of Biochemistry. In this system, an enzyme is grouped according to the reaction catalyzed and given an EC number, which is made up of four digits. Enzymes catalyzing the same reaction have the same EC number, independent of the origin and sequence. There are seven classes, numbered from 1 to 7, and a number of sub-classes. For instance, class 1 includes all oxidoreductases, enzymes catalyzing redox reactions, such as alcohol dehydrogenase (EC 1.1.1.1)

$$\text{ethanol} + \text{NAD}^+ \rightleftharpoons \text{acetaldehyde} + \text{NADH} + \text{H}^+ \tag{9.37}$$

All classes, with example reactions, are listed in Table 9.3.

Table 9.3: Enzyme classification.

Class	Catalyzed reaction	Enzyme example
EC 1 Oxidoreductases	Redox reactions AH (red) + B $\rightleftarrows$ A (ox) + BH (red)	Dehydrogenases
EC 2 Transferases	Transfer of functional group from a substrate to another AB + C $\rightleftarrows$ A + BC	Transaminase, kinase
EC 3 Hydrolases	Hydrolytic cleavage of covalent bond AB + H_2O $\rightleftarrows$ AOH + BH	Lipase, peptidase, phosphatase
EC 4 Lyases	Nonhydrolytic cleavage of covalent bond RCOCOOH $\rightleftarrows$ RCHO + CO_2	Decarboxylase
EC 5 Isomerases	Intramolecular rearrangement ABC $\rightleftarrows$ ACB	Isomerase, mutase
EC 6 Ligases	Covalent joining of two molecules with the concomitant hydrolysis of ATP A + B + ATP $\rightleftarrows$ AB + ADP + P_i	Synthases
EC 7 Translocases	Moving ions or molecules across membranes	Transporters

Further reading

Arnold, F.H. (2018). Directed evolution: Bringing new chemistry to life. *Angew Chem Int Ed Engl* 57:4143–4148.

Bell, E.L., Finnigan, W., France, S.P., Green, A.P., Hayes, M.A., Hepworth, L.J., Lovelock, S.L., Niikura, H., Osuna, S., Romero, E., Ryan, K.S., Turner, N.J. and Flitsch, S.L. (2021). Biocatalysis. *Nature Rev Methods Primer* 1:46.

Blake, C.C., Koenig, D.F., Mair, G.A., North, A.C., Phillips, D.C. and Sarma, V.R. (1965). Structure of hen egg-white lysozyme. A three-dimensional Fourier synthesis at 2 angstrom resolution. *Nature* 206:757–761.

Briggs, G.E. and Haldane, J.B. (1925). A note on the kinetics of enzyme action. *Biochem J* 19:338–339.

Buller, R., Lutz, S., Kazlauskas, R.J., Snajdrova, R., Moore, J.C. and Bornscheuer, U.T. (2023). From nature to industry: Harnessing enzymes for biocatalysis. *Science* 382:eadh8615.

Chapin, J.C. and Hajjar, K.A. (2015). Fibrinolysis and the control of blood coagulation. *Blood Rev* 29:17–24.

Chowdhury, R. and Maranas, C.D. (2020). From directed evolution to computational enzyme engineering – A review. *AIChE Journal* 66:e16847.

Fischer, E. (1894). Einfluß der configuration auf die wirkung der enzyme. *Ber Dtsch Chem Ges* 27:2985–2993.

Hartley, B.S. and Kilby, B.A. (1954). The reaction of p-nitrophenyl esters with chymotrypsin and insulin. *Biochem J* 56:288–297.

Hedstrom, L. (2002). Serine protease mechanism and specificity. *Chem Rev* 102:4501–4524.

Hofstee, B.H. (1952). On the evaluation of the constants V_M and K_M in enzyme reactions. *Science* 116:329–331.

Huang, J. and Lipscomb, W.N. (2004). Products in the t-state of aspartate transcarbamylase: Crystal structure of the phosphate and N-carbamyl-L-aspartate ligated enzyme. *Biochemistry* 43:6422–6426.

Koshland, D.E. (1958). Application of a theory of enzyme specificity to protein synthesis. *Proc Natl Acad Sci U S A* 44:98–104.

Kursula, P. (2014). The many structural faces of calmodulin: A multitasking molecular jackknife. *Amino Acids* 46:2295–2304.

Lineweaver, H. and Burk, D. (1934). The determination of enzyme dissociation constants. *J Am Chem Soc* 56:658–666.

Vocadlo, D.J., Davies, G.J., Laine, R. and Withers, S.G. (2001). Catalysis by hen egg-white lysozyme proceeds via a covalent intermediate. *Nature* 412:835–838.

10 The striker

When it comes to protein folding, two very pertinent questions can be asked: *Why* do proteins fold and *how* do proteins fold?

Why a protein folds is simply because the folding process reduces Gibbs free energy ($G_{folded} - G_{unfolded} < 0$) of the native protein relative to the unfolded polypeptide chain. As mentioned before, the driving force of folding is the hydrophobic effect and that folding leads to a large number of noncovalent interactions that stabilizes the folded and native conformation of the protein.

Around the beginning of 1960s, Christian Anfinsen and colleagues showed in a series of groundbreaking experiments that it was possible to regain the native folded state and activity of an unfolded enzyme in the absence of any external factors. They used urea to unfold ribonuclease A in the presence of β-mercaptoethanol, a reducing agent that break disulfide bonds. When urea was removed first and thereafter β-mercaptoethanol, all enzyme molecules folded into the same structure and the activity was recovered to 100% (Figure 10.1). However, when both urea and β-mercaptoethanol were removed at the same time, only a fraction of the activity was recovered. Thus, a large fraction of the enzyme molecules folded into an incorrect and inactive structure. Ribonuclease A contains eight cysteines that form four disulfide bonds in the native protein, but when both denaturant and reductant were removed at the same time, the disulfide bonds were reformed but not correctly.

At the time, the interpretation of Anfinsen's experiments was very straightforward: the folded structure and the folding pathway were somehow encoded in the amino acid sequence and that a certain amino acid sequence always gave a certain folded structure. However, later it has turned out that this conclusion was not completely correct.

10.1 Levinthal's paradox

The other question, *how* a protein folds, is much more difficult to answer. If we take a 100 amino acid long polypeptide and assume that each bond connecting the amino acids can have three different states, there are $3^{99} = 1.7 \cdot 10^{47}$ possible conformations the polypeptide could attain. As pointed out initially by Cyrus Levinthal, even if the polypeptide can sample 10^{13} conformations per second, it would take more than $5 \cdot 10^{26}$ years to try them all. As proteins fold within seconds or even faster, it was concluded that it is not possible to find the correct folding by a random search of the conformation space. This is Levinthal's paradox.

Therefore, Levinthal reasoned that folding cannot occur by randomly probing all possible conformations a polypeptide chain can attain to find the one with lowest energy. However, the conformation space can be reduced by assuming that initially short local secondary structures form, which subsequently form larger and larger

https://doi.org/10.1515/9783111350684-010

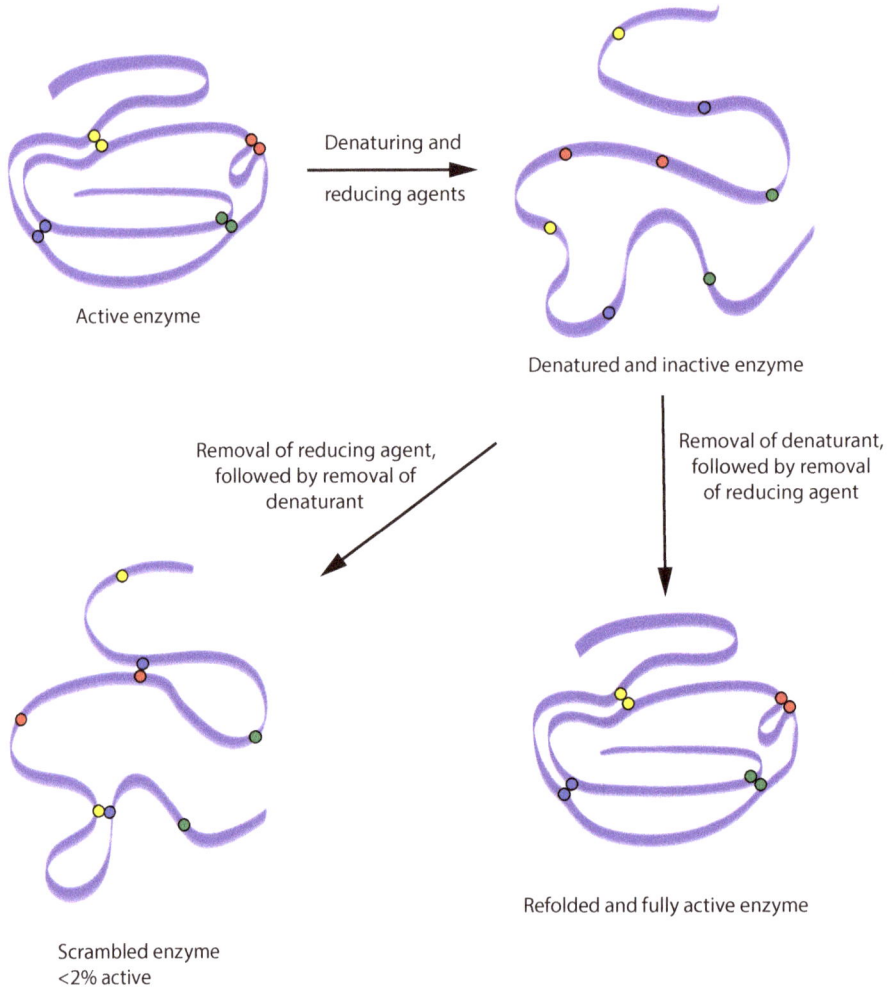

Figure 10.1: Anfinsen's classical experiment. Ribonuclease A was denatured by urea in the presence of the reducing agent β-mercaptoethanol that broke the four disulfide bonds present. The denatured and reduced enzyme lost all activity. When the reducing agent was removed before the urea, the four disulfide bonds reformed in all possible ways but only a few enzyme molecules reformed the disulfides as in the native enzyme. The activity of the "scrambled" enzyme was only a few percent of the native ribonuclease A. However, when the order was reversed, urea was removed before β-mercaptoethanol, all four disulfides were reformed correctly and the activity was fully regained.

structural elements to finally yield the conformation with lowest energy. Thus, instead of a random search of all possible conformations, the native structure is attained by a guided search of the conformational space.

10.2 *Why* do proteins fold?

The reason *why* proteins fold is that the free energy of the native state is less than that of the unfolded state

$$\Delta G = G_{\text{folded}} - G_{\text{unfolded}} < 0 \tag{10.1}$$

An unfolded protein makes numerous interactions with water whereas in the folded protein those noncovalent interactions are exchanged for intramolecular interactions. Hydrophobic residues tend to pack together and hydrogen bond donors and acceptors form intramolecular bonds, as those stabilizing α-helices and β-sheets. The energy contribution from each noncovalent bond in both the folded and unfolded states is small but due to the large number of interactions the total interaction energy is very large, several thousands of kilojoules per mole. Still folded proteins are only marginally stable as the free energy of unfolding is in the order of 20–60 kJ/mol (5–15 kcal/mol). This tiny difference corresponds to the breaking of one or two hydrogen bonds. The folded protein balance on a very narrow ridge and not very much is needed to push it over the edge.

The contribution to the change in free energy can be expressed as

$$\Delta G = \Delta G_{\text{polar}} + \Delta G_{\text{nonpolar}} + \Delta G_{\text{van der Waals}} \tag{10.2}$$

where ΔG_{polar} is the change in free energy due to hydrogen and ionic bonds. In the unfolded protein, polar groups form bonds with water molecules, whereas in the folded state some of those bonds are exchanged for intramolecular bonds with other polar groups. The contribution to free energy from interaction between polar groups and water molecules probably does not differ much between unfolded and folded states. Polar groups on the surface of the folded protein may interact with other polar groups, but due to the screening effect of the water molecules the energy contribution is small. Interactions between interior polar groups contribution to the stability of the folded protein, at least when these interactions occur in a nonpolar environment where the dielectric constant is much lower than that in water (~2 versus ~80). At the same time, a polar group that does not interact with any other groups, is a strong destabilizing factor.

$\Delta G_{\text{nonpolar}}$ represents the driving force of folding, the hydrophobic effect. Although the folding reduces the conformation space to a single conformation, the released caged water molecules reduces the entropy more than the folding increases it. Experimentally it has been found that each 1Å^2 buried in the folded state contribute somewhere around 100 J/mol (25 cal/mol).

As van der Waals interactions are strongly distance dependent, and packing in the folded state is much more compact, $\Delta G_{\text{van der Waals}}$ is negative and therefore contributes to stabilize the folded conformation.

10.3 *How* do proteins fold?

One way to describe *how* a protein folds is to describe the folding process as a downward path in a funnel-shaped energy landscape where the bottom represents the native structure as well as the conformation with lowest energy. The top of the funnel represents free energy of the unfolded protein molecules, with its unlimited conformations; the width of the funnel is related to the entropy. Downward the funnel is narrower, representing fewer and fewer conformations and therefore also reduced entropy (Figure 10.2).

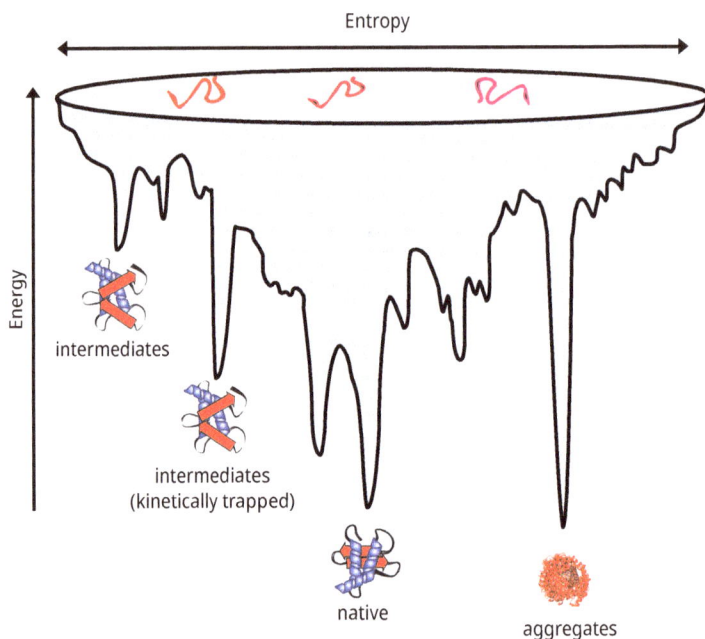

Figure 10.2: A rugged folding funnel. The top of the funnel represents the conformational space of the denatured state. Each polypeptide chain follows its own way down the energy landscape. Some of these paths lead to a local energy minimum and a trapped intermediate conformation, whereas others lead to the global minimum and the native conformation. Adapted with permission after A.I. Bartlett and S.E. Radford (2009) An expanding arsenal of experimental methods yields an explosion of insights into protein folding mechanisms. *Nature Struct Molec Biol* 16:582–588.

So-called two-state proteins are either folded or unfolded; no transition state intermediates are observed. Folding of a two-state protein proceeds down a smooth funnel-shaped landscape with no significant kinetic traps. Other proteins may encounter a rugged energy landscape with valleys and barriers, which may hinder the polypeptide to fold to the native conformation or lead to accumulation of misfolded intermediates, as happened in Anfinsen's experiments. A trapped intermediate may need to unfold before it can find a new and better way down the energy landscape.

At the beginning of the folding process short regions with predominantly hydrophobic amino acids form a secondary structure element that may be present in the final native structure. This model, the framework model, is challenged by the nucleation condensation model. In the nucleation condensation model, a small folding nucleus is formed at the transition state and formation of secondary and tertiary structures occur simultaneously on the downhill side of the energetic barrier. If misfolded, the structure element must unfold and move uphill the energy landscape before refolding. On the way downhill, such structure elements come together and interact to build larger and larger structural elements.

The energy landscape theory implies that the folding process can start from any of the unfolded molecules. Therefore there is not a single pathway but many pathways downhill the energy landscape. Thus, folding can be seen as several parallel processes of a large ensemble of polypeptide chains that all strive to reach the bottom of the landscape funnel. Ideally, all folding paths should converge at the bottom of the funnel and yield the same native conformation. However, some polypeptides can be trapped in deep energy valleys in misfolded conformations due to nonnative interactions within the polypeptide or between different polypeptides. These misfolded intermediates may unfold and follow another path down the energy landscape and finally attain the native state. It is also possible that they are unable to unfold and therefore end up in a misfolded structure that need to be degraded to recycle the amino acids. It is also possible that misfolded intermediates form stable oligomers or higher aggregates with even lower free energy than the native state. Such aggregates have been linked to neurological disorders such as Alzheimer's disease.

The foldon model is an alternative explanation to *how* proteins fold. It is suggested that short stretches of 20–30 residues, called foldons, fold into a native or native-like structure and that folding occurs through a stepwise formation of cooperative foldon units. The process follows a single pathway energetically downhill towards the final native state. Also, in this model it is possible that intermediates can be trapped in local energy minima. Unfolding follows the same trajectory but in the opposite direction (Figure 10.3).

It is important to realize that a protein is either native and functional or denatured and nonfunctional. No proteins are "half-native" and "half-functional." It is likewise important to understand that there is an equilibrium between the folded and unfolded states; a protein that is folded at one time may be unfolded at another time. Like in all reactions, there is an activation barrier ($\Delta G^{\#}$) that determines the rate of the folding reaction whereas the change in Gibbs free energy ($\Delta G = G_{folded} - G_{unfolded}$) determines the equilibrium.

Most knowledge of protein folding comes from experimental and theoretical studies on isolated proteins. The real cellular milieu is much more complex, with a multitude of proteins, organelles and small molecules that could and probably will affect the folding process. Thus, the cell volume is crowded. Crowding or the excluded volume effect increases the factual concentration of dissolved molecules. This may affect

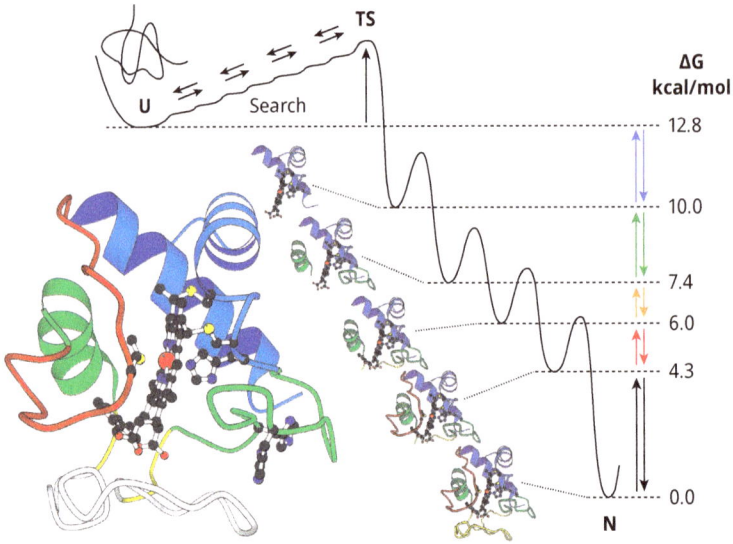

Figure 10.3: The foldon model illustrated with the trajectory of cytochrome c (cyt c). The stepwise folding process follows a single pathway energetically downhill towards the final native state (N). U: unfolded state; TS: transition state. Adapted with permission after W. Hu, Z.-Y Kan, L. Mayne and S. W. Englander (2016) Cytochrome c folds through foldon. *Proc Natl Acad Sci* 113: 3809–3814.

the stability of both native and nonnative conformations as well as increase the folding rate. At the same time crowding may reduce diffusion rates that could make it more difficult for subunit in multisubunit proteins to encounter each other and associate. It is also possible that the presence of other proteins leads to intermolecular interactions that influence the folding process, which in the worst cases lead to nonfunctional proteins.

10.4 Co-translational folding

Protein synthesis is linear and the N-terminal region is synthesized first and it is also much slower than folding, to synthesize a 300-residues protein takes around 3 min for a bacterium whereas folding often occurs in milliseconds. The folding process may start as soon as the polypeptide chain emerges from the exit tunnel of the ribosome or even before inside the tunnel. Therefore it is possible that the initial folding leads to faulty interactions and misfolded intermediates. However, Nature has developed systems to evade this (Figure 10.4).

Heat shock proteins (Hsp) are a family of proteins that are upregulated by cellular stress. They were first discovered in cells stressed by heat, hence the name, and have since been found to be essential for all organisms. Many members of the Hsp family

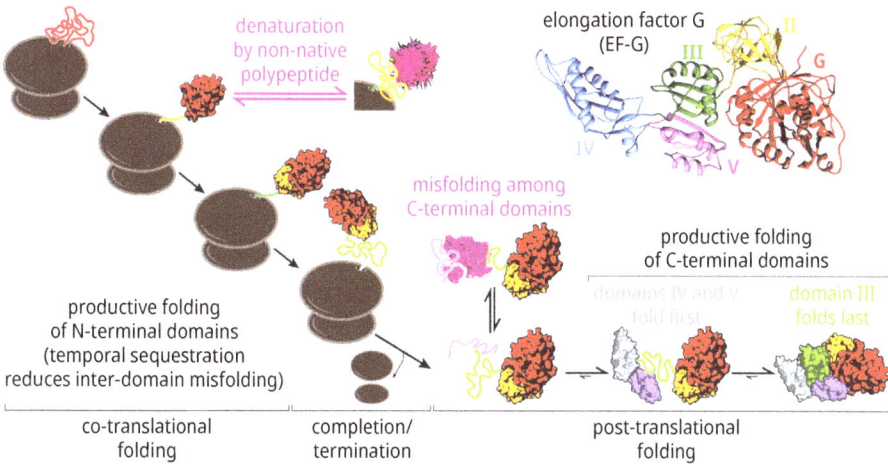

Figure 10.4: Chaperone assisted co-translational folding of elongation factor G. In the presence of the chaperone (trigger factor), the two N-terminal domains I and II (red and yellow) fold correctly. However, without the trigger factor protein, domain II causes the initial folded domain I to unfold. When domain III (green) emerge, folding but not translation is disrupted as the folding of domain III require stabilizing contacts with domains IV and V. This results in an accumulation of unfolded polypeptide during post-translation folding that has a propensity to misfold. Reprinted from N. Rajasekaran and C. M. Kaiser (2022) Co-translational folding of multi-domain proteins. *Front Mol Biosci* 9:869027. Licensed by CC BY (creativecommons.org/licenses).

perform a chaperone function. For instance, Hsp70 and Hsp90 both assist other proteins to fold correct, by binding extended peptide segments or partly folded proteins to prevent misfolding or aggregation. The process is energy-dependent as Hsp-bound ATP is consumed during the process.

The names of Hsps are related to their molecular weights; Hsp60, Hsp70 and Hsp90 are heat shock proteins with molecular weights around 60, 70 and 90 kDa, respectively.

Chaperonins are another type of folding helpers; they give a misfolded polypeptide a second chance to fold correctly. Chaperonins form a double ring structure, resembling two tires stacked on top of each, with a central cage where a misfolded polypeptide is assisted to refold correctly. The bacterial GroEL and the eukaryotic chaperonin-containing TCP-1 (CCT), also called T-complex 1 ring protein (TRiC), have similar ring structures. GroEL requires the co-chaperonin GroES that functions as a lid to the cage (Figure 10.5). There are two general views (not mutual exclusive) on how chaperonins function. One view is that the cage reduces the conformation space accessible to the polypeptide and prevents aggregation and thereby facilitates proper folding. In the other view, the chaperonin has an active role by forming interactions with the polypeptide that guides the folding process. Energy is required for the refolding process; ATP is hydrolyzed during the process.

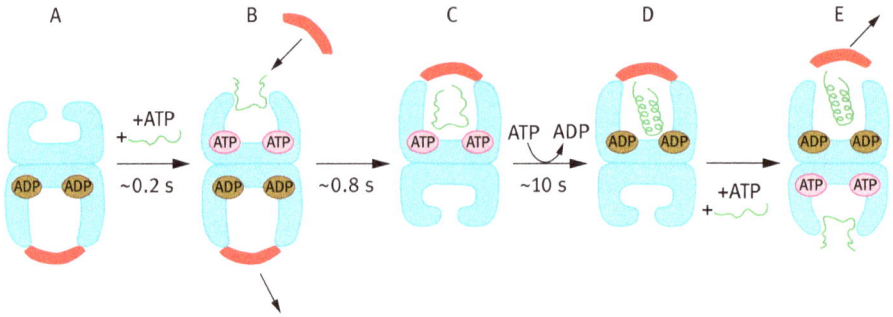

Figure 10.5: The reaction cycle of a chaperonin. The reaction cycle begins with the GroEL-GroES-ADP complex (A). ATP and a misfolded polypeptide (green) bind to the open ring (B). ATP binding induces conformational changes enabling GroES (red) binding, producing a stable folding complex (C). The misfolded polypeptide refolds and ATP hydrolyses (D). GroES and the folded polypeptide dissociate and ATP and a misfolded polypeptide bind to the opposite ring (E). Adapted with permission from A.L. Horwich, A.C. Apetri and W.A. Fenton (2009) The GroEL/GroES cis cavity as a passive anti-aggregation device. *FEBS Lett* 583:2654–2662.

If the chaperonin fails and the polypeptide is still misfolded, it may enter the cage again for another try to fold properly.

Another type of folding helpers is protein disulfide isomerase (PDI) and peptidyl-prolyl *cis-trans* isomerase (PPI). PDI is present in the endoplasmatic reticulum in eukaryotic organisms and in the periplasmic space in bacteria. PDI catalyzes formation and breakage of disulfide bonds, which aids a protein to fold and find proper arrangements of disulfide bonds.

Nearly all peptide bonds in a folded protein are in a *trans* conformation. It is estimated that only 0.05% of nonproline peptide bonds are *cis* while around 6% X-Pro peptide bonds have a *cis* conformation. In an unfolded polypeptide, the ration between *cis* and *trans* is around 1:4. As the activation energy for a *cis* to *trans* conversion is very high (80 kJ/mol), this conversion will slow down folding. PPI catalyzes the isomerization of a *cis* bond to a *trans* bond (and reverse), thereby speeding up the folding process considerably.

10.5 Proteins with two distinct structures

In the 1980s, Stanley Prusiner suggested that prion proteins are the causative agent of some nervous system diseases. Later it has been shown that Creutzfeldt-Jakob disease, Mad Cow disease (bovine spongiform encephalopathy), sheep scrapie and other related neuronal disorders are caused by prions. The diseases are caused by the conversion of the normal cellular prion protein (PrP^C) to the disease-causing isoform PrP^{Sc}. The presence of PrP^{Sc} stimulates the conversion of PrP^C to PrP^{Sc}.

The hamster PrP^C is synthesized as a 254-amino acid precursor protein. In the endoplasmatic reticulum, a 22-amino acid residue signal sequence at the N-terminal and a 23-amino acid peptide at the C-terminal are removed. PrP^C is glycosylated and a glycosylphosphatidylinositol anchor at the C-terminal is attached. The final 209-amino acid protein contains several repeated sequences (hexapeptides and octapeptides) at the N-terminal region. The lipid anchor attaches Prp^C to the cell membrane. The membrane-bound PrP^C is internalized with time (half-life ca. 5 h) and degraded by the lysozyme. The structure of PrP^C is mainly α-helical with two short antiparallel β-strands (Figure 10.6).

A   B

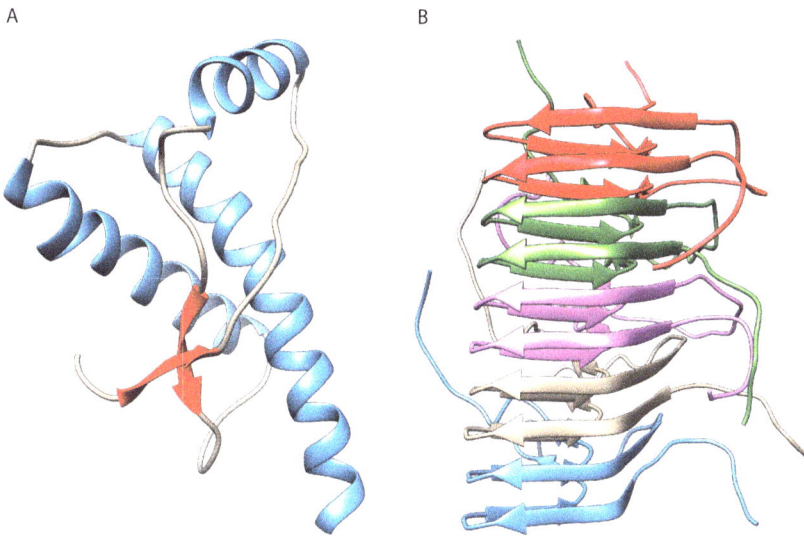

Figure 10.6: Prion proteins. (A) The structure of human PrP^C was determined by solution NMR (pdb: 1QLX). The structure is mainly α-helical (blue) but there are two short antiparallel β-strands (red). (B) Side view of the five central molecules of the fungal HET-s fibril determined by solid state NMR (pdb: 2RNM).

Due to the propensity to aggregate and form fibrils, it has not been possible to obtain a high-resolution structure of PrP^{Sc}. However, spectroscopic studies have indicated that PrP^{Sc} contains 40–50% β-sheet and 50–60% coils or turns. Several models of PrP^{Sc} have been proposed, often based on structures of other misfolding prone proteins like amyloid-β and the fungal HET-s protein. Although not proven experimentally, it is likely that fibrils of PrP^{Sc} have a similar β-solenoid fold as the HET-s fibril.

Amyloid-β peptides are generally believed to be a major cause of Alzheimer's disease. These peptides are released in the extracellular volume in the brain after proteolytic cleavage of the membrane-bound amyloid precursor protein (APP). The native functional role of APP is not well understood but it has been proposed that APP is involved in cell signaling, long-term potentiation and cell adhesion.

Humans express several alternatively spliced isoforms of APP. The most abundant neuronal isoform is a 695 amino acid long polypeptide, consisting of several functional domains. The protein undergoes extensive modifications such as glycosylation, phosphorylation, sialylation and sulfation. It is also degraded by proteases of the secretase family. Most of the extracellular domain is removed by α-secretase and β-secretase. The amyloid-β peptides ($A\beta_{1-40/42}$) are generated by γ-secretase proteolysis within the membrane-spanning region if APP first is cleaved by β-secretase (Figure 10.7).

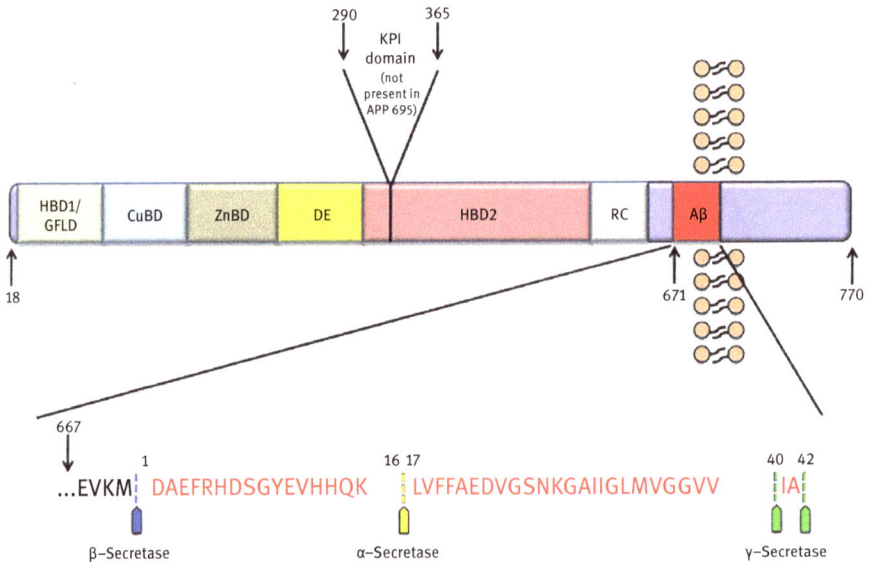

Figure 10.7: Domain structure of amyloid precursor protein. The amyloid precursor protein contains several functional domains. A heparin binding and growth factor-like domain (HBD1/GFLD), a copper binding domain (CuBD), a zinc binding domain (ZnBD), an acidic region (DE), a Kunitz-type protease inhibitor domain (KPI), another heparin binding domain (HBD2), a random coil region (RC) and the amyloid-β domain (Aβ). The KPI domain is not present in the predominant neuronal amyloid precursor protein (APP$_{695}$). The insert displays the sequence of the human amyloid b peptide with the sites of proteolysis by α-, β- and γ-secretase. Reprinted from O. Lazarov and M.P. Demars (2012) All in the family: How the APPs regulate neurogenesis. *Front Neurosci* 6:81. Licensed by CC BY (creativecommons.org/licenses).

The sequence and experimental data indicate that the domain corresponding to $A\beta_{1-42}$ in the full-length APP is α-helical. However, when released into the extracellular volume, $A\beta_{1-42}$ changes conformation, from α-helical to β-strands. These β-strands aggregate and form amyloid fibrils, which are typical of the plaques found in brains of Alzheimer patients (Figure 10.8).

It appears that many polypeptides have a propensity to aggregate and form amyloid fibrils, as long as the solvent conditions are appropriate. Even the highly soluble and non-aggregation-prone SH3 domain, nucleates and forms amyloid fibrils at acidic pHs. The structure of SH3 amyloid fibrils consists of aggregated β-strands, similar to those of $A\beta_{1-42}$.

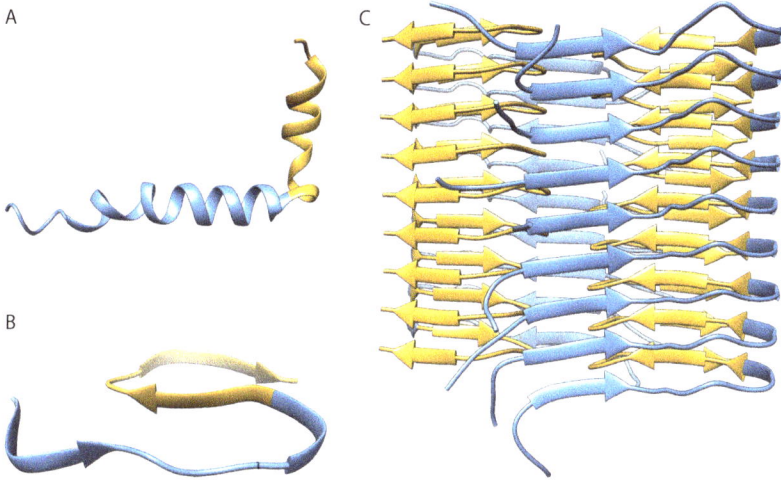

Figure 10.8: Structure of amyloid beta$_{1-42}$. (A) A soluble structure of Aβ_{1-42}. The membrane spanning (gold) and extracellular (blue) α-helices are indicated. (pdb: 1IYT). (B) Structure of one Aβ_{1-42} peptide extracted from the structure of an amyloid fibril. (pdb: 2MXU). (C) Structure of a fibril core consisting of a dimer of Aβ_{1-42} molecules (pdb: 5KK3).

10.6 Intrinsic disordered proteins

Some proteins or regions of certain proteins are disordered, and it is not until they bind to their targets, they become structured, as illustrated in Figure 10.9. These regions often have a sequence bias for small and hydrophilic amino acids as well as proline residues. As the binding surface is not developed until it binds to the target, it is possible to adjust the contact area that can increase both binding strength and specificity. A disorder protein (or region) involved in signaling can interact with different receptors and thereby transmit the signal to several pathways. An unstructured region can also function as a dynamic linker between two domains. In some sense, a disordered region in a protein provides increased versatility.

10.7 Moonlighting

Moonlighting proteins represent a group of proteins in which a single protein has more than one biochemical or biophysical functions that are not the result of alternative splicing; a single polypeptide chain performs two or more functions. Such moonlighting proteins are found in all organisms, from man to bacteria.

The crystallins, the major proteins of the eye's transparent lens, are a diverse group of proteins responsible for the optical properties of the lens. Epsilon- and tau-crystallins

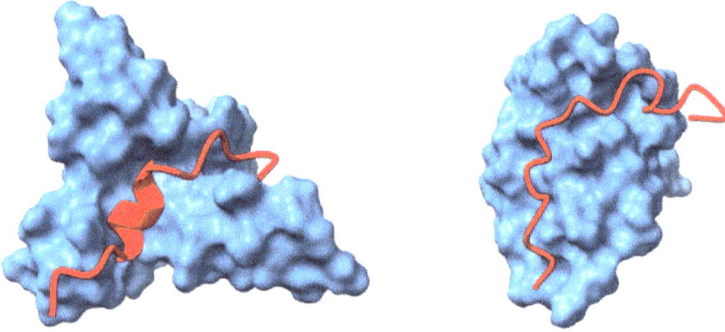

Figure 10.9: Malleability of an intrinsic disordered protein. When the N-terminal transactivation domain of p53 (red) binds to the target protein (blue; pdb: 2MDZ and 2RUK) it folds to fit the binding surface on the target.

of the lens of ducks and other birds are in fact lactate dehydrogenase and enolase, each encoded by a single-copy gene. Thus, both proteins fulfill their role as catalyzing agents in glycolysis and at the same time perform a structural function in the eye.

It has been suggested that during evolution moonlighting enzymes often have been recruited from metabolic enzymes and stress proteins. Hexokinase doubles as transcription regulator in yeast and several other glycolytic enzymes, such as phosphofructokinase, triosephosphate isomerase, aldolase, glyceraldehyd-3-phosphate dehydrogenase and enolase, also act as plasminogen- and fibronectin-binding proteins. The bacterial chaperonin Hsp60 that prevent protein misfolding is also an adhesin that attaches the bacterium to the target cell.

The human glucose-6-phosphate isomerase (GPI) catalyzes the interconversion of glucose-6-phosphate and fructose-6-phosphate in the glycolysis. However, extracellularly GPI functions as a neurotrophic factor promoting survival of neurons and when secreted from tumors GPI becomes an autocrine motility factor.

Gene sharing has some obvious advantages. It is economical to use the same gene to produce a single protein with several function, as it makes the genome smaller and thus cheaper (in energy cost) to replicate during cell division. There is no need to "invent" a new protein with that particular function, a process that is slow and costly.

The human opiomelanocortin protein is an example on how to economize by other means. This protein is expressed as a 285-amino acid long prepro-polypeptide, with a 44-amino acid long signal peptide, mainly in the anterior pituitary gland. Removal of the signal peptide yields a pro-polypeptide with 241 amino acids that undergoes extensive tissue-specific proteolytic processing by prohormone convertases and other proteases. Processing in the pituitary gland generates adrenocorticotropic hormone that regulates cortisol and androgen production and β-lipotropin, a precursor of β-endorphin. Further tissue-specific processing generates several different hormones as illustrated in Figure 10.10.

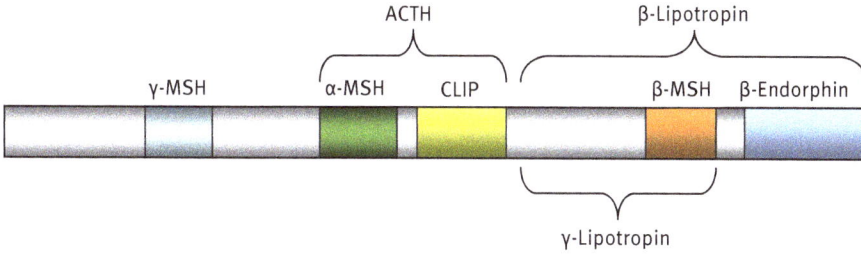

Figure 10.10: The pro-opiomelanocortin protein. The pro-polypeptide is a precursor for several melanotropins, lipotropins and endorphins. Proteolytic cleavage in the pituitary gland generates adrenocorticotropic hormone (ACTH) and β-lipotropin. Further tissue-specific processing generates melanotropins (α-MSH, β-MSH and γ-MSH), corticotropin-like intermediary peptide (CLIP) and β-endorphin. Reprinted from M.S. Menditatta, Y. Yang, A.E. Balazs, A.S. Willis, C.M. Eng, L.P. Karaviti and L. Potocki (2011) Early onset obesity and adrenal insufficiency associated with a homozygous POMC mutation. *Int J Pediatr Endocrinol* 2011:5. Licensed by CC BY (creativecommons.org/licenses).

Further reading

Anfinsen, C.B. (1973). Principles that govern the folding of protein chains. *Science* 181:223–230.

Checler, F. and Vincent, B. (2002). Alzheimer's and prion diseases: Distinct pathologies, common proteolytic denominators. *Tr. Neurosci* 25:616–620.

Finkelstein, A.V., Bogatyreva, N.S., Ivankov, D.N. and Garbuzynskiy, S.O. (2022). Protein folding problem: Enigma, paradox, solution. *Biophys. Rev* 14:1255–1272.

Hu, W., Kan, Z.Y., Mayne, L. and Englander, S.W. (2016). Cytochrome c folds through foldon-dependent native-like intermediates in an ordered pathway. *Proc Natl Acad Sci USA* 113:3809–3814.

Jeffery, C.J. (2016). Protein species and moonlighting proteins: Very small changes in a protein's covalent structure can change its biochemical function. *J Proteomics* 134:19–24.

Levinthal, C. (1969). How to fold graciously. In: Mössbauer spectroscopy in biological systems: Proceedings of a meeting held at Allerton House. Monticello, Illinois, Debrunner, P., Tsibris, J.C.M. and Munck, E. Urban-Champaign, University of Illinois Press: 22–24.

Motojima, F. (2015). How do chaperonins fold protein?. *Biophysics (Nagoya-shi)* 11:93–102.

Müller, U.C., Deller, T. and Korte, M. (2017). Not just amyloid: Physiological functions of the amyloid precursor protein family. *Nat Rev Neurosci* 18:281–298.

Nassar, R., Dignon, G.L., Razban, R.M. and Dill, K.A. (2021). The protein folding problem: The role of theory. *J Mol Biol* 433:167126.

Rajasekaran, N. and Kaiser, C.M. (2022). Co-translational folding of multi-domain proteins. *Front Mol Biosci* 9:869027.

Wolynes, P.G. (2015). Evolution, energy landscapes and the paradoxes of protein folding. *Biochimie* 119:218–230.

11 Full-time

There are many different biological machines; some move cells or whole organisms, whereas others convert energy from one form to another, translate genetic information or transport cargo from one place in the cell to another.

Like any engine, these biological machines also require fuel to run. Some are driven by hydrolysis of ATP, like muscle contraction, cargo transport and mitosis. Motors attached to membranes, such as the ATP synthase and the bacterial flagellar motor, utilize a concentration difference of protons over the membrane.

11.1 Rotary motors

F_1F_0-ATPase or ATP synthetase is a ubiquitous biological rotary motor conserved throughout all organisms. It is found in the mitochondrial inner membrane, the thylakoid membrane of plants and the plasma membrane of bacteria. It is composed of two functional motors: a membrane-spanning F_0 motor and a membrane-extrinsic F_1 motor. Independent of the origin of the ATP synthases, their overall compositions and structures are similar, although the subunit composition of the two complexes can differ. The F_1F_0-ATPase uses energy from a proton gradient and the membrane potential, the proton motive force, to produce ATP from ADP and P_i ($H_2PO_4^-$).

The membrane-embedded F_0 is composed of 8–15 α-helical c-subunits, forming the c-ring, and the a-subunit. The number of c-subunits in the c-ring differs between species. For instance, the c-rings of vertebrates contain eight c-subunits and that of *Escherichia coli* 10 c-subunits.

The F1 motor consists of three globular heterodimers, each composed of an α-subunit and a β-subunit (Figure 11.1). The three heterodimers surround the γ-subunit (the stalk). The stalk extends below the heterodimers and binds to the F_0 motor. In bacteria and chloroplasts, the binding is augmented by the ε-subunit (in mitochondrial F_1F_0-ATPase the analogous subunit is called δ).

The peripheral stalk couples F_0 and F_1 to each other. It consists of two identical and α-helical b-subunits, which form an amphiphatic coiled-coil dimer that spans the periphery of the whole complex. In some organisms the two b-subunits are not identical but rather homologous. The transmembrane region of the dimer binds to the membrane-embedded a-subunit. The other end of the peripheral stalk attaches to one of the three α-subunits of the knob. The δ-subunit provides additional coupling between the b-subunits and the N-terminal region of the α-subunit.

The c-ring and the attached γ- and ε-subunits form the rotor, the moving part of the F_1F_0-ATPase. When the c-ring of the F_0 motor rotates, the central shaft (the γ- and ε-subunits) transfers the torque to the F_1 motor.

https://doi.org/10.1515/9783111350684-011

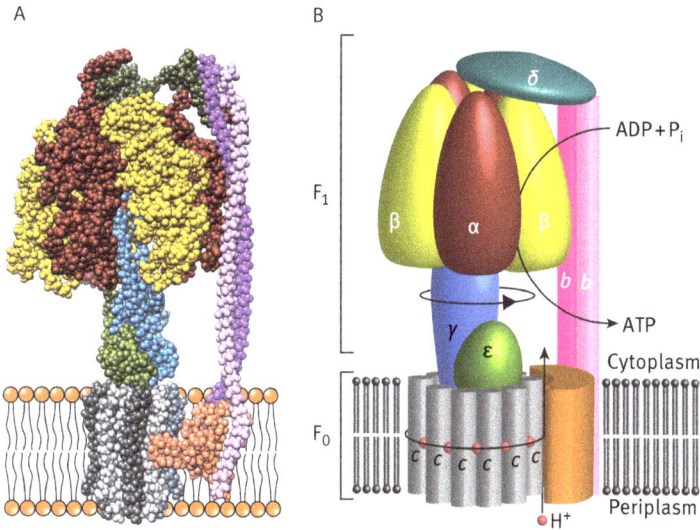

Figure 11.1: The ATP synthase. (A) The structure of *E. coli* F_1F_0-ATPase determined by cryo-electron microscopy (pdb: 5T4O). (B) Schematic structure of the F_1F_0-ATPase. The F_0 motor consists of the c-ring (gray) and the a-subunit (brown). The ATP-producing F_1 motor consists of three heterodimers, each composed by an α-subunit (dark red) and a β-subunit (yellow), a central shaft composed by the γ-subunit (blue) and the ε-subunit (green). The peripheral stalk (pink) connects the F0 motor with the F1 motor through interactions between b-subunit (pink) and the a-subunit in the membrane and between one of the α-subunit (dark red) and b-subunits (pink), augmented by the δ-subunit (dark green). The same color scheme was used in both A and B. Reprinted with permission from M. Sobti, C. Smits, A.S.W. Wong, R. Ishmukhametov, D. Stock, S. Sandin and A.G. Stewart (2016) Cryo-EM structures of the autoinhibited E. coli ATP synthase in three rotational states. *eLife* 5:e21598. Licensed by CC BY 4.0 (creativecommons. org/licenses).

Each c-subunit has a negatively charged aspartate or glutamate residue (Asp_{61} in *E. coli* c-subunits) in the middle of the C-terminal α-helix. This side chain is exposed on the external circumference of the c-ring. Two c-subunits and the a-subunit form a passage that protons can enter, as illustrated in Figure 11.2. The proton neutralizes the negatively charged carboxy group in the interface between the c-ring and the a-subunit. This causes the side chain to move away from the a-subunit and generates a rotation of the c-ring (clockwise when viewed from F_0 toward F_1). The generated rotation places a charged carboxy group at the end of the passage, which allows another proton to enter the passage and neutralize the charge, generating another rotational substep. By these rotational substeps, the protons are carried around until they reach a second site on the a-subunit where the neutralize carboxy group is reionized and the proton is released through another narrow passage into the opposite side of the membrane. As the shaft (the γ- and ε-subunits) is attached to the c-ring, the torque generated by the proton gradient is transmitted to the F_1 motor.

A

B

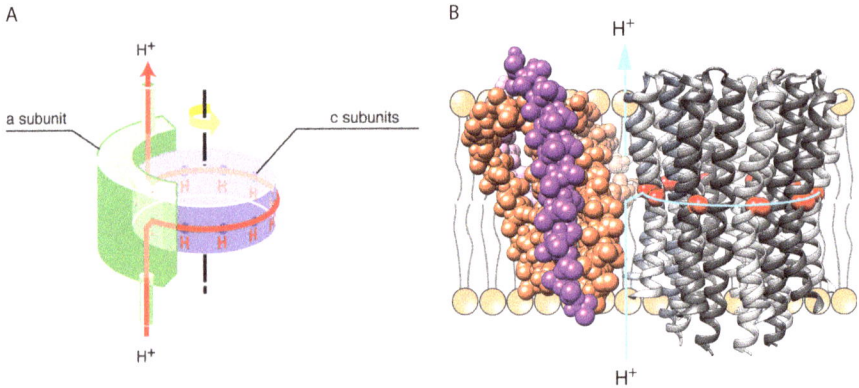

Figure 11.2: The F_0 motor. (A) Schematic illustration of the rotation of the c-ring. A proton enters through a narrow passage between the a-subunit (green) and two c-subunits and neutralizes a negatively charged carboxy group on the interface between the a- and c-subunits. This induces a counterclockwise rotation of the c-ring that allows another proton to enter and rotate the rotor another substep. After a full turn, the protons reach a site where the carboxy group is reionized and the protons released through another narrow passage into the opposite side of the membrane. Reprinted with permission from J.E. Walker (2013) The ATP synthase: the understood, the uncertain and the unknown. *Biochem Soc Trans* 41: 1–16. (B) The path of protons (light blue) through the a-subunit (brown) and the c-ring (gray) illustrated on the structure of *E. coli* F_1F_0-ATPase (pdb: 5T2O). The carboxy group in the middle of the C-terminal α-helix in each c-subunit is indicated in red.

The β-subunits of F_1 catalyzes the synthesis of ATP in a process where each β-subunit cycles through four conformational states due to the rotation of the asymmetric central shaft (the γ-subunit). The empty β-subunit ($β_E$) converts to a half-closed state ($β_{HC}$) by a ~30° rotation of the shaft (Figure 11.3). This conversion is necessary for binding of ADP and P_i to $β_{HC}$. As the rotation continues, the half-open state converts to a closed state ($β_{DP}$) and ATP forms from the bound ADP and P_i ($β_{TP}$). The synthesized ATP releases from $β_{TP}$ as it opens and converts back to the empty state (βE). Each clockwise (as viewed from beneath) 360° rotation of the shaft leads to the formation of three ATP molecules, one from each of the three β-subunits.

Since a 360° rotation of the F_0 c-ring requires the translocation of an equal number of protons as there are c-subunits in the c-ring, the direct energy cost is the number of c-subunits divided by 3. The bovine mitochondrial F_1F_0-ATPase (and probably the F_1F_0-ATPase of all other vertebrates and invertebrates) has a c-ring with 8 c-subunits, which gives a proton to ATP ratio of 2.7. Since most of the ATP is used outside the mitochondria, ATP must translocate from the mitochondrial matrix, where it is produced, to the cytoplasm and ADP must enter the mitochondria to support further synthesis of ATP. The adenine nucleotide translocator catalyzes the electrogenic ATP/ADP exchange across the mitochondrial inner membrane, driven by the mitochondrial membrane potential, a component of the proton motive force.

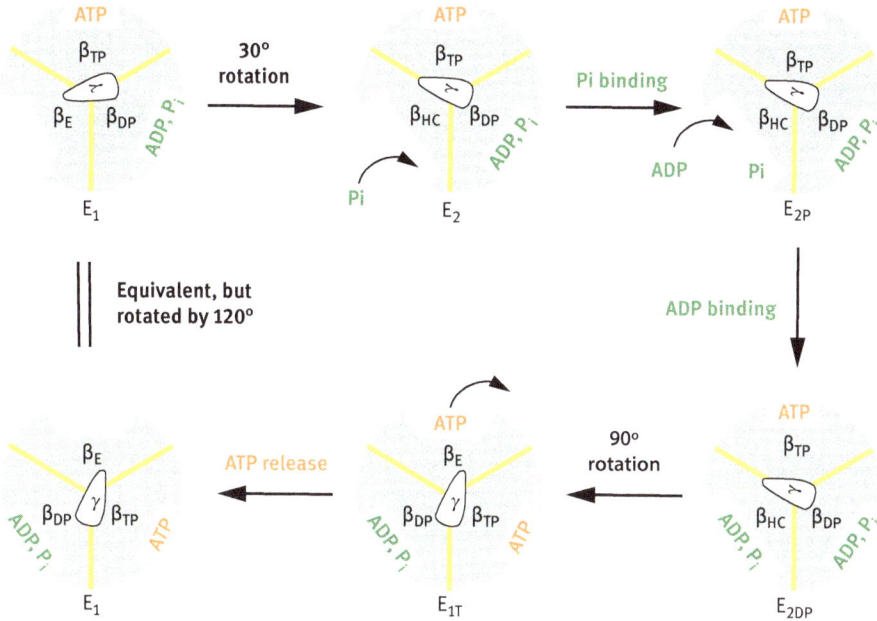

Figure 11.3: Proposed model for synthesis of ATP by the F_1 motor. The catalytic sites of the β-subunits are labeled $β_E$ (empty), $β_{HC}$ (half-closed), $β_{DP}$ (ADP and P_i bound) and $β_{TP}$ (ATP bound). The rotation of the c-ring rotates the γ-subunit ~30° and converts $β_E$ to $β_{HC}$, which binds ADP and Pi. As the rotation continues, $β_{HC}$ closes ($β_{DP}$) and ATP forms and $β_{TP}$ converts back to $β_E$ and ATP releases. This constitutes a third of a 360° rotation. Adapted with permission from Y.Q. Gao, W. Yang and M. Karplus (2005) A structure-based model for the synthesis and hydrolysis of ATP by F1-ATPase. *Cell* 123: 195–205.

The structure of the ~30 kDa human ADP/ATP translocase 1 (SLC25A4) contains six transmembrane α-helices that form a barrel with a deep cone-shaped depression in the mitochondrial inner membrane. When the depression faces the matrix, ATP can bind and induce an eversion that releases ATP in the intermembrane space. After that, cytoplasmic ADP binds to the depression, induces the conformational change and translocates into the matrix.

A H^+/P_i symporter catalyzes replenishment of the mitochondrial P_i but at an energy cost as a proton is co-transported with every P_i molecule. Therefore, ~3.7 protons are required for the production of each mitochondrial ATP molecule rather than ~2.7 protons.

Since the c-ring of spinach chloroplast F_1F_0-ATPase contains 14 c-subunits, synthesis of each ATP molecule requires ~4.7 protons. However, this ATP synthesis occurs on the outside of the chloroplast where it is used and does not require translocation of ADP and P_i across the chloroplast membrane.

When the membrane potential is small, the F_1F_0-ATPase can run in the reverse direction, and function as a proton pump. Thus, instead of synthesizing ATP, hydrolysis of ATP reverses the rotation of the c-ring and transports protons the opposite way. Bacteria use this to pump protons over the plasma membrane to the periplasmic space to create a proton gradient.

The vacuolar (H^+)-ATPases (V_1V_0-ATPases) are the major ATP-dependent proton pumps in eukaryotes. They are located in a number of cellular membranes of eukaryotes and play important roles in many physiological processes. The structure of the V_1V_0-ATPase is similar to that of F_1F_0-ATPase; a peripheral V_1 domain that catalyze hydrolysis of ATP and a membrane-spanning V_0 domain that translocates protons across the membrane into the lumen of organelles such lysosomes, endosomes and vacuoles. The c-ring of the V_0 domain consists of several different but homologous subunits that form a hexameric ring structure.

In Archaea, ATP is produced by A_1A_0-ATPases. Like the F_1F_0-ATPases, they use a proton gradient to drive the synthesis of ATP but can also use sodium ions. The A_1A_0-ATPases can also work in the reverse direction and can work as an ATP-driven ion pump. The general structure of the archaeal ATPases are similar to other rotary ATPases but has several distinct structural features such as an extended central shaft and two peripheral stalks.

11.1.1 Bacterial flagellar motor

The bacterial flagellar motor has a similar construction as that of ATPases; a torque generating rotor and stabilizing stators. This rotary motor rotates the bacterial flagella and allows bacteria to swim. An *E. coli* cell has six to eight motors, each capable to rotate the flagellum up to 18,000 rpm (300 Hz) and able to switch between clockwise and counterclockwise rotation nearly instantly. When all motors rotate counterclockwise, the cell swims rather straight, but when one (or more) motors switch to clockwise rotation, the cell tumbles and makes a turn.

The *E. coli* motor is anchored to the peptidoglycan through the stators, which also convert the proton flux into rotation of the rotor. The proton flux generates rotation of the cytoplasmic c-ring, assembled from the FliG (26–34 copies), FliM (~34 copies) and FliN (>100 copies) mounted on the MS ring (membrane and supramembrane ring). The rod attaches to the c-ring and spans the peptidoglycan and the outer membrane. The L and P rings are believed to function as bearings for the rod. On the extracellular side, the rod connects to the curved hook and the flagellar filament, as shown in Figure 11.4.

A chemotactic signal induces phosphorylation of CheY. Phospho-CheY interacts with the N-terminal of FliM of the c-ring and changes the counterclockwise rotation of the motor to a clockwise rotation and the bacteria tumble instead of swimming in a straight line. This allows the bacteria to "sense" a gradient and swim toward attractant molecules or avoid repellent molecules.

Each stator complex consists of 4 MotA and 2 MotB proteins and forms two ion channels that allow flow of proton from the periplasmic space into the cytoplasm. Up to 12 stator complexes form a concentric ring around the MS ring. Interaction between the C-terminal domain of FliG and the cytoplasmic domain of MotA generates the rotational torque. A conserved aspartate residue in MotB, which could provide a proton binding site during proton transfer, is essential for motor function.

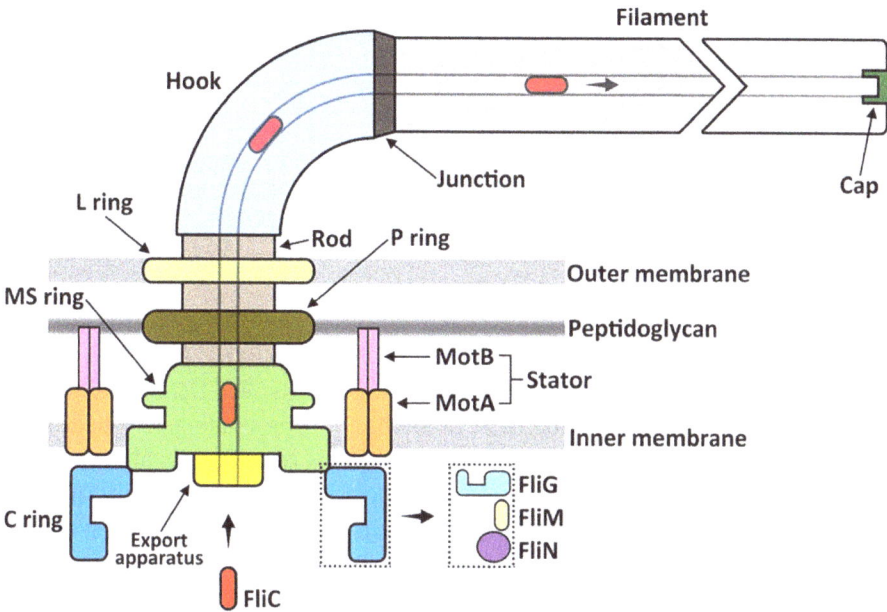

Figure 11.4: A schematic illustration of the bacterial flagellar motor. Reprinted from R. Xue, Q. Ma, M.A. Baker and F. Bai (2015) A delicate nanoscale motor made by Nature – The bacterial flagellar motor. *Adv Sci* 2:1500129. Licensed by CC BY 4.0 (creativecommons.org/licenses).

11.2 Muscle contraction

The force-generating structure in skeletal muscle is the sarcomere. The sarcomere contains overlapping thin (actin) and thick (myosin) filaments, as illustrated in Figure 11.5. Since actin filaments are polar and attach to both sides of the Z line, filaments on each side of the Z line points in opposite directions. The bipolar thick filaments form a staggered assembly of myosin molecules and associate with the A band at the center of the sarcomere. This gives the sarcomere a striated appearance. In addition to actin and myosin, several other proteins are present in the sarcomeres, such as the giant proteins titin and nebulin as well as Z line proteins like palladin, zyxin, N-WASP and α-actinin.

Force is generated when the thin filaments slide over the thick filaments and the sarcomere shortens due to the activity and conformation change of the motor domain of myosin. Hydrolysis of ATP fuels the muscle contraction.

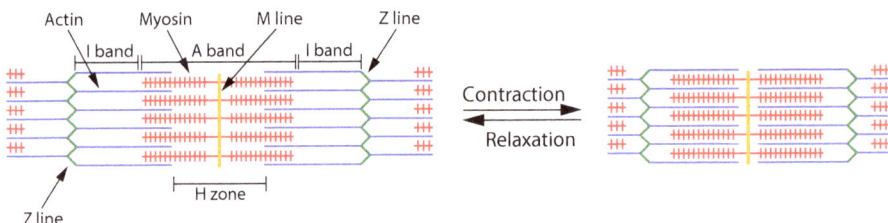

Figure 11.5: Schematic drawing of a sarcomere, the force-generating unit of striated skeletal muscle tissue. Force is produced when the sarcomere shortens as a result of contraction. Adapted with permission from A.C. Wareham (2011) Muscle. *Anaesth Intens Care Med* 12: 249–252.

11.2.1 Myosin

The myosin protein family is very large and contains at least 40 different but homologous members. They all share the basic properties of actin binding, ATP hydrolysis and force transduction. At the N-terminal all myosins contain a highly conserved motor domain with a lever-arm followed by a tail at the C-terminal region. The neck domain (lever-arm) contains one to six IQ motifs (with the amino acid sequence IQxxxRGxxxR where x can be any amino acid) with affinity for the myosin essential and regulatory light chains or calmodulin. The tail is the least conserved region. Some myosins contain a coiled-coil region that is used for dimerization, whereas others contain domains, such as SH3 (src homology domain 3) and PH (pleckstrin homology domain) that interact with specific adaptor proteins.

The thick filaments of the skeletal muscle sarcomere consist of myosin II. Myosin II is a very large protein, with a molecular mass of ca. 520 kDa. It is composed of two 220 kDa heavy chains and four 17–22 kDa light chains. The N-terminal motor domains contain an actin-binding site and a catalytic ATP-binding site. Myosin binding to actin filaments enhances hydrolysis of ATP. The neck domain has sites for the myosin essential and regulatory light chains. The α-helical C-terminal tail domains wrap around each other and form a ~150 nm long coiled-coil region with the characteristic heptad repeats (Figure 11.6).

The tail exhibits additional repetitive sequences. Four heptads form a 28-residue long repeat unit with alternating positive and negative charged zones along the entire coiled-coil. Two myosin dimers can interact and form an antiparallel or parallel dimer of dimers. The interaction depends on the charge distribution along the coiled-coil tail. Due to the alternating charges along the molecule, axial staggering between adjacent myosin II molecules in the filament occurs at multiples of 98 residues or 14.3 nm. The

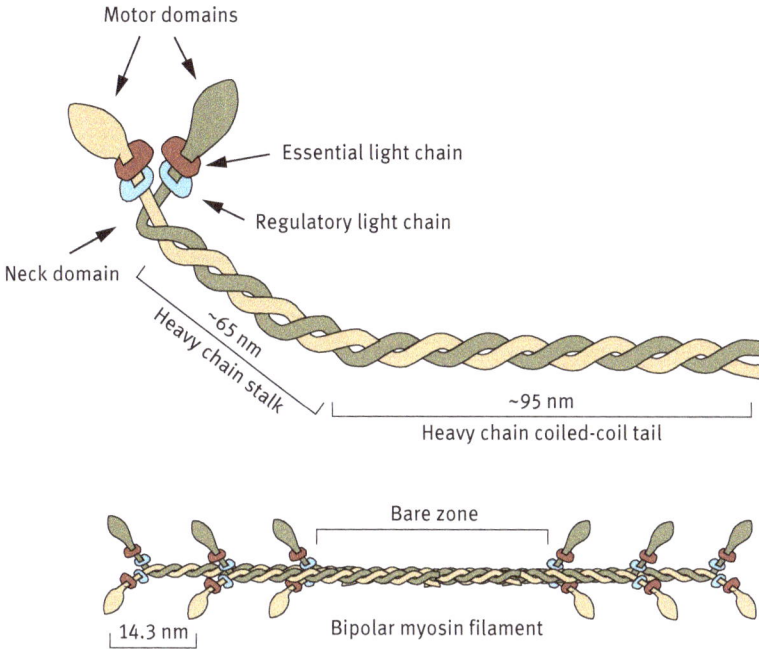

Figure 11.6: Schematic structure of myosin II.

bipolar thick filaments in striated skeletal muscle consist of several hundreds of myosin II molecules. There is a bare zone in the middle of the ca. 1,600 nm long thick filament.

When myosin II is digested with trypsin, two fragments are released: light mero-myosin (LMM) and heavy meromyosin (HMM) as illustrated in Figure 11.7. Further proteolysis of heavy meromyosin by papain produces two more fragments: S1 and S2. The S1 fragment, with the actin-binding site, has become very useful experimentally, as it can be used to determine the polarity of actin filaments. When S1 binds to actin fila-ment it does so at an angle, creating arrowheads along the filament with the arrow pointing toward the pointed end.

11.2.2 The motor domain

The motor domain is made up of the N-terminal, the upper and lower 50 kDa and the converter subdomains. These four subdomains are linked by three structural ele-ments, called switch II, the relay and the SH1 helix (Figure 11.8). The cleft between the upper and lower 50 kDa subdomains constitutes the actin binding site. The binding site for ATP is at the bottom of the actin-binding cleft where a P-loop and switches I and II are found. The P-loop is a conserved nucleotide phosphate binding motif and a common feature of members of the P-loop NTPase domain superfamily. The motif,

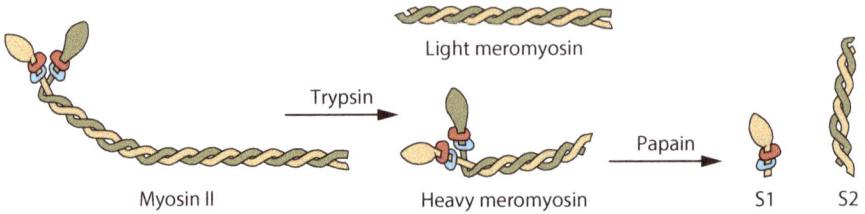

Figure 11.7: Proteolytic digestion of myosin II. Trypsin proteolysis of myosin II produces two fragments: light meromyosin and heavy meromyosin. Further proteolysis of heavy meromyosin by papain produces myosin S1 and S2 fragments. S1 is used extensively to determine the polarity of actin filaments. When bound to actin filament, S1 fragments create arrowheads along the filament pointing toward the pointed end, the slow growing end of the filament.

also known as a Walker A motif, has the pattern GxxxxGK[T/S] where x is any residue. A Walker B motif (hhhh[D/E] where h is a hydrophobic residue) is usually present in proteins with a P-loop. The P-loop binds the β- and γ-phosphate groups of ATP or GTP, whereas the magnesium cation is bound to the Walker B motif. ATP does not enter the binding site through the large cleft but rather via a pocket on the side of the protein. The phosphate moiety of ATP molecule enters the pocket first. After hydrolysis, ADP blocks the exit of the liberated phosphate group (P_i). It is not fully clear how P_i can exit the binding site before ADP as required. It has been suggested that P_i uses a "backdoor" to exit the binding site.

11.2.3 The cross-bridge cycle

In the absence of bound nucleotide, the myosin motor domain binds strongly to actin filaments, forming what is called a rigor complex. Binding of ATP opens the actin binding cleft and leads to the dissociation of the actomyosin complex, as illustrated in Figure 11.9. Cleft opening is connected with closure of switch I onto the γ-phosphate of ATP, followed by closure of switch II. This induces a repriming transition of the converter subdomain and the lever arm moves from a "down" position to an "up" position. In the "up" position, hydrolysis of ATP occurs, which stabilizes the pre-power stroke state and entails cleft closure and stereospecific binding to actin. This induces conformational transitions in the motor domain coupled to movement of the lever arm as P_i leaves the nucleotide binding site. The force-generating power stroke occurs when first P_i and the ADP are released from the motor domain. By this, the system is back in the rigor state and is ready to bind ATP and initiate the cross-bridge cycle again. Thus, the bipolar myosin filaments pull the actin filaments from both sides and thereby shorten the sarcomere and contract the muscle.

The conformational transitions in the motor domain during the cross-bridge cycle are proportionately small. The upper 50 kDa and N-terminal subdomains change very

Switch I
ATP binding site
Upper 50 kDa domain
P-loop
Lever arm
Regulatory light chain
Essential light chain
Actin binding cleft
Converter
Lower 50 kDa domain
Switch II
N-terminal domain
Relay
SH1 helix

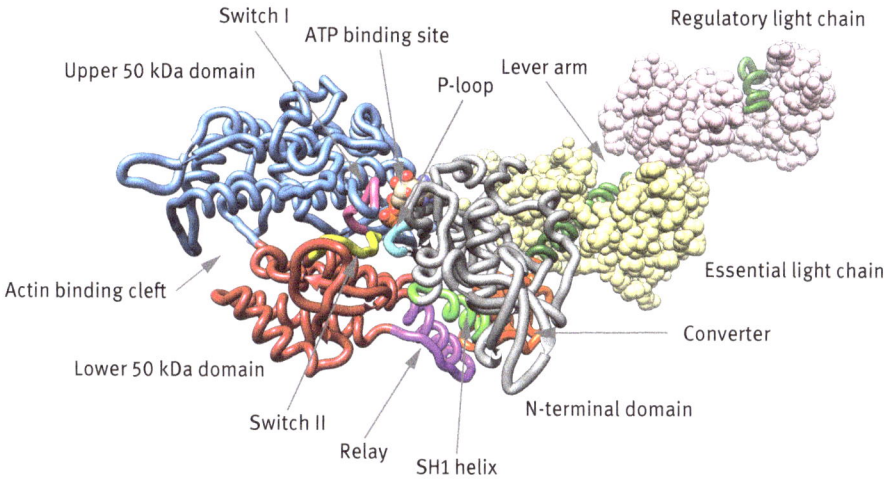

Figure 11.8: The myosin II motor domain. The motor domain contains four subdomains: the N-terminal subdomain (gray), the upper (blue) and lower (red) 50 kDa subdomains and the converter subdomain (orange). The lever arm (green) is linked to the converter. The essential (khaki) and regulatory (pink) light chains bind to the lever arm. The functional important loops are indicated: the P-loop (cyan), relay (purple), switch I (hot pink), switch II (yellow), SH1 helix (lime green). A space filling model of ADP is present in the ATP binding site. Pdb: 1QVI.

little when going from one state to another. The changes in the lower 50 kDa subdomain are much more pronounced, but still the transitions are not very large. The conformational changes are most noticeable in the SH1 helix, which unfolds more or less completely. As the SH1 helix is linked with the converter subdomain, any changes in the SH1 helix affect the converter. As can be seen in Figure 11.10, the lever arm's position in the detached state is completely different from the position in the pre-power stroke state. The small conformational changes in the lower 50 kDa subdomain and switch II, relay and SH1 helix are amplified by the converter and leads to a ~60° rotation of the lever arm.

11.2.4 Role of calcium ions in skeletal muscle contraction

Calcium ions have a very important role in muscle contraction. The calcium concentration of a relaxed muscle cell is around 10^{-7} M. A nerve impulse causes a very fast release of calcium ions from the sarcoplasmatic reticulum that surrounds each myofiber. The increased calcium ion concentration activates the troponin complex by calcium-binding to troponin C and a concomitant conformation change. This causes a rearrangement of the troponin complex and tropomyosin, which exposes the myosin-binding sites on the thin filaments. The muscle fiber is prepared for contraction.

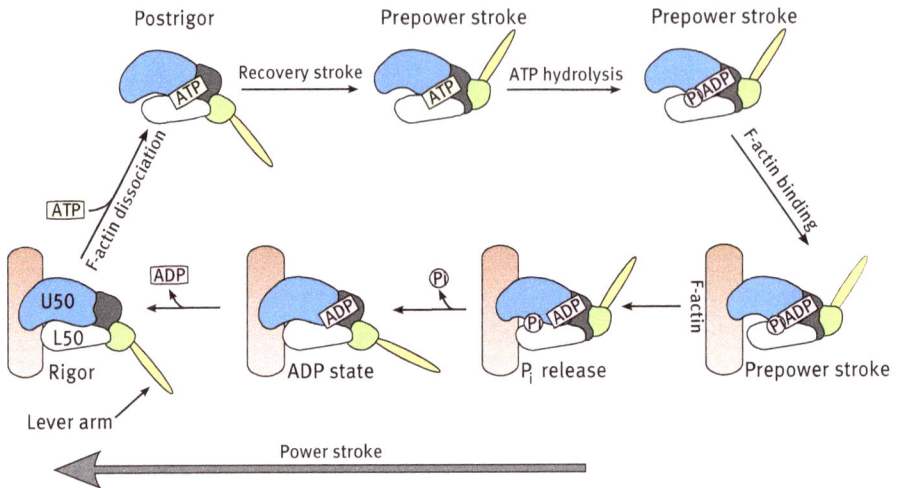

Figure 11.9: The cross-bridge cycle. ATP binding dissociates the rigor actomyosin complex and moves the lever arm (yellow) from a "down" position to an "up" position, effectively making the myosin motor domain ready to pull the actin filaments and contract the muscle. As ATP hydrolyses the motor domain binds to actin (brown). Actin binding induces conformational changes in the motor domain that is transmitted to the lever arm as P_i leaves the active site. Then, Pi-release followed by release of ADP induce a series of conformational changes leading to the power stroke and force production. The subdomains of the motor domain are color coded: upper 50 kDa (U50, blue), lower 50 kDa (L50, gray), N-terminal (black), converter (green) and lever arm (yellow). Adapted from V.J. Planelles-Herrero, J.J. Hartman, J. Robert-Paganin, F.I. Malik and A. Houdusse (2017) Mechanistic and structural basis for activation of cardiac myosin force production by omecamtiv mecarbil. *Nat Commun* 8: 190. Licensed by CC BY 4.0 (creativecommons.org/licenses).

11.2.5 Myosin domain architecture

Independent of type or origin, all myosins have a conserved motor domain that interact with actin filaments and hydrolyze ATP and they all have motile functions. Several of the myosin classes can be divided into subclasses with distinct tissue expression patterns. For instance, the human genome contains 14 different myosin II genes, where some are expressed in skeletal or cardiac muscle, whereas other are only expressed in nonmuscle tissues. All myosin IIs form bipolar filaments, but the organization and regulation of the contractile system differs. Smooth muscle and nonmuscle myosin IIs are regulated by phosphorylation of the regulatory light chain. When unphosphorylated these myosins are present in an inactive folded conformation where the tail interacts with the motor domain. Phosphorylation of the regulatory light chain promotes filament formation and activates the ATPase of these myosins.

The major structural differences are located to the tail region that varies in length and domain architecture. Only myosin II, V and XVIII have the typical coiled-coil re-

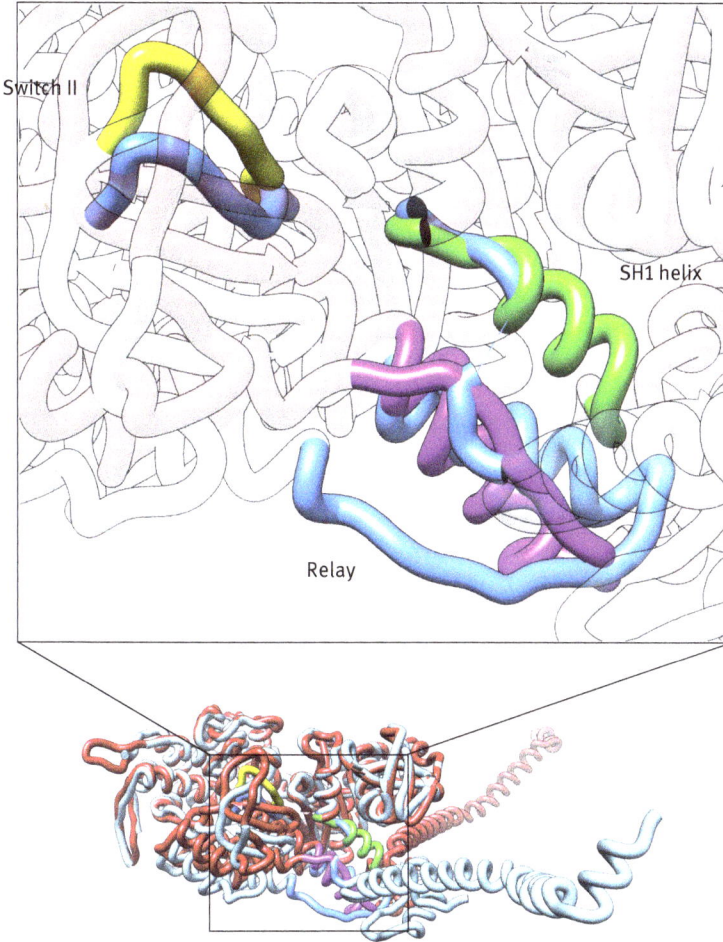

Figure 11.10: Two states of the motor domain of myosin II. The structure of the prepower stroke state (red) was superimposed on the actin-detached state (blue). The upper panel highlights the conformational changes in switch II (yellow and blue), relay (purple and blue) and SH1 helix (green and blue). Pre-power stroke state: pdb: 1QVI and detached state: pdb: 1L2O.

gions in their tails required for dimer formation. Other myosins are either monomeric or form dimers due to other structural elements or associated proteins.

Myosin V and VI have globular cargo-binding domains at their C-terminals and transport cargos along actin filaments. Myosin V has six IQ motifs in the neck region, while myosin VI only has one IQ motif. Myosin VI has a 22-amino acid insert close to the ATP binding pocket that makes myosin VI to move cargo in the opposite direction (toward the pointed end) to all other myosins (that move toward the barbed end). Tails of other myosins contain Rho-GTPase-activating domains (myosin IX), lipid-binding PH domains (myosin I and X) or FERM (protein 4.1-ezrin-radixin-moesin) do-

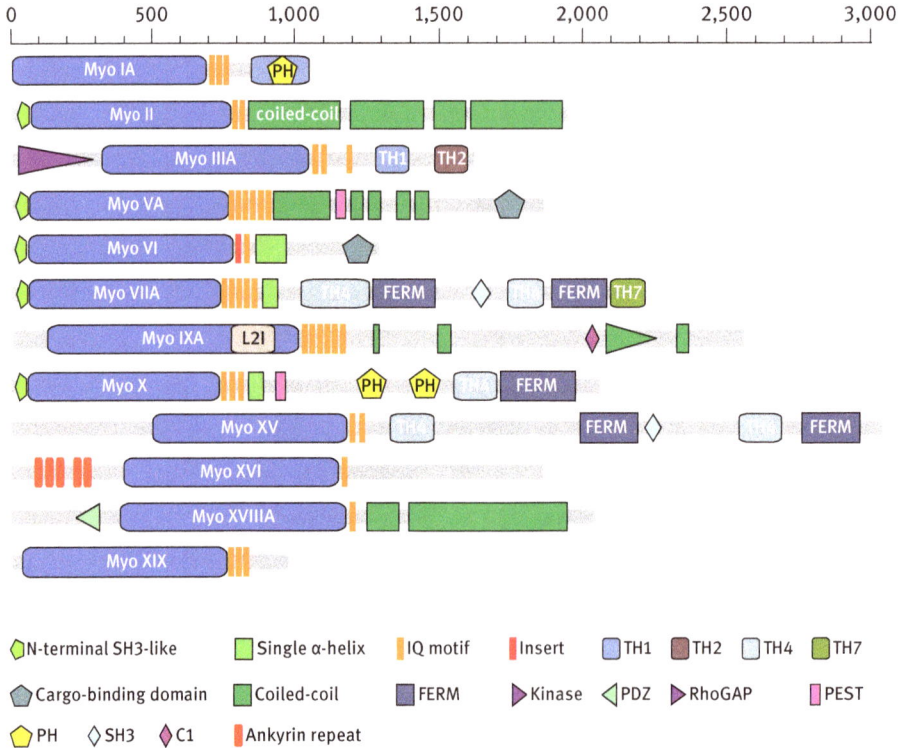

Figure 11.11: Domain organization of human myosins. The myosin isoform is given in the motor domain (blue box). The key to domain names and symbols is given at the bottom of the figure, except for the motor domain that is colored in blue. The abbreviations used are: TH1-TH7, myosin tail homology domain 1–7; FERM, protein 4.1-ezrin-radixin-moesin domain; PDZ, PSD95-Dlg1-ZO1 protein domain; RhoGAP, Rho-GTPase-activating protein domain; PEST, region rich in proline, glutamic acid, serine and threonine; PH, pleckstrin homology domain; SH3, src homology domain 3; C1, protein kinase C conserved region 1; L2I, long insert in the motor domain. The ruler at the top indicates the number of amino acid residues. Adapted after F. Odronitz and M. Kollmar (2007) Drawing the tree of eukaryotic life based on the analysis of 2,269 manually annotated myosins from 328 species. *Genome Biol* 8: R196. Licensed by CC BY 2.0 (creativecommons.org/licenses).

mains (myosin VII, X and XV) to mention some of the domains that can be found in the myosin tails (see Figure 11.11).

11.2.6 Myosin-dependent cargo transport

Some of the myosins transport cargo inside the cell along actin filaments. Myosin V transport cargo toward the barbed end (plus end) of actin filaments and deliver it at the cell cortex. Myosin VI, on the other hand, is the only myosin that transports cargo in the opposite direction toward the pointed end (minus end) and the cell center. Both

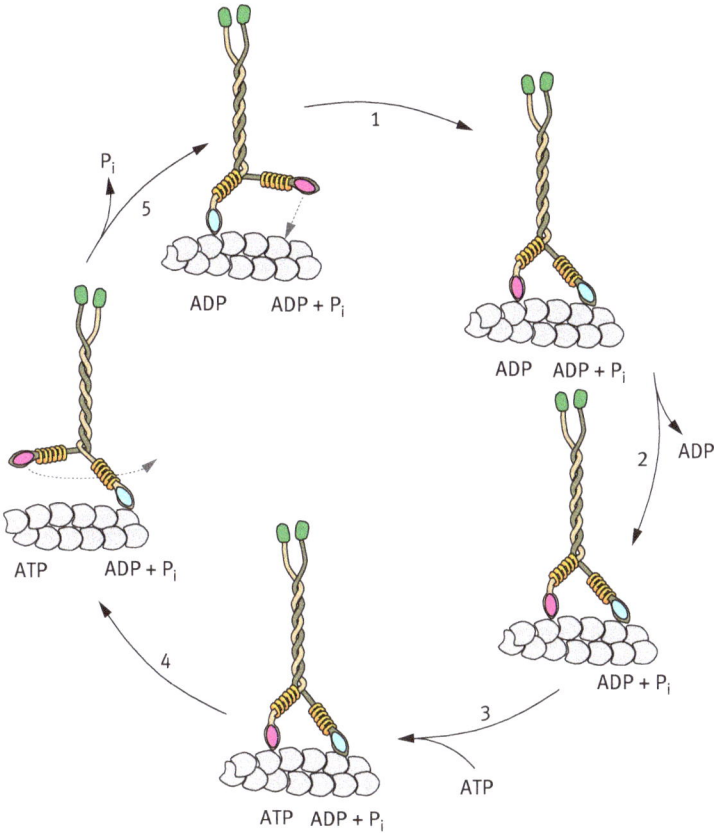

Figure 11.12: Myosin V hand-over-hand walk along an actin filament. The nucleotide in the binding pocket is indicated below the actin filament. (See text for details.).

myosin V and VI "walk" along the actin filament in a hand-over-hand manner, as illustrated in Figure 11.12.

After hydrolysis of bound ATP, the leading motor domain attaches to actin (step 1 in Figure 11.12). The trailing motor domain releases bound ADP readily in contrast to the leading motor domain (step 2). Consequently, binding of ATP to the empty nucleotide binding pocket (step 3) detaches the trailing motor domain from actin (step 4). ATP binding induces a conformational change that together with the power stroke rotates the trailing motor domain (step 5). The trailing motor domain attaches to the filament and becomes the new leading motor domain, whereas the previous leading motor domain becomes the new trailing motor domain (step 1) and the cycle repeats. For each cycle one molecule ATP is consumed and myosin V moves 36 nm toward the barbed end of the actin filament.

11.3 Transport along microtubules

Reliable intracellular transport of vesicles and many other types of cargo is fundamental for cellular function, cell survival and morphogenesis. As the processivity of
myosin-dependent cargo transport is limited, it is only used for deliveries over short
distances. For long-distance transport, particularly in polarized cells like neurons and
epithelial cells, dynein and kinesin are preferred due to their high processivity. Both
dynein and kinesin haul cargo along tracks of microtubule protofilaments, though in
opposite directions. Kinesin is an anterograde haulier that transports cargo toward
the plus end of microtubules, from the cell center toward the cell periphery, whereas
the retrograde dynein hauls cargo toward the minus end, from the cell periphery to
the cell center.

   Dynein was discovered by Ian Gibbons in 1965, and soon it was evident that dynein is the force-generating motor in beating cilia and flagella. Around 20 years later,
Richard Valle and colleagues isolated dynein from brain tissue and showed that this
dynein can move on microtubule tracks. To distinguish between the two types of dyneins, the one present in cilia and flagella was called axonemal dynein and the other
cytoplasmic dynein. At the same time, Ronald Vale, Thomas Reese and Michael Sheetz
isolated a motor protein from squid axons, which they named kinesin.

11.3.1 Dynein

All dyneins are multisubunit proteins; they all contain one to three heavy chains of
> 500 kDa and a number of intermediate chains (45–110 kDa) and light chains (8–55
kDa). Most of these accessory chains bind to the ~160 kDa N-terminal domain. The ~380
kDa C-terminal comprises the motor domain with six AAA + (ATPase associated with
diverse cellular activities) modules and a 15 nm long antiparallel coiled-coil stalk with a
globular microtubule-binding structure. The presence of Walker A and B motifs in the
AAA + module is crucial for ATP binding and hydrolysis, respectively. In dyneins, only
AAA1-4 have functional Walker A motifs and thus bind ATP. The Walker B motif of
AAA2 lacks the glutamate residue necessary for ATP hydrolysis and does not catalyze
hydrolysis of ATP. AAA3 and/or AAA4 probably do not hydrolyze ATP. Therefore, it is
hydrolysis by the AAA1 site that fuels the dynein motor. The tail domain is required for
heavy chain self-association and interaction with regulatory proteins.

   In the human genome there are 13 genes for axonemal heavy chains and two for
cytoplasmic heavy chains. The general domain structure of a dynein heavy chain is
illustrated in Figure 11.13.

   With an empty AAA1 site, the dynein microtubule binding domain (MTBD) is tightly
bound to the microtubule (step 1 in Figure 11.14). ATP binding to the AAA1 site rapidly
detaches the MTBD from the microtubule and induces several conformational changes
(step 2). The open ring conformation of the AAA modules, changes into a closed ring

A

B

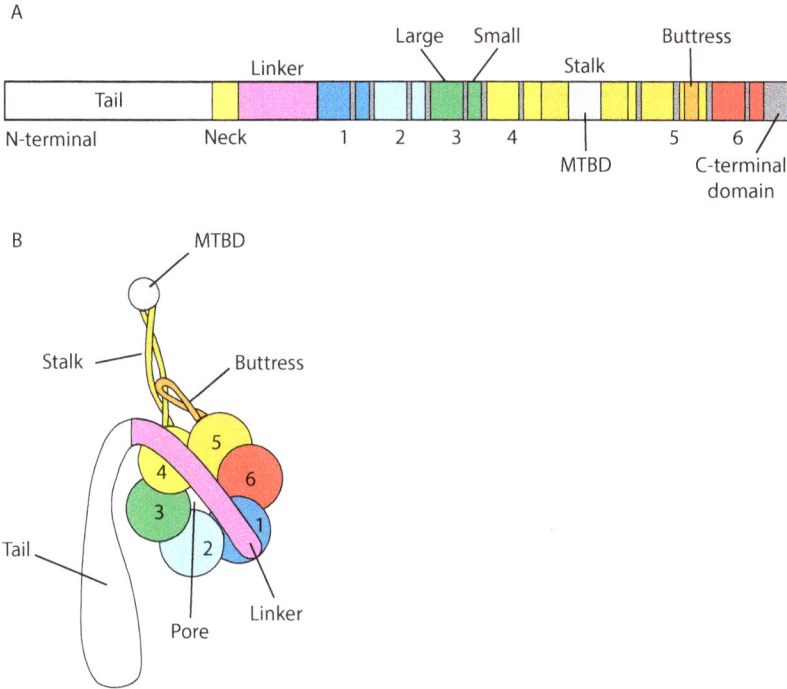

Figure 11.13: Domain organization of dynein heavy chain. A and B) The motor domain consists of a ring of six AAA ATPase modules (labeled 1–6), each with a large and small subdomain, a linker, a coiled-coil stalk with a microtubule binding domain (MTBD) and the buttress. The N-terminal tail domain is responsible for self-association. Adapted from H. Schmidt, E.S. Gleave and A.P. Carter (2012) Insights into dynein motor domain function from a 3.3-Å crystal structure. *Nat Struct Mol Biol* 19: 492–498.

conformation and the linker adopts a bent conformation when it is detached from its docking site at AAA5 and instead forms a contact with AAA3/2. Also, AAA4 and AAA5 move closer, which causes the buttress to pull one of the helices in the stalk. This induces a weak binding state of the MTBD and allows the MTBD to find a new binding site closer to the minus end of the microtubule. ATP hydrolysis converts the MTBD to a strong binding state (step 3). As P_i dissociates, the linker reverts to its former straight conformation (step 4). This is believed to be the power stroke and the main step when force is conveyed to the attached cargo. Before ATP can initiate a new cycle, ADP must dissociate from AAA1.

Axonemal dynein, together with accessory proteins, causes the beating of cilia and flagella by sliding along the microtubule doublets (see Figure 7.10 for the structure of the axoneme). The outer arms consist of three or two dynein heavy chains (depending on the organism) forming heterotrimers or heterodimers. The inner dynein arms consist of one or two heavy chains. As the microtubule doublets in the axo-

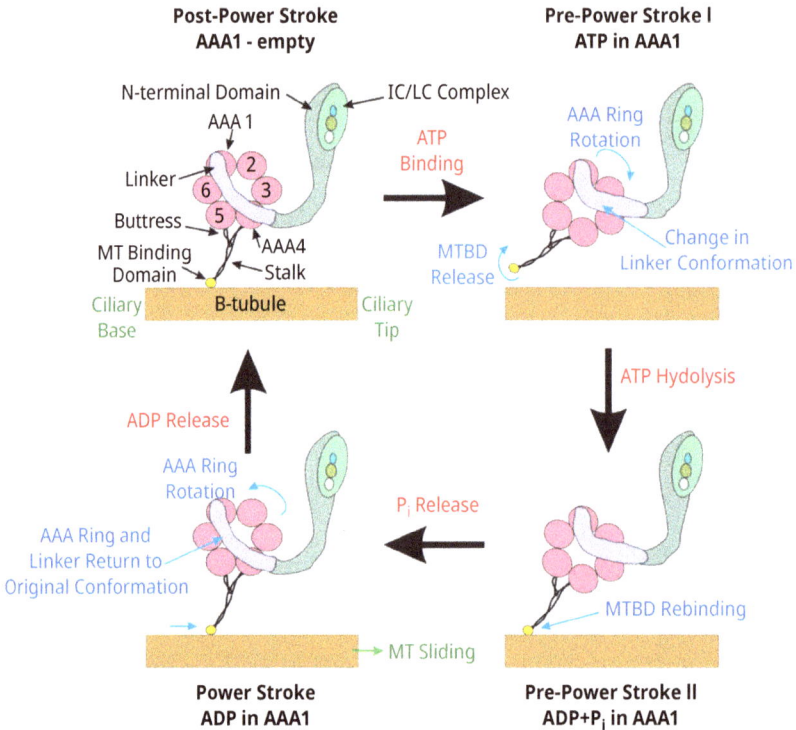

Figure 11.14: Model of the power stroke cycle of a dynein motor domain. The dynein microtubule binding domain (MTBD) moves toward the minus end of the microtubule with an attached object to the N-terminal domain. The attached object can be the A-tubule of one microtubule doublet of an axonemal dynein or a cargo transported by a cytoplasmic dynein dimer. (See text for details.) Reprinted with permission from S.M. King (2018). Turning dyneins off bends cilia. *Cytoskeleton (Hoboken)*. 75:372–81.

neme is attached to the basal body; the dynein sliding bends the microtubules and thus also the whole flagella, thereby creating the beating.

The cytoplasmic dyneins, dynein-1 and dynein-2 (also known as intraflagellar transport dynein), are homodimers of their respective heavy chains. Cytoplasmic dynein-1 has important roles in many different cellular processes as a transporter of vesicles, organelles, RNAs and proteins to specific locations in the cell, whereas dynein-2 carries cargo along microtubule in the axoneme. Cytoplasmic dynein is made up by dimers of heavy chains, light intermediate chains and intermediate chains. The intermediate chain dimers provide a scaffold for binding light chain dimers. In addition to the dynein multisubunit complex, several other proteins associate with the dynein complex and create functionally distinct transporters. Dynactin, a 1.2 MDa complex of more than 20 polypeptides, is required for most dynein functions in the cell. It is known to increase the processivity of dynein, probably by providing extra tether to the microtubule through the dynactin subunit p150^{Glued}.

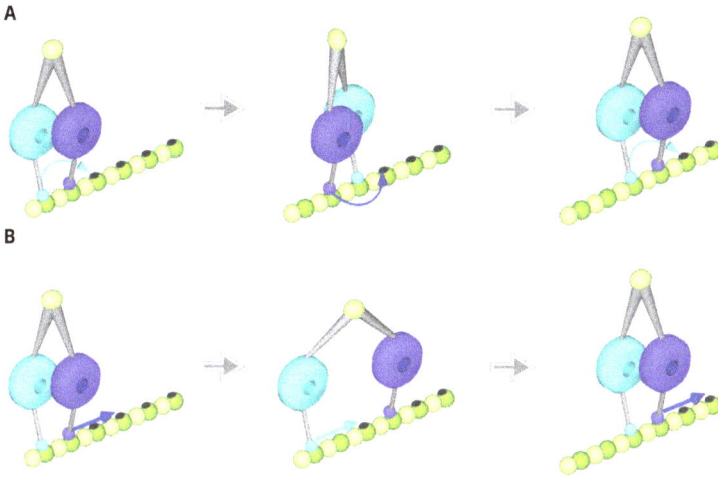

Figure 11.15: Model of cytoplasmic dynein walking along a microtubule. (A) Hand-over-hand model. The trailing microtubule binding domain (MTBD) detaches from the microtubule when ADP in the motor domain is exchanged for ATP and moves in front of the leading MTBD. Upon ATP hydrolysis and release of P_i, it is rebound strongly to the microtubule. The former leading motor domain, which now is the trailing motor domain, goes through the same process and overtakes the new leading motor domain. (B) Inchworm model. The leading motor domain is always in front of the trailing motor domain. Again, when ADP is exchanged for ATP, the MTBD detaches and can slide and advance along the microtubule without complete dissociation from the microtubule due to a weak interaction with the protofilament. Reprinted with permission from K. Shibata, M. Miura, Y. Watanabe, K. Saito, A. Nishimura, K. Furuta and Y.Y. Toyoshima (2012). A single protofilament is sufficient to support unidirectional walking of dynein and kinesin. *PLoS One*. 7:e42990.

When both motor domains of cytoplasmic dynein contain bound ADP, both MTBD are strongly bound to the microtubule. Exchange of ADP for ATP in the leading motor domain (the one closest to the minus end of the microtubule) induces conformational changes in the AAA + ring that is conveyed to the MTBD and causes dissociation from the microtubule. The movement in the motor domain and Brownian motion results in a search for a new binding site for the MTBD on the microtubule. The conversion weak-to-strong binding is favored when the stalk is angled backward and the MTBD finds a binding site. Strong rebinding to the microtubule induces ATP hydrolysis, release of P_i and straightening of the linker. This leads to tension in the connection with the lagging motor domain. Exchange of ADP for ATP in the lagging motor domain induces dissociation of the lagging MTBD from the microtubule. The tension between the two motor domains is relieved when the lagging motor domain takes a step forward toward the leading motor domain and the dynein dimer has moved one step toward the microtubule minus end. However, it is also possible that the ADP in the leading motor domain exchanges for ATP again. In this case, the tension between the two motor domains causes the leading MTBD to take a step backward to relive the tension after detaching from the microtubule. Thus, the dynein dimer makes a shuffle dance but does not move.

The motor domains of the dynein dimer walk are largely independent of each other. Similar to myosin and kinesin, dynein can use a hand-over-hand walking model but also an inch-worm type of walk, as illustrated in Figure 11.15. In this type of walk, the leading motor domain detaches and take one (or even several steps) forward while the lagging motor domain remains fixed in position. Then the lagging motor domain can take a step either to come closer to the leading motor domain (inch-worming) or walk past it (hand-over-hand). The average step size is 10–12 nm, but it can vary between 8 and 32 nm in both directions. Dynein may also move from one microtubule to another one in the microtubule lattice.

11.3.2 Kinesin

Like myosin, kinesin belongs to the P-loop NTPase superfamily. Hydrolysis of ATP fuels the motor activity of kinesin that allows anterograde transport of cargo along microtubules. Kinesin consists of a motor domain (often called "head"), a neck region that connects the head with the coiled-coil stalk, followed by the tail, as depicted in Figure 11.16. The head contains the catalytic site and the microtubule binding site, whereas the cargo usually binds to the tail region.

There are 45 mammalian genes coding for proteins of the kinesin superfamily (also known as KIFs). Due to alternative mRNA splicing, the number of expressed KIFs may be even larger. Phylogenetic analysis indicates that the kinesin superfamily comprises 15 families, termed kinesin-1 to kinesin-14B.

Kinesin 1 is composed of two kinesin heavy chains and two kinesin light chains. The N-terminal motor domain is followed by a linker region, called the neck, a coiled-coil stalk and a tail, to which the light chains associate (Figure 11.16). Kinesin 2 form heterotrimers, between KIF3A, KIF3B and kinesin-associated protein (KAP) or homodimers of KIF17. All other members of the 15 families form homo- or heterodimers, with the exception of kinesin 3 that exists as monomers or homodimers and kinesin 5 that forms homotetrameric motors.

Depending on the position of the motor domain, the families can be grouped in three types. Most kinesins have the motor domain in the N-terminal region (N-kinesins); some have a motor domain in the middle of the protein (M-kinesins) or located to the C-terminal region (C-kinesins), as depicted in Figure 11.17. N-kinesins transport cargo toward the microtubule plus end, whereas C-kinesins drives minus end-directed transport. M-kinesins depolymerize microtubule at both plus and minus ends.

Kinesins 1, 2 and 3 are mainly engaged in transport of vesicles and organelles. The other kinesins are involved in chromosome positioning, congression and segregation (kinesins 4, 7, 8, 10 and 13), spindle organization (kinesins 5, 6, 12 and 14) and for

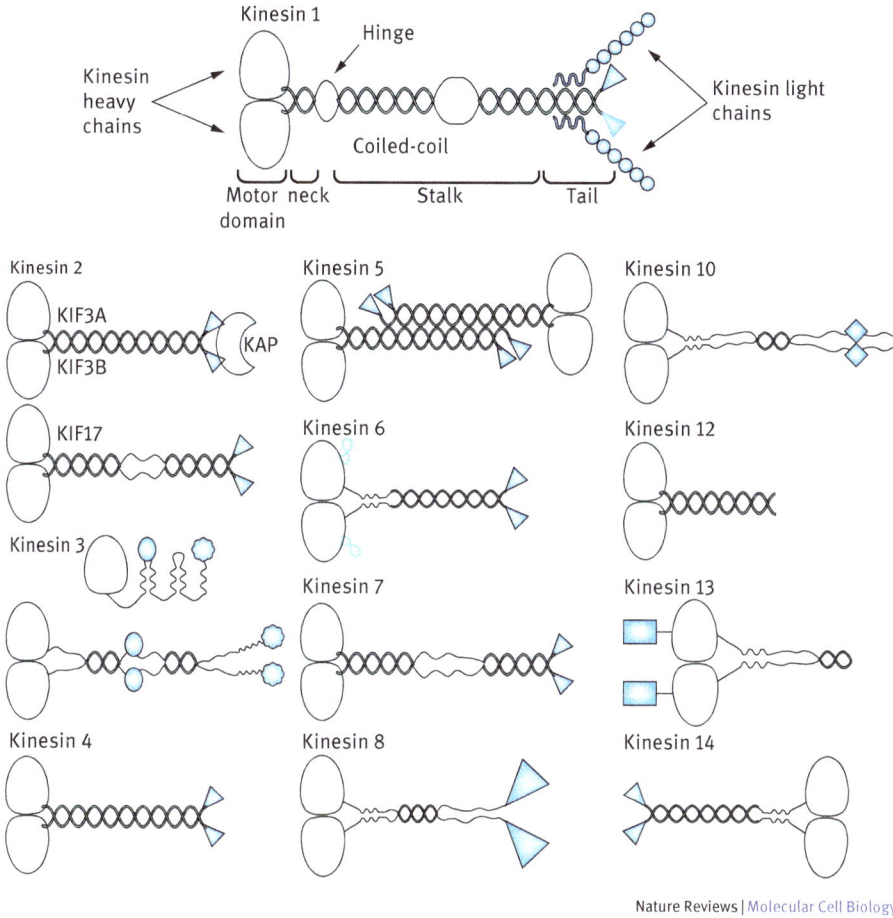

Nature Reviews | Molecular Cell Biology

Figure 11.16: Schematic structure of kinesins. Adapted with permission from K.J. Verhey and J.W. Hammond (2009) Traffic control: regulation of kinesin motors. *Nat Rev Mol Cell Biol* 10:765–777.

kinetochore-microtubule attachment (kinesins 7, 13 and 14). Kinesin 11 has a role in signal transduction, whereas the function of kinesin 9 is unclear.

As both ATP and ADP (i.e. actually Mg-ATP and Mg-ADP) can bind to kinesin with similar affinities (in the low micromolar range), they compete for binding. Therefore the kinesin motor will only be active when the concentration of ATP is much higher than that of ADP, which it is usually. This is very different from myosins that bind ATP much stronger than ADP.

The rather wide nucleotide binding pocket has a rigid P-loop at the bottom. The phosphate moiety of ATP (or ADP) enters the pocket first and docks on the P-loop, in the same way as nucleotides bind in myosin. Binding induces a conformational change in regions flanking the catalytic site. Switch I and switch II move closer to the

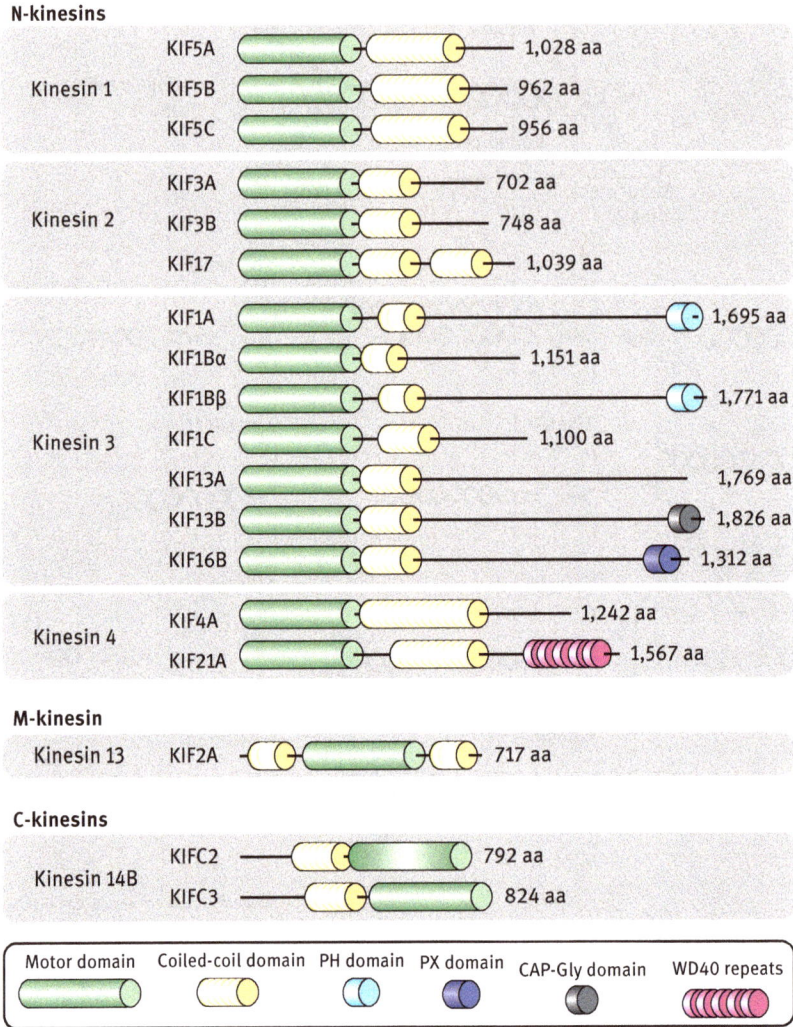

Figure 11.17: Domain structure of major kinesins. Adapted with permission from N. Hirokawa, Y. Noda, Y. Tanaka and S. Niwa (2009) Kinesin superfamily motor proteins and intracellular transport. *Nat Rev Mol Cell Biol* 10: 682–696.

nucleotide and a salt bridge forms between a conserved arginine (Arg203 in kinesin 1) and a glutamate (Glu236 in kinesin 1) side chains in switch I (SS<u>R</u>SH) and switch II (DLAG<u>SE</u>), respectively. The switch closure positions several side chains, two water molecules and a magnesium ion to allow a nucleophilic attack on the γ-phosphate and hydrolysis of ATP. The magnesium ion has a central role in the hydrolytic reaction by organizing the water molecules that are essential for catalysis (Figure 11.18).

Depending on the state of the bound nucleotide, kinesin cycles between strong and weak binding to microtubules. In the weakly bound state, kinesin can detach from the microtubule. Kinesin-ADP binds weakly to the microtubule, but the binding causes release of bound ADP that converts kinesin from a weak binder to a strong binder. The empty binding pocket can now accommodate ATP. Kinesin-ATP also binds strongly to microtubules. Hydrolysis of bound ATP to ADP and P_i, and the concomitant conformational change allows release of the phosphate from the binding pocket. The release of phosphate converts the motor domain from a strong binder to a weak binder and leads to detachment of the motor domain.

A short sequence of the neck, the neck linker, changes conformation depending on the nucleotide state of the motor domain. With ATP or ADP•P_i present in the binding pocket, the linker can dock into its motor domain, whereas with ADP or an empty binding pocket the linker cannot dock. When the motor domain detaches from the microtubule its neck linker is undocked. It is proposed that the neck linker functions as a lever arm to generate force when ATP binds (cf. the lever arm in myosin).

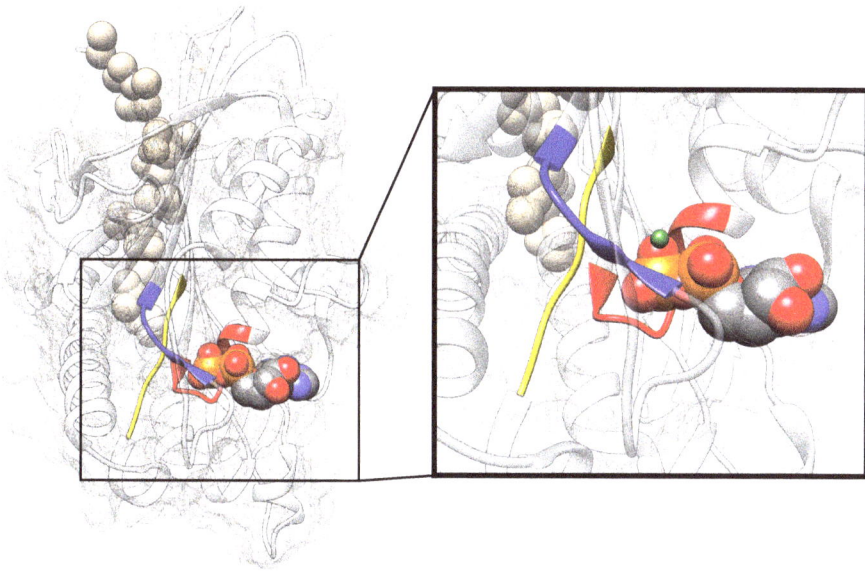

Figure 11.18: Structure of the human kinesin 1 motor domain. The figure shows the position of the P-loop (red), switch I (blue) and switch II (yellow) in the motor domain. The binding pocket contains a space-filling model of ADP-AlF$_4^-$ (an ATP-hydrolysis transition-state analog). The green sphere in the binding pocket indicates the magnesium ion. Aluminum fluoride and water molecules were omitted. The neck linker (tan) on the back side of the structure is docked against the motor domain (pdb: 4HNA).

The neck linker also senses force. When kinesin is exposed to a backward force the neck linker undocks and inhibits ATP binding. A forward force tends to dock the neck linker and promotes nucleotide binding.

This ability to sense direction of force can be used to coordinate the ATPase cycle of two coupled motor domains in a walking dimer. In the dimer, the leading head senses a backward load due to the lagging head, which promotes undocking of the neck linker and microtubule-activated release of ADP that convert the leading head to a strong binder. At the same time the lagging head senses a forward load that promotes neck linker docking with ADP in the binding pocket, favoring detachment.

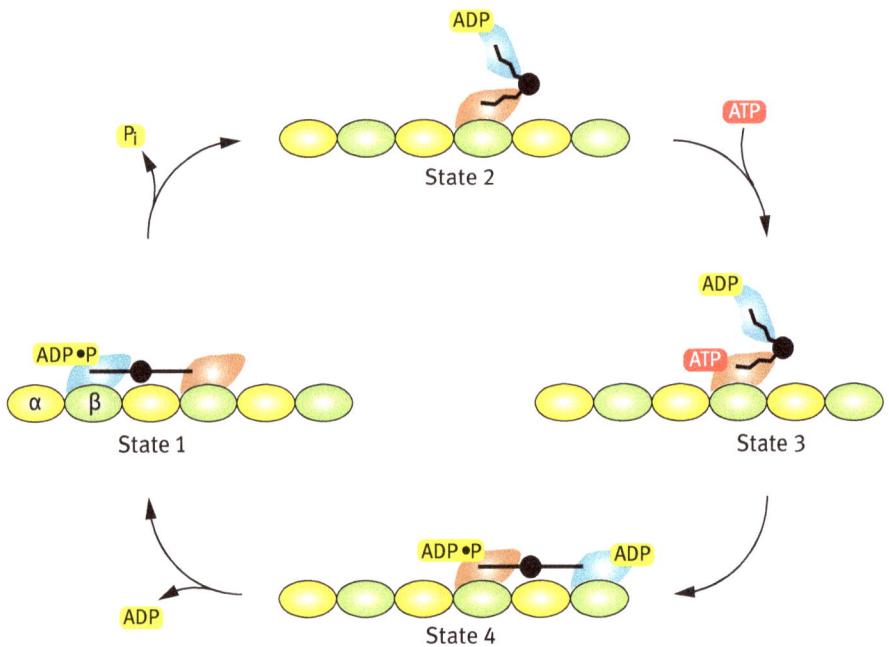

Figure 11.19: Kinesin walking on a microtubule. The distance between binding sites for kinesin on the microtubule is 8 nm (the distance of an αβ-tubulin dimer). The position of the neck linker (black line) depends on the nucleotide state of the heads. (See text for details.).

Dimeric kinesin walks along microtubules in a hand-over-hand manner, just as myosins do. At the start of kinesin's chemomechanical cycle, both the lagging head with $ADP \cdot P_i$ and the leading empty head binds strongly to the microtubule and the neck linker of the lagging head is docked, as illustrated in Figure 11.19 (state 1). Release of P_i from the lagging head weakens the interaction of that head with the microtubule and leads to detachment of the head and undocking of the neck linker (state 2). As the leading empty head binds (state 3) and hydrolyzes ATP (state 4), the neck linker docks into the head. The detached head, with ADP in the binding pocket, diffuses toward the plus end to a new site on the microtubule. Attachment to the microtubule induces a quick release of the bound ADP and the former lagging head becomes the new leading head, strongly bound to the microtubule (state 1). In one hand-over-hand cycle, the lagging head moves 16 nm forward to the next free binding site on the microtubule. At the same time, the center of the dimer moves 8 nm, which is the size of an αβ-tubulin dimer.

The processivity of kinesin is high. A single dimeric kinesin may take more than one hundred 8 nm steps and transport cargo nearly 1 mm, before detaching from the microtubule. Transport speed appears to depend on what is transported. In axons, vesicles can be transported up to 400 mm per day, whereas protein transport is much slower, <8 mm per day.

11.4 Nonmotile P-loop NTPase

The P-loop NTPase superfamily includes not only motor proteins but also many different proteins that hydrolyze ATP or GTP. For instance, SCOPe defines more than 25 different protein families of P-loop NTPases, such as heterotrimeric GTPases and elongation factors. One large and important group is the small GTPases.

Small monomeric GTPases transduce signals from receptors to targets by alternating between an active GTP-bound state and an inactive GDP-bound state (Figure 11.20). Thus, hydrolysis of bound GTP inactivates the GTPase and turns off the signal. However, both the catalytic efficiency and the exchange rate of GDP for GTP of these GTPases are not very good. Therefore, they are regulated by accessory proteins that increase the catalytic efficiency or the dissociation rate of bound GDP. GTPase-activating proteins (GAPs) stimulate GTP hydrolysis and guanine exchange factors (GEFs) increase the dissociation rate of GDP. The cellular location, cytoplasmic or membrane-bound, of a small GTPase is regulated by guanine dissociation inhibitors (GDIs). Binding of a GDI to a small GTPase in the GDP-bound state not only inhibits nucleotide exchange but also prevents membrane localization of the small GTPase.

Based on function and amino acid sequence, the small GTPases are divided into five groups: Ras (regulates cell proliferation, differentiation and survival), Rho (controls actin reorganization), Rab (regulators of vesicle transport and membrane trafficking), Ran (involved in transport across the nuclear membrane) and Arf (regulates vesicle transport and membrane trafficking as well as actin remodeling). It is obvious that the small GTPases and their control are of uttermost importance for survival. For instance, mutations in Ras are found in ~30% of all human cancers.

The architecture of all small GTPases is very similar. They form compact globular structures and all share a guanine-binding domain (G-domain) with five α-helices and a β-sheet with six strands. The G-domain, also known as the effector lobe, constitutes the N-terminal half of the catalytic domain. The guanine nucleotide binding pocket is lined with a P-loop, switch I and switch II, precisely as in myosin and kinesin. The other half of the catalytic domain forms the allosteric lobe.

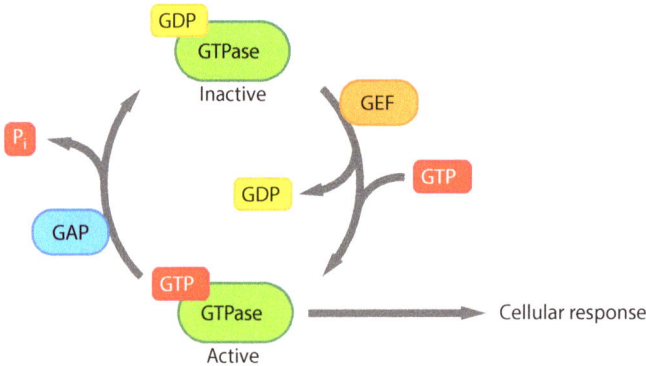

Figure 11.20: The small monomeric GTPase activity cycle.

The human genome contains three Ras genes, encoding H-Ras, K-Ras and N-Ras as well as several splice variants. Ras is a 22 kDa protein, with 189 (H-Ras and N-Ras) or 188 (K-Ras) amino acids. The catalytic domain of the isoforms (residues 1–166) is highly conserved (~90% identity), whereas the hypervariable region at the C-terminal (residue 167–189/188) shows a very low sequence identity (<10%).

To turn off a GTPase, the bound GTP must hydrolyze. Although small GTPases have an intrinsic catalytic activity, it is much too slow; the half-life of GTP bound to Ras is more than 30 min. A GAP can increase the hydrolytic activity more than five orders of magnitude.

Exchange of GDP for GTP (or vice versa) induces conformation changes in the structure, mainly in switch II but also in switch I. The P-loop as well as other parts of GTPase are more or less identical whether GTP or GDP is bound (Figure 11.21). The difference in structure, as measured by root-mean-square deviation (RMSD) between the catalytic domain of H-Ras with bound GppNHp (a nonhydrolyzable GTP analog) and GDP is 1.58 Å but if switch I and switch II are exclude from the comparison the RMSD is only 0.66 Å. When GTP binds switch II moves closer to the nucleotide and forms together with switch I and the P-loop the closed conformation of the binding pocket that catalyzes slow hydrolysis of the bound GTP (Figure 11.21).

GTP is placed correctly in the binding pocket due to hydrogen bonds to several residues. In H-Ras, Gly60 in switch II and Thr35 in switch I form hydrogen bonds with the γ-phosphate moiety of GTP, as illustrated in Figure 11.22. At the entrance of the pocket, there is an NKxD motif that provides an aspartate residue that can form hydrogen bonds with the guanine base but not with an adenine base and thereby creates specificity for GTP. At the bottom of the binding pocket, there is a magnesium ion bidentately coordinated to two nonbridging oxygens of the β- and γ-phosphate moieties of GTP, Ser17 in the P-loop, Thr35 in switch I and two water molecules. After hydrolysis of GTP, Gly60 and Thr35 cannot form hydrogen bonds with GDP, which

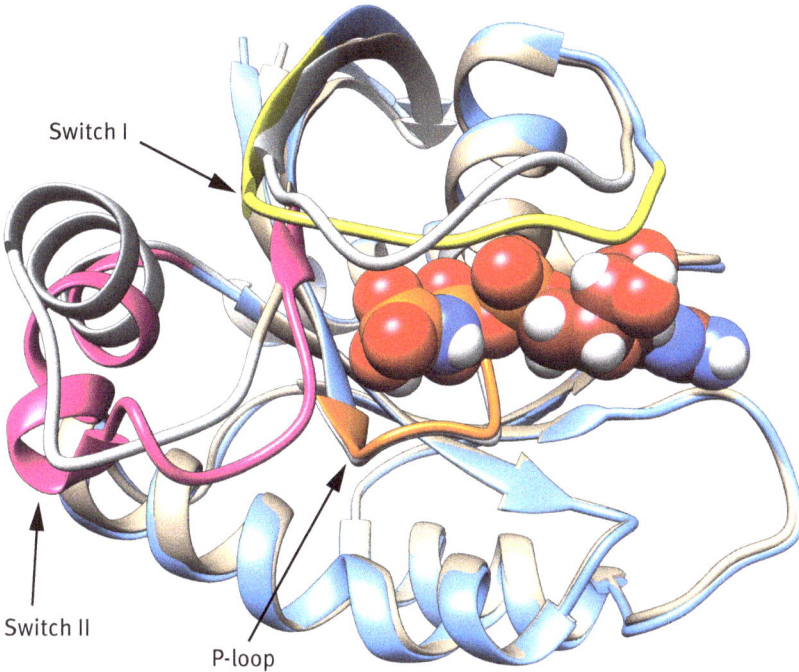

Figure 11.21: Two conformations of a GTPase. The crystal structure of human H-Ras with bound GppNHp (a nonhydrolyzable analog of GTP) in the binding pocket (blue, pdb: 5P21) was superimposed on the structure of human Ras with bound GDP (tan, pdb: 4Q21). In Ras-GTP, the P-loop (orange), switch I (yellow) and switch II (pink) are colored. The corresponding sequences in Ras-GDP are colored in gray.

results in the opening of the pocket and allows GDP to dissociate. Hydrolysis also causes Gln_{61} to move away from the catalytic site.

The Gln61 is critical for both intrinsic slow hydrolysis and GAP-mediated fast hydrolysis of GTP. The human Ras-specific GAP, p120GAP, interacts with H-Ras through the P-loop and regions with the two switches. In the structure, a positively charged arginine residue of p120GAP interacts with the α- and β-phosphate groups of the GTP in the binding pocket of H-Ras. Placed in the binding pocket, the positive so-called arginine finger can neutralize negative charges during the catalysis. p120GAP may also influence the switch II region, making it more rigid and a hydrogen bond between the arginine backbone carbonyl and Gln61 in switch II positions Gln61 correctly. In this position, the side chain of Gln61 forms hydrogen bonds and extracts hydrogen from a water molecule. The resulting hydroxyl ion can attack the γ-phosphoanhydrid bond. Mutation experiments have indicated that Gln61 and its ability to extract a hydrogen from a water molecule is absolutely essential for GAP-mediated GTP hydrolysis.

It is remarkable how often Nature reuse a theme in different contexts. The P-loop and the two switches form a catalytic site that is perfectly suited for hydrolysis of NTPs and this structure is reused in all P-loop NTPases. The amino acid sequences of

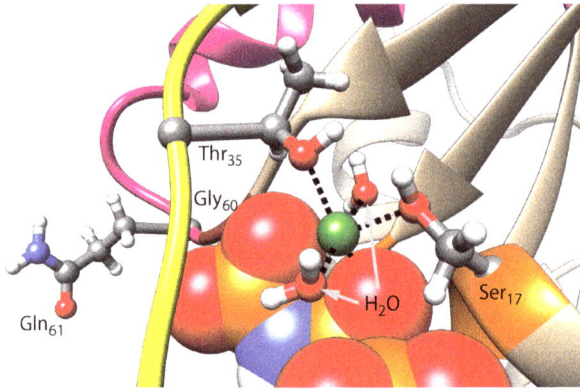

Figure 11.22: The binding pocket of Ras with GppNHp (a nonhydrolyzable analog of GTP). Ser17, Thr35 and two water molecules position the magnesium ion (green sphere) between the γ- and β-phosphate groups in the nucleotide. Gln61 extracts hydrogen from the water molecule, making it more nucleophilic, that makes an attack on the bond between the γ- and β-phosphate groups. The P-loop (orange) switch I (yellow) and switch II (pink) are color coded as in Figure 11.21.

these enzymes are very different but still the catalytic mechanism is the same independent of whether ATP or GTP is the substrate. Another example of a reused theme is the catalytic triad present in all serine proteases.

Although the functions of myosin and kinesin differ from that of small GTPases, still there are several similarities. Binding of ATP to myosin causes a small conformation change that is amplified by the converter region and the lever arm that eventually leads to a power stroke. Binding of GTP to the binding pocket of Ras induces a conformational change that recruits a GAP and converts a "bad" enzyme to a "good" enzyme, a "power activity." Dissociation of ADP from the binding pocket of myosin requires binding to actin filaments and GDP dissociation depends on a GEF protein.

Further reading

Bhabha, G., Johnson, G.T., Schroeder, C.M. and Vale, R.D. (2016). How dynein moves along microtubules. *Tr Biochem Sci* 41:94–105.

Bourne, H.R., Sanders, D.A. and McCormick, F. (1991). The gtpase superfamily: Conserved structure and molecular mechanism. *Nature* 349:117–127.

Boyer, P.D. (1998). Atp synthase–past and future. *Biochim Biophys Acta* 1365:3–9.

Cherfils, J. and Zeghouf, M. (2013). Regulation of small gtpases by gefs, gaps, and gdis. *Physiol Rev* 93:269–309.

Diez, M., Zimmermann, B., Borsch, M., Konig, M., Schweinberger, E., Steigmiller, S., Reuter, R., Felekyan, S., Kudryavtsev, V., Seidel, C.A. and Graber, P. (2004). Proton-powered subunit rotation in single membrane-bound f0f1-atp synthase. *Nat Struct Mol Biol* 11:135–141.

Geeves, M.A. (2016). Review: The atpase mechanism of myosin and actomyosin. *Biopolymers* 105:483–491.

Gibbons, I.R. and Rowe, A.J. (1965). Dynein: A protein with adenosine triphosphatase activity from cilia. *Science* 149:424–426.

Grüber, G., Manimekalai, M.S.S., Mayer, F. and Müller, V. (2014). Atp synthases from archaea: The beauty of a molecular motor. *Biochim Biophys Acta* 1837:940–952.

Huxley, H.E. (1957). The double array of filaments in cross-striated muscle. *J Biophys Biochem Cytol* 3:631–648.

Lymn, R.W. and Taylor, E.W. (1971). Mechanism of adenosine triphosphate hydrolysis by actomyosin. *Biochemistry* 10:4617–4624.

Maxson, M.E. and Grinstein, S. (2014). The vacuolar-type h(+)-atpase at a glance – More than a proton pump. *J Cell Sci* 127:4987–4993.

Mishra, A.K. and Lambright, D.G. (2016). Invited review: Small gtpases and their gaps. *Biopolymers* 105:431–448.

Mueller, M.P. and Goody, R.S. (2016). Review: Ras gtpases and myosin: Qualitative conservation and quantitative diversification in signal and energy transduction. *Biopolymers* 105:422–430.

Nirody, J.A., Sun, Y.-R. and Lo, C.-J. (2017). The biophysicist's guide to the bacterial flagellar motor. *Advances in Physics: X* 2:324–343.

Walker, J.E. (2013). The atp synthase: The understood, the uncertain and the unknown. *Biochem Soc Trans* 41:1–16.

Xue, R., Ma, Q., Baker, M.A. and Bai, F. (2015). A delicate nanoscale motor made by nature – The bacterial flagellar motor. *Adv Sci (Weinh)* 2:1500129.

12 Extra time

Structural information is necessary to fully understand the function and mechanism of a protein. This requires techniques that can determine the three-dimensional structure at atomic or close to atomic resolution. Many experimental methods are used to obtain structural information of proteins, but only some give a detailed view of the backbone and the position of amino acid side chains.

There are three major techniques for protein structure determination that give atomic resolution: X-ray crystallography, nuclear magnetic resonance spectroscopy (NMR) and cryogenic electron microscopy (cryo-EM). In 2023, 85% of the structures in the database at the Research Collaboration Structural Bioinformatics Protein Data Bank (RCSB PDB) have been determined by X-ray crystallography. However, an increasing number of structures are determined by cryo-EM; of all protein structures added to RCSB PDB during 2023, around 30% were determined by cryo-EM.

These methods have their special merits as well as disadvantages. X-ray crystallography requires little hands-on time and has no size limitation but requires that the protein can form crystals, which is a major obstacle. NMR gives a dynamic view of the structure but has a size limit of around 25 kDa and the analysis of data is time consuming. Cryo-EM is excellent for visualizing protein complexes and membrane proteins. Although cryo-EM mainly has been used to determine structures of large proteins or protein complexes, it is possible to resolve structures of proteins smaller than 50 kDa by cryo-EM.

Whatever method is used for collecting structural data, a properly folded and pure protein is a prerequisite. Thus, the protein needs to be purified from its source, perhaps not to homogeneity but at least to 95% purity. Otherwise, it will be difficult to distinguish whether collected structural data originates from the protein of interest or from contaminating proteins. Impurities in the protein sample may also prevent crystallization. The purity requirement of cryo-EM is less strict as a purity level of around 50–80% can give acceptable data. It must also be possible to keep the protein in solution at very high concentrations; this is particularly important for proteins to be investigated by NMR.

Many other methods give structural information about proteins. Mass spectrometry analysis can pinpoint post-translational modified residues. Intrinsic fluorescence spectroscopy can give information on the milieu around aromatic residue. Far-ultraviolet (UV) circular dichroism gives an estimate of the different secondary structures in a protein. Small-angle X-ray scattering (SAXS) can be used to image molecules and particles that are around 1 nm or larger. Although these and many other methods give valuable and useful structural information, they do not give atomic resolution.

However, in 2020 AlphaFold2 enters the scene, which led to a paradigm shift. For well-ordered single-chain proteins, it was now possible to predict the structures with very high accuracy, with nearly atomic resolution.

https://doi.org/10.1515/9783111350684-012

12.1 Protein structure prediction

Ever since the mid-1970s many approaches have been applied to predict a protein's native structure from its amino acid sequence. However, prediction of protein structures is a formidable task for several reasons. Numerous residues need to be positioned correctly in relation to all other residues and bonds, both covalent and noncovalent bonds, need to be optimized.

Most prediction approaches have involved ab initio methods (energy calculation), homology comparison with known structures or threading, which combines homology comparison and molecular modeling. Although predictions have improved significantly with time, it was not until 2020 that predictions became good enough for truly structural analysis.

Since 1994, the improvement of structure predictions has been monitored by the Protein Structure Prediction Center. The center organizes a competition, Critical Assessment of Protein Structure Prediction (CASP), where participants (computer prediction programs) predict the structures from only the amino acid sequences of proteins for which the experimental structures are not yet publicly available.

12.2 AlphaFold

Deep learning is a machine learning process that uses neural networks to learn from data and to process information like the human brain. A deep learning model is trained on a large dataset of labeled data. By processing the data, the model learns to identify patterns in the data. When trained, the model is able to make predictions on new data.

AlphaGo, created by Alphabet's DeepMind, was taught to play the ancient board game of Go, by deep learning. In 2016, AlphaGo became the first artificial computer program to beat human players in Go. AlphaGo was superseded by AlphaZero and later by AlphaMu. AlphaMu was able to learn how to fully master Go, chess and shogi (Japanese chess) after training on a large set of previously played games.

The team at DeepMind continued their development and created AlphaFold, a protein structure prediction model. In CASP13 (2018), AlphaFold outperformed most other prediction methods and caused a great improvement in folding prediction.

Two years later, AlphaFold2 entered the stage and made a historic breakthrough. The results in CASP14 showed that AlphaFold2 was outstanding compared to the other prediction models. Further, it was clear that AlphaFold2 could predict structures with remarkable precision; comparable to experimentally determined structures (Figure 12.1).

Later, AlphaFold2 has been used to predict the structures of more than 200 million proteins. Although the prediction accuracy usually is very high, the predicted structures should be considered as very good models. As some predicted models deviate significantly from experimentally determined structures, it is evident that classical experimental methods (X-ray crystallography, NMR or cryo-EM) still are needed.

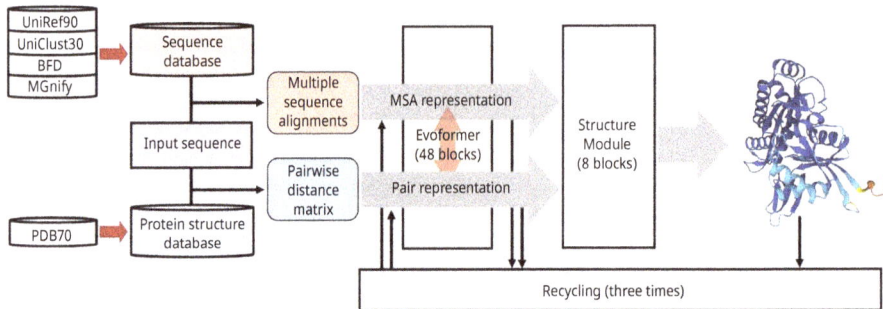

Figure 12.1: Schematic architecture of AlphaFold2 (AF2). There are three modules in AF2. The input module takes the amino acid sequence of the target protein, and generates a pairwise distance matrix (PDM) representation and a multiple sequence alignment (MSA) representation.

The PDM provides a database of structural information (i.e., distance constraints) derived from known structures, with vast examples of how different amino acid sequences fold. The PDM is analyzed to identify patterns and relationships between sequence and structure, which are then used to predict the structure of the target protein. The MSA contains sequences of related proteins that share a common ancestry. By comparing the amino acid residues across these sequences, it is possible to identify pairs of residues that have co-evolved, which indicates that they probably are close to each other in the structure. This information provides additional distance constraints that improve the predicted structure.

The second module is a machine learning algorithm, which takes the MSA and PDM representations and passes them iteratively through the deep learning module, the Evoformer. This module extracts the best alignments and pairwise interactions and passes the information to the next module.

The third module is the structure module, which creates the 3D atom coordinates of the target protein from the abstract representation of the protein structure determined by the Evoformer. Adapted with permission from Z. Yang, X. Zeng, Y. Zhao and R. Chen (2023) AlphaFold2 and its applications in the field of biology and medicine. *Sig Transduct Target Ther* 8:115. Licensed by CC BY (creativecommons.org/licenses).

Despite its usefulness, AlphaFold2 has limitations. For accurate prediction, a multiple sequence alignment (MSA) of homologous sequences is required. Therefore, if none or a few homologous sequences are available, the accuracy of the prediction may be inferior. Since the neuronal network is trained on available experimental determined structures, AlphaFold2 would likely predict structures similar to those present in the databases. Further, the predicted structure is a static snapshot that does not appraise the structural flexibility and conformational heterogeneity of proteins, such as the active site of an enzyme. Another limitation is that some proteins are multimeric, interact with other proteins or bind small ligands and thereby change conformation. In such cases, it can be expected that AlphaFold2 prediction would be less reliable.

Adenylate kinase (AK) is an illustrative example of this limitation. AK is an important enzyme in cellular energy metabolism. AK catalyzes the reversible reaction 2 ADP $\rightleftarrows$ AMP + ATP, thereby involved in AMP metabolic signaling and ATP supply. During the catalysis cycle, AK undergoes considerable conformation changes, from an open state, which allows the substrates to enter the active sites, to a closed state during the reaction. Structural studies have shown that the conformation changes mainly involve a "lid" that covers the active site in the closed state. When the AK sequence was submitted to AlphaFold2, the closed conformation was predicted with very good accuracy (rmsd 0.8 Å), whereas the prediction of the open conformation was not very good (rmsd ~6 Å). As seen in Figure 12.2, the major mismatch occurs in the lid region.

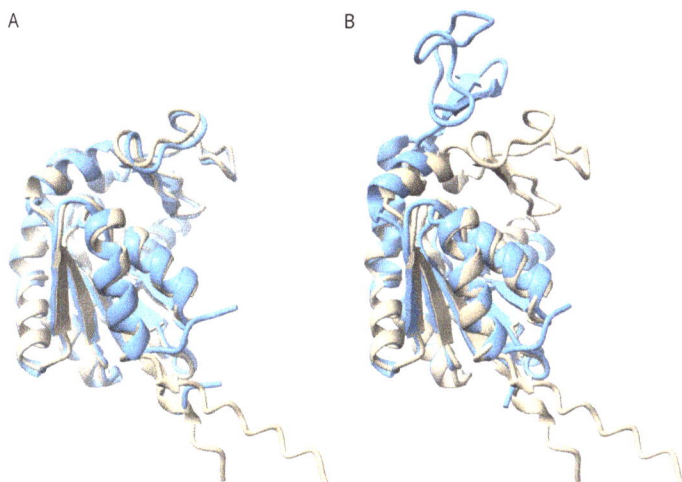

Figure 12.2: Human adenylate kinase. (A) The closed state of adenylate kinase (pdb: 6ZJB). (B) The open state of adenylate kinase (pdb: 6ZJED). Both the closed and open conformations of adenylate kinase (blue) are superimposed on the structure predicted by AlphaFold2 (tan).

Even with these known limitations, the development of AlphaFold2 was a gigantic step forward in understanding the folding and structures of proteins. AlphaFold2 is perhaps the long-awaited Holy Grail of protein folding.

In the footsteps of AlphaFold, several other similar predicting models have emerged that have certain advantages compared to AlphaFold but also some disadvantages; such as increased speed but reduced accuracy.

AlphaFold-Multimer is a model trained specifically for multimeric proteins. This model predicts the multimeric interfaces of multimeric proteins with known stoichiometry, while maintaining the high accuracy of each polypeptide.

12.3 Molecular dynamic simulation

If the positions of all atoms in a protein and surrounding solvent are known, the forces exerted on each atom by all other atoms can be calculated, if there is a force field that describes the potential of all interactions. These forces can then be used in Newton's equations of motion to predict the position and velocity of each atom in the system. The forces on each atom are then calculated, but for the new spatial position of each atom.

These new forces are then used to update the prediction of the position and velocity of each atom. Repeating the process over and over again generates a trajectory of simulation of the protein's structure over time. Such a molecular dynamic (MD) simulation can provide insight into the folding, stability and structural transitions of a protein.

The time steps in an MD simulation must be very short, usually in the femtosecond (10^{-15} s) range. Many biochemical processes, like structure transitions in a protein, occur in the microsecond (10^{-6} s) range or slower. Thus, an MD simulation requires many million time steps. This, together with the many thousands of atoms that must be included in the calculation of each time step, makes the calculations extremely computationally demanding. Therefore, all-atom MD simulations of protein folding or protein dynamics have mostly been used to simulate very fast processes, in the microsecond range.

Improvement of algorithms and introduction of course-grained force fields as well as specialized hardware and computers have made it computationally possible to simulate biochemical processes occurring in the millisecond (10^{-3} s) range and even longer.

The villin headpiece (e.g., a 35 residue subdomain of the headpiece, HP35) has been a popular protein in several MD simulations, due to its small size and fast folding. The unstructured HP35 folds into a native-like structure in ~5 microseconds, which is similar to experimental measurements. The simulation also indicated that many nonnative interactions were formed during the simulation, and that these nonnative interactions prevented the formation of a hydrophobic core and further folding. Once these nonnative interactions were broken, by thermal fluctuations, the native state formed rapidly.

Proteins are highly dynamic; the structure is not static but changes when they interact with other proteins or biomolecules. MD is a versatile tool; it can be used to create useful predictions of protein function, give support to and improve analysis of experimental results and much more.

12.4 X-ray crystallography

Structure determination by X-ray crystallography starts with the crystallization of the protein. Unfortunately, it is not possible to predict the conditions at which a particular protein will form crystals or even if it will crystallize. Therefore, protein crystallization is a matter of trial and error, but based on a large body of existing empirical knowledge.

Hanging drop vapor diffusion crystallization is one common method to grow crystals. For this, a drop of protein is mixed with a buffered precipitant solution and placed in a sealed container (often a titer plate) with a reservoir of the precipitant solution. With time, as water vapor leaves the drop, the concentrations of both protein and reagent increase and a crystal may start to grow. The initial screening is usually followed by optimization of the crystallization conditions to obtain well-ordered and larger crystals.

With crystals in hand, one major obstacle has been overcome!

When the crystal is exposed to a beam of X-rays, the waves diffract or scatter due to interactions with electrons in the sample. Scattered waves along a certain direction are in phase and give strong scattering, whereas others are out of phase and cancel the scattering.

As the crystal contains a regular lattice of identical molecules, the electron density is repeated at regular intervals. This repeating unit is called the unit cell.

The intensity and phase of each spot in the diffraction patterns are related to the electron density of the unit cell. Thus, by Fourier transformation of the intensity and relative phase of each spot the electron density map can be constructed (Figure 12.3).

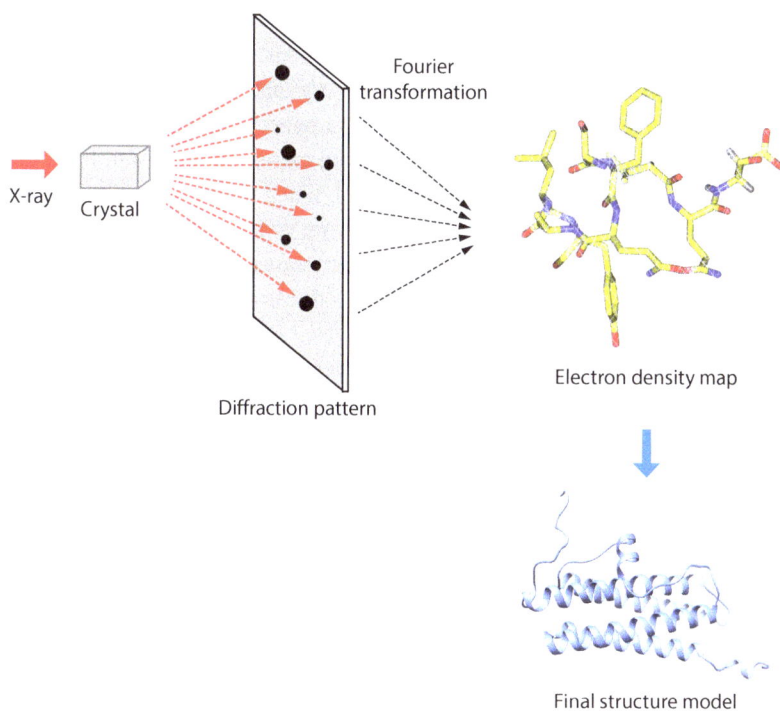

Figure 12.3: Workflow for solving the structure by X-ray crystallography. A single crystal is exposed to a beam of X-rays. The waves diffract due to interactions with the electrons in the crystal and produce a diffraction pattern. An electron density map is constructed by Fourier transformation of the intensity and phase of each spot in the diffraction pattern. A structure model is then fitted in the electron density map.

The diffraction pattern contains the intensity but not the phase. Therefore, the phase must be determined by other experiments. The phase can be obtained by multiwavelength anomalous dispersion (MAD) or multiple isomorphous replacement (in cases where a similar structure exits in the PDB database). MAD experiments are usually performed with proteins in which all methionines have been exchanged for selenomethionines. The chemical properties of methionine and seleno-methionine are similar but in contrast to sulfur, selenium is an effective anomalous scatterer.

With intensities and phases, an initial electron density map can be calculated. Chemical knowledge (such as covalent bond lengths) and solvent flattening are then used to improve the electron density map. From the improved electron density map, new phase angles are calculated, which are used to refine the map and the structural model. The process is repeated to improve agreement between the structural model and the experimental diffraction intensities until no further refinement is obtained.

It should be noted that the structure determined by X-ray crystallography is a static image of the protein, as a predicted structure by AlphaFold. It gives little information on parts of the protein that are not "frozen" in the structure. The final structure may lack some parts of the structure, such as loops or other mobile segments of the protein, as these parts do not give sufficient electron densities.

It is not always possible to grow large, high-quality crystals required for X-ray crystallography. Serial femtosecond crystallography (SFX) overcomes this limitation by using tiny crystals (micrometer or smaller) together with extremely short and bright X-ray pulses generated by X-ray free electron laser (XFEL). When the pulses are generated by a synchrotron the technique is called serial synchrotron crystallography (SSX).

In both SFX and SSX, large numbers of the microcrystals (>100,000) are delivered to the X-ray beam in liquid suspension. Each pulse produces a diffraction pattern, as in single crystal ("classic") X-ray crystallography. However, to obtain a complete data set, the diffraction patterns from very many crystals are collected and combined. The collected data is then analyzed to determine the electron density and the three-dimensional structure of the protein.

Time-resolved SFX takes advantage of the extremely short pulses, which makes it possible to take snapshots of the protein in different states and enables the study of fast dynamic processes at atomic resolution.

12.5 Nuclear magnetic resonance (NMR) spectroscopy

The basis of NMR spectroscopy is that an atom nucleus with half-integer spin quantum number can be excited by a radio-frequency pulse if placed in a static magnetic field. The resonance frequency or chemical shift depends on the identity and the electron distribution around the nucleus. As each nucleus in a folded protein has a distinct electronic environment, it also has a distinct chemical shift. A protein has thousands of resonances, and in a one-dimensional spectrum it would be impossible to discern all chemical shifts due to incidental overlaps and it is also possible that two protons have the same chemical shift. By multidimensional experiments, overlapping resonances can be decreased.

Proteins contain virtually only hydrogen, carbon, nitrogen and oxygen. The predominant isotopes of carbon (^{12}C) and oxygen (^{16}O) have spin 0 and that of nitrogen

(^{14} N) has spin 1. These nuclei are therefore NMR silent or give broad signals. To apply multidimensional NMR, it is therefore necessary to label proteins with suitable ½-spin nuclei, such as ^{15} N or ^{13}C.

Isotopic labeling of proteins is most conveniently done including ^{15} N-labeled ammonium chloride and ^{13}C-glucose as the only nitrogen and carbon source in the a defined culture media used for the expression system. Hence all hydrogens, carbons and nitrogens in the final expressed and isolated protein will be isotopically labeled as well as NMR-active.

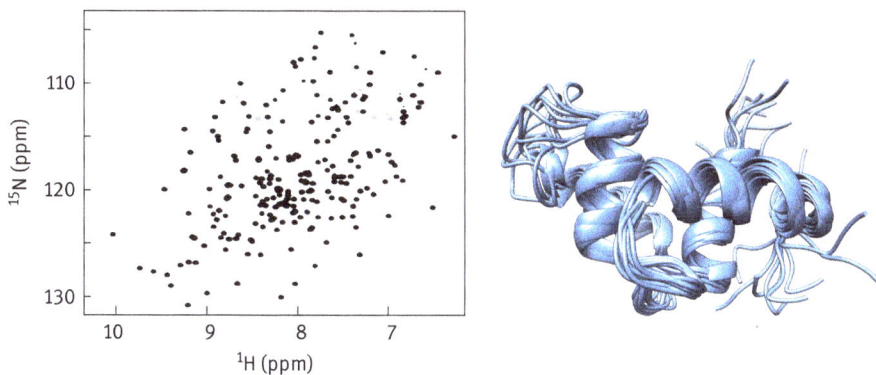

Figure 12.4: HSQC spectrum and an ensemble of structures.

Two types of NMR experiments are required for the structure determination. One set of experiments is required for assigning each chemical shift to atoms in specific amino acids in the protein sequence. This usually involves collecting a ^{1}H-^{15}N HSQC (heteronuclear single quantum correlation) spectrum (Figure 12.4). A HSQC spectrum is a "finger-print" of the protein and reports the correlation of the nuclear spins of the H^N (the amide proton) and N in the peptide bond. This correlation is present in every amino acid residue except for proline residues and the N-terminal residue. As each peak in the spectrum is correlated to a specific amino acid, the number of observable resonances should be the same as the number of amino acids in the protein sequence (minus the number of prolines and the N-terminal residue. A HSQC spectrum with dispersed peaks, such as in Figure 12.4, indicates that the protein is folded. In the next stage, several different types of triple-resonance heteronuclear spectra are used for sequential assignment of each cross-peak to a specific amino acid in the protein sequence.

In the second set of experiments, internuclear distances are obtained by the nuclear Overhauser effect (NOE). In this case, correlation through-space is established between nuclei that are close to each other instead of correlation through-bond. Only nuclei that are separated by less than ~5 Å give rise to a NOE signal and the intensity varies with distance. NOE spectroscopy (NOESY) is used to determine the NOE and hence distances between nuclei. Other NMR experiments measure local (by J-couplings) and global bond (by residual dipolar couplings) angles as well as longer distances (by paramagnetic relaxation).

These distances and angles are then used to compute a model of the structure that is consistent with the measured data. As the measured distances are not precise, several structural models are usually consistent with the data. Therefore, structures determined by NMR are presented by an ensemble of structures that all fit the data, as shown in Figure 12.4.

Structure determination by NMR requires that the protein is not too large; there is a practical limit of around 25–30 kDa, and that the protein is soluble at high concentrations and does not aggregate. It is though possible to determine the structure of much larger proteins, but this requires highly sophisticated labeling schemes and NMR experiments.

In contrast to X-ray crystallography, NMR measures properties in a reasonable native-like aqueous milieu; still protein structures determined by both methods are very similar. With NMR it is also possible to determine order parameters and dynamics at certain positions in the amino acid sequence of the protein as well as conformational changes and ligand binding.

12.6 Cryogenic electron microscopy (cryo-EM)

Transmission electron microscopy (TEM) has been used for a long time to observe various specimens at very high resolution. However, to use classical TEM imaging, water must be removed from specimens, but the removal may change the structure or lead to aggregation. Furthermore, the high vacuum and the intense electron beam used in TEM eventually destroy biomolecules.

By quickly freezing the specimen in such a way that the water forms a disordered glass rather than crystalline ice, it can be imaged by EM. In such a vitrified sample, the water is disordered and does not diffract the electron beam as crystalline ice would, and the structure of the specimen is retained.

In cryogenic electron microscopy (cryo-EM) several hundred thousands of noisy two-dimensional images of randomly oriented protein particles are collected. Using computer methods, these particle images are sorted according to their orientation and similar ones are grouped together and averaged, generating high-resolution 2D images of the different orientations. The different 2D images are then used to generate a high-resolution 3D structure of the specimen.

Although cryo-EM has only resolved atomic details in a few cases so far, the method has produced many startling structures of large protein complexes such as ribosomes, photosystems and many viruses (Figure 12.5).

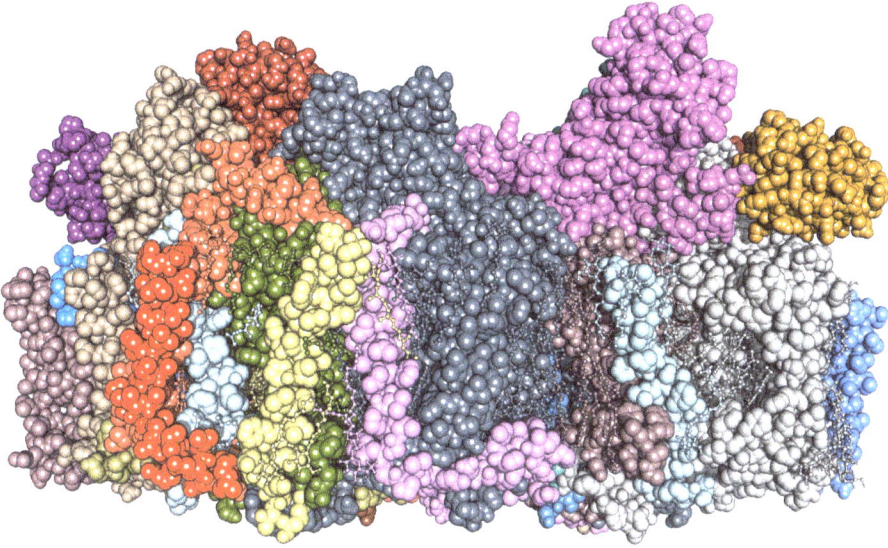

Figure 12.5: Cryo-EM image of photosystem II of a cyanobacterium. The structure in the image contains 42 protein subunits (space-filled) and chlorophyll and other molecules (ball-and-stick) associated with the complex. The attained resolution was 1.93 Å. (pdb: 7N8O).

Cryo-electron tomography (cryo-ET) combined with cryo-focused ion beam-milling (cryo-FIB) has evolved as a technique to visualize the internal structures of cells and macromolecules with high resolution in their native state. By cryo-FIB, it is possible to prepare thin lamella samples (a few hundred nanometers thick) of vitrified specimens that can be imaged using cryo-ET or high-resolution TEM.

Also in cryo-ET, a large number of 2D images are collected, but at different angles by tilting the sample. The collected images are then computationally processed and combined to generate a 3D rendering of the specimen (Figure 12.6).

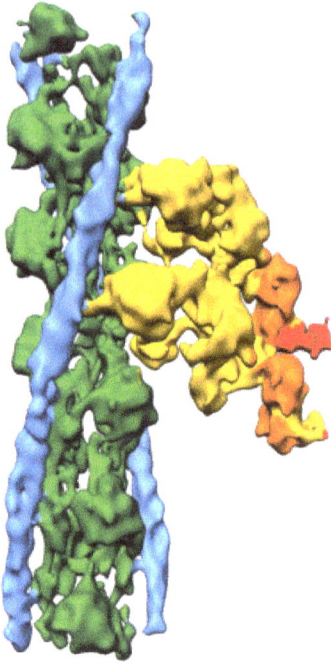

Figure 12.6: Surface view of the actomyosin structure imaged by cryo-electron tomography. Actin filament (green) with bound myosin (yellow), myosin essential light (orange) and myosin regulatory light chains (red). Estimated resolution 10.2 Å. From Z. Wang, M. Grange, T. Wagner, A.L. Kho, M. Gautel and S. Raunser (2021). The molecular basis for sarcomere organization in vertebrate skeletal muscle. *Cell*. 184:2135–50 e13. (emd-12289 and pdb: 7NEP).

12.7 Mass spectrometry

Mass spectrometry (MS) can be used to determine certain aspects of protein structure. By ionizing proteins or peptides (as well as other molecules) MS can determine the mass-to-charge ratio (m/z) of the ions with high accuracy. Thus, it is possible to determine whether an amino acid residue or peptide has been modified in some way, for instance by phosphorylation or cross-linking to another residue or peptide. Further, MS can be used to sequence peptides and proteins, from both ends. Therefore, MS is a powerful tool that also can be used for the study of protein structure.

Loops on the surface of a native protein are more vulnerable to proteases than peptide bonds in the interior. Therefore a short protease treatment may release such loops. After the isolation of these peptides, MS can be used to determine the sequence of the peptide and, hence, give valuable information on what residues are located on the surface.

Cross-linking combined with MS analysis, can identify residues that are close in space but far apart in sequence. Since a cross-linker has a specific length, only residues within this specific distance will be cross-linked. In such a workflow, a specific cross-linker is added to the protein, and after a certain time the reaction is quenched. Protease digestion then gives a mixture of linear peptides and cross-linked peptides, which can be enriched by liquid chromatography (LC). LC-MS/MS (tandem mass spectrometry) is then used to fragment and analyze the cross-linked peptides, as illustrated in Figure 12.7. The analysis will identify intramolecular as well as intermolecular cross-linked residues, giving structural information.

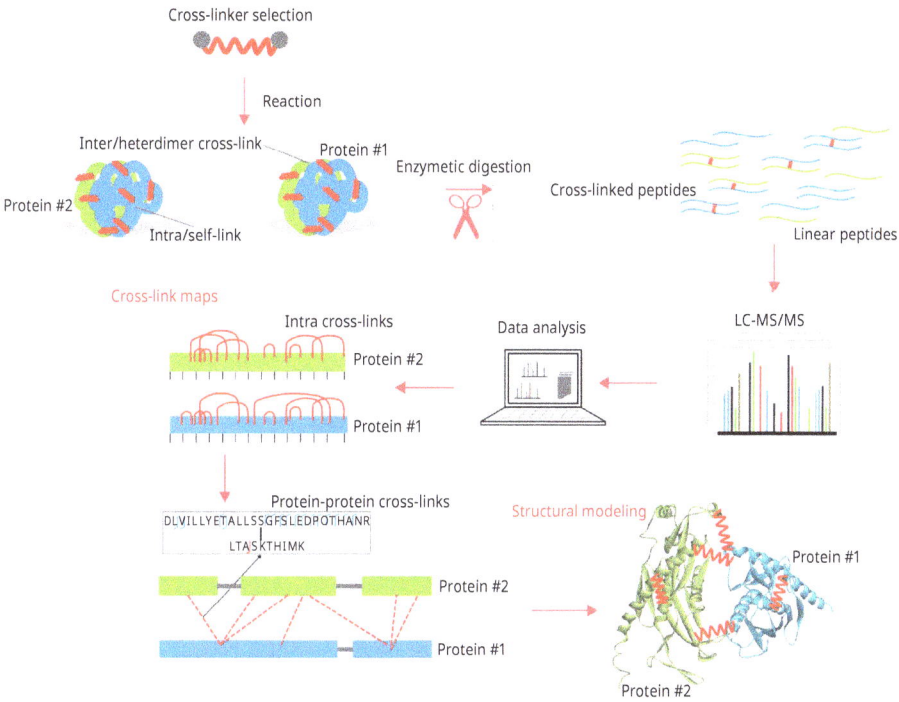

Figure 12.7: Typical workflow of cross-linking mass spectrometry (CLMS) experiments. Adapted after U. Kalathiya, M. Padariya, J. Faktor, E. Coyaud, J.A. Alfaro, R. Fahraeus, T.R. Hupp and D.R Goodlett (2021) Interfaces with Structure Dynamics of the Workhorses from Cells Revealed through Cross-Linking Mass Spectrometry (CLMS). *Biomolecules* 11, 382. Licensed by CC BY (creativecommons.org/licenses).

Exposed hydrogens in a protein are in continued exchange with hydrogens in the solvent. Since the exchange depends on the conformation and dynamics, the rate of exchange can give structural information. By using deuterium in the solvent, the exchange process can be followed experimentally by MS, as the mass of the protein increases due to deuteration (Figure 12.8).

Hydrogen/deuterium exchange (H/D exchange) MS is a very versatile tool that can be applied to many different analyses, as indicated in Figure 12.9.

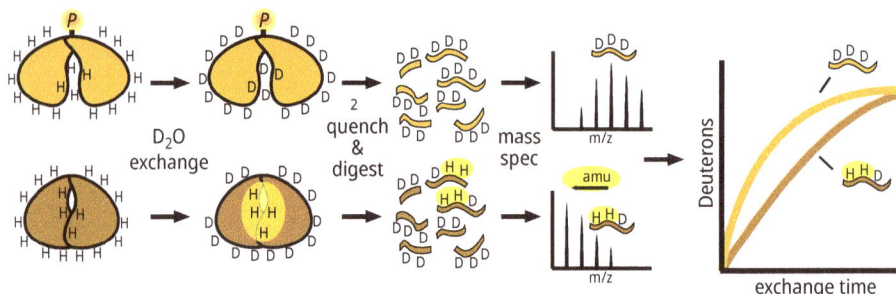

Figure 12.8: Hydrogen/deuterium exchange mass spectrometry. Incubating the protein in a deuterated solvent allows the exchange of available hydrogens. After quenching the exchange reaction and digestion, deuterated peptides can be distinguished from undeuterated peptides by MS, and thus give structural information. Copyright Eric S. Underbakke. Reproduced with permission.

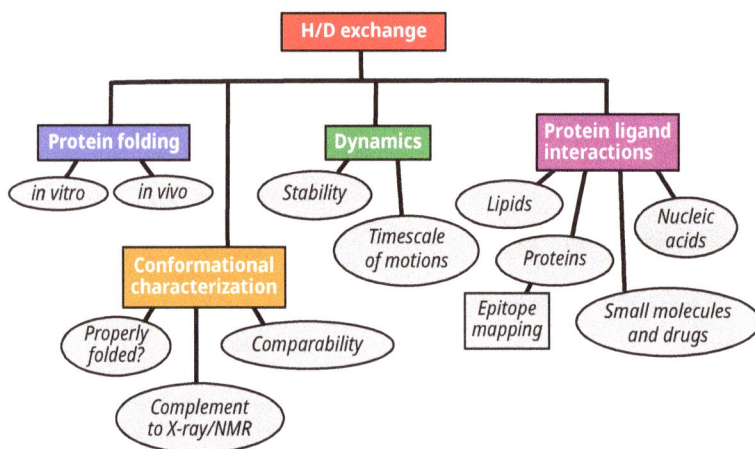

Figure 12.9: Different application of hydrogen/deuterium exchange (H/D exchange) mass spectrometry. Adapted with permission after J.R. Engen, T.E. Wales and X. Shi (2011) Hydrogen Exchange Mass Spectrometry for Conformational Analysis of Proteins. *Ency Anal Chem*, ISBN: 9780470027318.

12.8 What does the protein structure look like?

Independent of how the protein structure was determined, the general idea is that the structure will tell us something about the protein and its function. However, to study and explore the structure of a protein, it is necessary to visualize the structure in such a way that makes the analysis possible. Several molecular graphics programs can load the atomic coordinates from a pdb or mmCIF file (the mmCIF format is used for very large molecular structures) and display the structure. These graphics programs usually include a variety of displaying tools that can emphasize certain aspects of the structure as well as analytic tools that allow the measurement of distances and bond angles.

The most common way to display a protein structure is the cartoon or ribbon style. This style depicts α-helices and β-strands as spring-shaped ribbons and flat arrows with the arrowhead at the C-terminal of the β-strand, respectively. This type of display makes it easy to discern the overall fold, secondary structure elements and loops in the structure (Figure 12.10).

A ball-and-stick rendering is more useful for a detailed atomic view of the structure. Thus, it is possible to explore how a substrate is recognized and which residues in the active site of an enzyme that interact with the substrate. As the ball-and-stick drawing is scaled appropriately, it is possible to measure bond distances or bond angles. In a wire trace, only the backbone is displayed and each line represents the covalent bonds between atoms. Since a wire trace displays less details, it may be useful for comparing the folding of two proteins.

To understand how well a substrate fits in the active site or a ligand fits in a binding site, a space-filling molecular model is probably more convenient. In the space-filling model, each atom is displayed as a sphere with a size reflecting its van der Waals radius, making it easy to see if any atoms are too close to each other and therefore clash.

It is also possible to display only the surface of the protein. Such a topographic display can be used to color code the electrostatic potential or hydrophobicity on the protein surface.

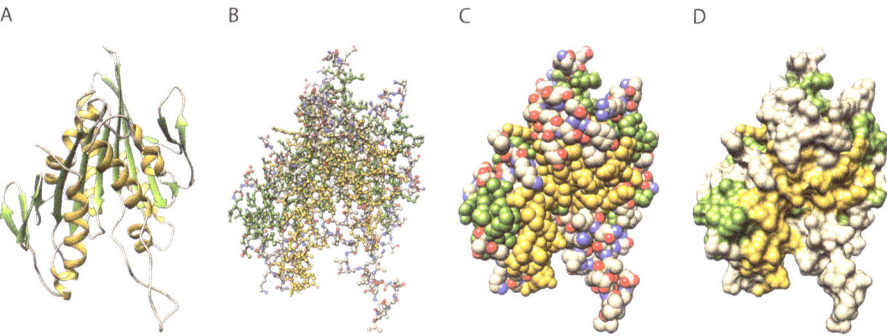

A   B   C   D

Figure 12.10: Different rendering of a protein structure model of kinesin. (A) The ribbon or cartoon rendering displays α-helices as spring-shaped ribbons and β-strands as arrows. (B) A ball-and-stick model displays all atoms (except hydrogens) as balls and bonds as straight lines. (C) In the space-filling model all atoms are displayed as spheres scaled to their van der Waals radius. (D) A surface representation displays a surface property, in this case, the secondary structure. The different rendered models display the same structure with the same size.
In the figure atoms and secondary structures are color-coded: α-helices: golden; β-strands: green; carbons: tan; nitrogens: blue; oxygens: red. (pdb: 1BG2

Further reading

Altman, R.B. (2023). A holy grail – The prediction of protein structure. *N Engl J Med* 389:15.

Berger, C., Premaraj, N., Ravelli, R.B.G., Knoops, K., Lopez-Iglesias, C. and Peters, P.J. (2023). Cryo-electron tomography on focused ion beam lamellae transforms structural cell biology. *Nat Methods* 20:499–511.

Biehn, S.E. and Lindert, S. (2022). Protein Structure Prediction with Mass Spectrometry Data. *Annu Rev Phys Chem* 73:1–19.

Bertoline, L.M.F., Lima, A.N., Krieger, J.E. and Teixeira, S.K. (2023). Before and after AlphaFold2: An overview of protein structure prediction. *Front Bioinform* 3:1120370.

Freddolino, P.L., Harrison, C.B., Liu, Y. and Schulten, K. (2010). Challenges in protein folding simulations: Timescale, representation, and analysis. *Nat Phys* 6:751–758.

Henderson, R. (2018). From electron crystallography to single particle cryoem (Nobel lecture). *Angew Chem Int Ed Engl* 57:10804–10825.

Hollingsworth, S.A. and Dror, R.O. (2018). Molecular dynamics simulation for all. *Neuron* 99:1129–1143.

Hu, Y., Cheng, K., He, L., Zhang, X., Jiang, B., Jiang, L., Li, C., Wang, G., Yang, Y. and Liu, M. (2021). NMR-based methods for protein analysis. *Anal Chem* 93:1866–1879.

Ikeya, T., Guntert, P. and Ito, Y. (2019). Protein structure determination in living cells. *Int J Mol Sci* 20:2442.

Jumper, J. and Hassabis, D. (2022). Protein structure predictions to atomic accuracy with AlphaFold. *Nature Meth* 19:11–12.

Klukowski, P., Riek, R. and Guntert, P. (2022). Rapid protein assignments and structures from raw nmr spectra with the deep learning technique artina. *Nat Commun* 13:6151.

Meisburger, S.P., Thomas, W.C., Watkins, M.B. and Ando, N. (2017). X-ray scattering studies of protein structural dynamics. *Chem Rev* 117:7615–7672.

Murata, K. and Wolf, M. (2018). Cryo-electron microscopy for structural analysis of dynamic biological macromolecules. *Biochim Biophys Acta Gen Subj* 1862:324–334.

Rajan, A., Freddolino, P.L. and Schulten, K. (2010). Going beyond clustering in MD trajectory analysis: An Application to villin headpiece folding. *PLoS ONE* 5:e9890.

Ribeiro, J.V., Bernardi, R.C., Rudack, T., Stone, J.E., Phillips, J.C., Freddolino, P.L. and Schulten, K. (2016). Qwikmd – Integrative molecular dynamics toolkit for novices and experts. *Sci Rep* 6:26536.

Shen, M.Y. and Freed, K.F. (2002). All-atom fast protein folding simulations: The villin headpiece. *Proteins* 49:439–445.

Skolnick, J., Gao, M., Zhou, H. and Singh, S. (2021). AlphaFold 2: Why it works and its implications for understanding the relationships of protein sequence, structure, and function. *J Chem Inf Model* 61:4827–4831.

Teng, Q. (2013). Structural Biology. Practical NMR applications. 2nd ed., Springer US. New York, NY.

Terwilliger, T.C., Liebschner, D., Croll, T.I., Williams, C.J., McCoy, A.J., Poon, B.K., Afonine, P.V., Oeffner, R.D., Richardson, J.S., Read, R.J. and Adams, P.D. (2023). Alphafold predictions are valuable hypotheses and accelerate but do not replace experimental structure determination. *Nat Methods* doi.org/10.1038/s41592-023-02087-4.

Wlodawer, A., Minor, W., Dauter, Z. and Jaskolski, M. (2013). Protein crystallography for aspiring crystallographers or how to avoid pitfalls and traps in macromolecular structure determination. *FEBS J* 280:5705–5736.

13 Penalty shoot-out

Isolation of proteins is often a prerequisite for further structural or functional analysis. Some proteins can be isolated without much effort in large quantities and with high purity whereas others require elaborated procedures to obtain even the faintest amounts. Independent of the type of protein, any purification strategy follows the same principle, as illustrated in Figure 13.1. The initial step in a purification strategy is usually the extraction of a soluble fraction from the chosen tissue, followed by clarification where solid remnants and cell debris are removed. In the next step, the clarified protein solution should be concentrated and stabilized by adding appropriate salts and/or other additatives. The final steps involve removement of impurities to achieve a final pure protein.

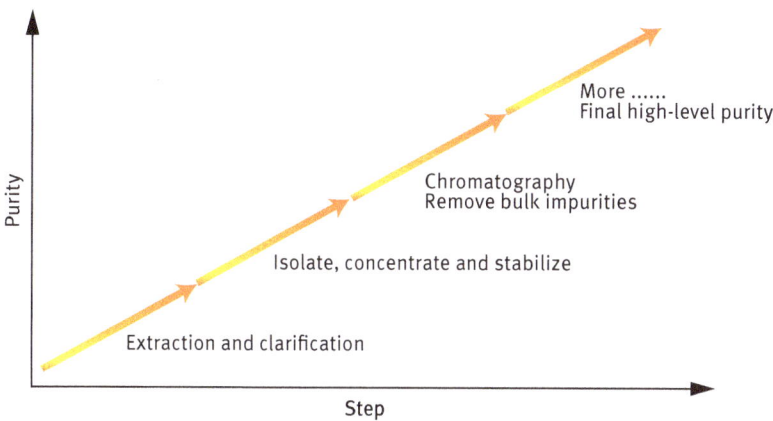

Figure 13.1: A common purification strategy.

Before embarking on a purification procedure, it is pertinent first to obtain as much information as possible about the protein to be purified. It is helpful to know the amino acid sequence, as size, isoelectric point, solubility and optical properties can be extracted from the sequence. For eukaryotic proteins, the localization in the cell is valuable to know; if the target protein is mitochondrial, many contaminants can be avoided by isolating mitochondria from other cell organelles. If the protein has a measurable activity, this can be used to follow the protein during the isolation; if not, an assay must be devised that distinguish the target protein from other proteins. The enzymatic activity can usually be used to follow the purification of an enzyme. Nonenzymatic proteins are more difficult to follow as an assay based on function or some other property is required.

Depending on the starting material, it can be necessary to consider ethics. This is particularly important if human or animal tissue is going to be used. Although recombinant proteins require access to the corresponding gene, cloning and expression in a suitable host, it has many advantages, making it very convenient as a starting material.

https://doi.org/10.1515/9783111350684-013

13.1 Buffers

Whenever working with proteins, it is important to keep pH constant and at a predetermined value; therefore, buffers are used. The charge of acidic and basic side chains depends on the pH of the solution. When pH changes, so may the charges of side chains, which may break noncovalent bonds. Breaking a single bond can affect both the stability and the folded state of a protein.

The ability to alter the net charge of a protein is often used in ion-exchange chromatography (IEX). A reduced pH can change the net charge of a protein from negative to positive.

There is no buffer that covers the whole pH range, but different buffers can be used for different pH ranges. A buffer component usually has a good buffering capacity 1 pH unit above and below its pK_a-value. Table 13.1 lists some commonly used buffers.

Table 13.1: Common buffer components.

Buffer component	pKa	pH range
Sodium acetate	4.75	3.6–5.6
MES(2-morpholinoethanesulfonic acid monohydrate)	6.15	5.5–6.7
Sodium or potassium phosphate	2.12, 7.21, 12.67	5.8–8.0
PIPES(piperazine-1,4-bis(2-ethanesulfonic acid))	6.82	6.1–7.5
MOPS(3-morpholinopropanesulfonic acid	7.2	6.5–7.9
HEPES(4-(2-hydroxyethyl))piperazine-1-ethanesulfonic acid)	7.55	6.8–8.2
Tris(Tris(hydroxymethyl)aminomethane)	8.08	7.0–9.0
TAPS(N-[tris(hydroxymethyl)methyl]-3-aminopropanesulfonic acid)	8.4	7.7–9.1
Boric acid	9.14, 12.74, 13.8	8.2–10.1
CAPS(3-(cyclohexylamino)-1-propanesulfonic acid)	10.4	9.7–11.1

It is important to realize that the ionic strength of a phosphate or citrate buffer is pH-dependent in contrast to monovalent buffers. If it is necessary to use a phosphate buffer and to keep the ionic strength constant at different pHs, the ionic strength can be controlled by including a neutral salt.

13.2 Preparation of extract

Whatever be the starting material, to extract the intracellular content, it must first be disrupted. The cell type determines which method is most suitable for homogenizing. "Soft" cells, such as red blood cells, can be lysed by osmosis, whereas "tough" cells, like yeast, require much harsher methods.

The cell wall of bacterial cells, particularly Gram-positive bacteria, can be disrupted enzymatically by lysozyme. By adjusting the ionic strength, the plasma membrane can be kept intact or lysed. If the cell suspension has a high ionic strength (high salt concentration), the plasma membrane stays intact and only periplasmic proteins will be released. Alternatively, sonication can be used to lyse bacteria as well as many other cell types.

Plant cells and yeast have a rigid cell wall that require other methods, such as bead mills or French press. In a bead mill, the cells are crushed by rotating small glass or metal beads together with the cells. In a French press, the cell suspension is pushed through tiny orifice with high pressure. Shear stress and decompression cause cell disruption.

Animal tissues and cells can be disrupted by mechanical knife homogenizers, such as various types of blenders.

Independent of the method used to disrupt the cells, the main objective is that most of the proteins should be freed from the cell compartment in a functional state. Thus, too harsh treatments should be avoided as the target protein can be harmed and can lose its structure and biological activity. The addition of a low concentration of a detergent may increase the release of proteins.

Before homogenization, cells or tissues are suspended in a suitable solvent, usually a buffer that may include salt and other additatives. If possible, it is practical to choose buffer, salt and additatives that are compatible with the next step in the workflow.

After cells have been homogenized, all cell remnants and debris must be removed to obtain a clarified solution. This can be done by centrifugation or filtration. A 20 min spin at 20,000 rpm is usually sufficient to get rid of all solid parts. For large volumes (liters and more), filtration is a better choice.

If the target protein is known to be present in a particular cell organelle, differential centrifugation can be used to fractionate this organelle from other cellular constituents. The homogenate is first centrifuged at low speed to sediment or pellet whole cells, membrane remnants and other debris. The pellet is removed and the supernatant is centrifuged at higher speed, and the process is repeated with higher and higher speeds. For details see Table 13.2.

With a clarified cell homogenate, the real purification can begin!

During the whole purification process, it is important to keep the activity of contaminating proteases as low as possible. As soon as the cells lyse, the detrimental proteases awake and start to degrade proteins. By keeping all solutions cold, for instance on ice, the protease activities are reduced. The addition of protease inhibitors or metal chelators, such as EDTA or EGTA, to the cell homogenate can also reduce degradation but may interfere with future purification steps. Speed is also an essential factor to consider; as the longer proteases are active the more damage they cause.

Table 13.2: Separation of eukaryotic subcellular fractions.

Sample	Time and G force	Content of pellet	Content of supernatant
Lysed eukaryotic cells	10 min @ 1000 x g	Whole cells, cell debris, nuclei	Cytosol[#]
Supernatant of previous row	20 min @ 20,000 x g	Mitochondria, chloroplasts, lysosomes, peroxisomes	Cytosol and the so-called microsome fraction[*]
Supernatant of previous row	60 min @ 80,000 x g	Microsome fraction, plasma membrane	Cytosol, ribosomes
Supernatant of previous row	180 min @150,000 x g	Ribosomes	Cytosol

[#] The cytosol is the soluble part of the cell content.
[*] The microsome fraction contains vesicles formed from the endoplasmatic reticulum and the Golgi apparatus when the cells are lysed.

13.3 Solubility and precipitation

The solubility of a protein depends on the amino acid residue present on the surface. Although hydrophobic residues predominantly are present in the interior, some are present on the protein surface. Proteins with many hydrophobic residues on the surface have lower solubility in aqueous solvents than proteins with hydrophilic and charged residues. This difference in solubility can be used for purification purposes.

The addition of salt reduces the hydration shell around a protein and thereby exposing any hydrophobic patches. With increasing salt concentration, more and more hydrophobic patches are exposed on the surface of the protein. Due to the hydropho-

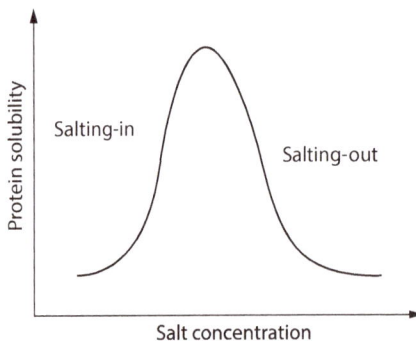

Figure 13.2: Salt precipitation. Salting-in: Addition of low concentrations of salt often increases the solubility of a protein. Salting-out: High salt concentrations lead to aggregation and precipitation of the protein.

bic effect, these patches bind to each other, which eventually cause the protein to aggregate and precipitate. This is known as salting-out (Figure 13.2).

A pH close to the isoelectric point of the protein (where the net charge is zero), reduces the repulsive forces and improves precipitation.

As proteins have different solubility properties and isoelectric points, salting-out can be used as a means to remove some contaminating proteins. However, the salt must not have any detrimental effects on the target protein, and if the target protein is precipitated, it must be possible to renature it.

The addition of low concentrations of salt usually increases the solubility, as the salt ions screen the charges on the protein surface and reduces charge interactions between proteins. This is known as salting-in.

The effectiveness of different salts to precipitate proteins follows the Hofmeister series. Due to the very good solubility in water and the lack of harmful effects on most proteins, ammonium sulfate is the most commonly used precipitation agent.

If the target protein can be renatured after ammonium sulfate precipitation, salting-out can be used to concentrate a protein solution and reduce the volume. The amount of solid ammonium sulfate that must be added to a solution to reach a certain concentration is tabulated in Table 13.3. For this, the precipitate is collected after centrifugation and redissolved in a small volume of a suitable buffer. Since the salt concentration will be very high after precipitation, it is often necessary to reduce the salt concentration.

Dialysis is a simple method to reduce the salt concentration or change the buffer. The protein solution is added to a dialysis bag that is submerged in a buffer with the desired composition. The dialysis bag is made of a semipermeable membrane that allows small molecules, like salt ions or buffer components, but not proteins to pass through the membrane. After equilibration, the solute concentrations in the dialysis bag and dialysis buffer will be the same. For example, dialysis of 10 mL protein solution with 4 M ammonium sulfate against 990 mL buffer reduces the final salt concentration 100 times to 40 mM. Repeating the dialysis once again, reduces the salt concentration another 100 times, now to 0.4 mM.

If the initial volume of the protein solution is around 10–15 mL or less, a desalting column is a quicker alternative to dialysis. Desalting columns function in essence as size exclusion chromatography, also called gel filtration, by separating according to molecular size. A desalting column contains a porous matrix. When the protein solution flows through the column, small molecules enter the matrix and are retarded and separated from large molecules that do not enter the matrix.

Table 13.3: The amount of solid ammonium sulfate required to bring a solution of known initial saturation to a desired final saturation at 0 °C.

%	Final percentage saturation at 0 °C																
	20	25	30	35	40	45	50	55	60	65	70	75	80	85	90	95	100
	g solid ammonium sulfate to add to 1000 ml of solution																
0	106	134	164	194	226	258	291	326	361	398	436	476	516	559	603	650	697
10	53	81	109	139	169	200	233	266	301	337	374	412	452	493	536	581	627
15	26	54	82	111	141	172	204	237	271	306	343	381	420	460	503	547	592
20	0	27	55	83	113	143	175	207	241	276	312	349	387	427	469	512	557
25		0	27	56	84	115	146	179	211	245	280	317	355	395	436	478	522
30			0	28	56	86	117	148	181	214	249	285	323	362	402	445	488
35				0	28	57	87	118	151	184	218	254	291	329	369	410	453
40					0	29	58	89	120	153	187	222	258	296	335	376	418
45						0	29	59	90	123	156	190	226	263	302	342	383
50							0	30	60	92	125	159	194	230	268	308	348
55								0	30	61	93	127	161	197	235	273	313
60									0	31	62	95	129	164	201	239	279
65										0	31	63	97	132	168	205	244
70											0	32	65	99	134	171	209
75												0	32	66	101	137	174
80													0	33	67	103	139
85														0	34	68	104
90															0	34	70
95																0	35

Initial percentage saturation of ammonium sulfate

Adapted from R.M.C. Dawson, D.C. Elliott, W.H. Elliot and K.M. Jones, eds., Data for Biochemical Research, 2nd edition, 1969, Oxford University Press, Oxford.

13.4 Chromatography

Chromatography separates components of a mixture by their relative attraction for a moving mobile phase and a contiguous stationary phase. In liquid chromatography, which is the main chromatography technique used for proteins, the mobile phase is a buffer with suitable additives and the stationary phase a solid but porous matrix with certain properties. Liquid chromatography can be divided into adsorption chromatography and partition chromatography. In adsorption chromatography, the protein is bound, adsorbed, to the matrix, whereas in partition chromatography, the protein partitions between the mobile and stationary phases.

The ideal matrix should be hydrophilic, should not bind proteins, should be chemical and physical stable, should be rigid to allow high flow rates of the mobile phase and, most important, must contain functional groups that interact and bind proteins. Several different types of matrices are available, made of various materials such as dextran, agarose, polyacrylamide or silica.

Functionalization of the matrix determines the property used to adsorb proteins: matrices for ion exchange chromatography (IEX) contain charged functional groups; matrices for hydrophobic interaction chromatography (HIC) contain hydrophobic functional groups and affinity chromatography (AC) uses matrices functionalized with ligands that bind a certain protein.

The basic principle for all adsorption chromatography is the same. The protein sample, dissolved in the mobile phase, is applied to the column packed with a matrix. A pump or gravity creates a continuous flow of mobile phase through the column. Some of the proteins in the sample will be adsorbed on the stationary phase, whereas all other proteins will flow through the column with the mobile phase and elute from the other end of the column. By changing the mobile phase is such a way that any adsorbed proteins are released from the matrix; they can also be eluted from the column and thus separated from those proteins that did not bind to the matrix.

A chromatography system, as illustrated in Figure 13.3, can be used for any type of liquid chromatography.

13.4.1 Ion exchange chromatography (IEX)

IEX exploits the fact that proteins are charged and that the charge can be manipulated by adjusting the pH of the buffer. A protein with positive net charge will be adsorbed on a cation exchanger, whereas a negatively charged protein will flow through the column and elute separately. Adsorbed proteins can be unbound and eluted from the column by including a large excess of salt in the elution buffer that competes with the protein for binding to the matrix. Sodium chloride is the mainly used salt in IEX, but any salt that reduces the binding of the protein to the matrix can be used. It would

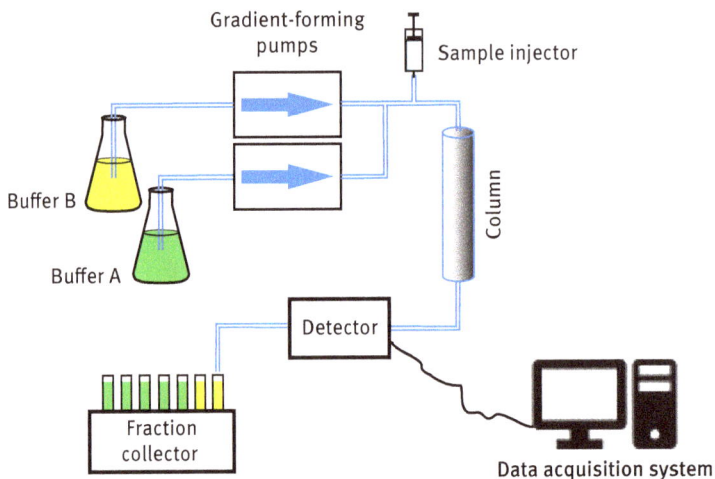

also be possible to elute an adsorbed protein by changing the pH and thereby change the net charge of the bound protein.

It should be noted that during the adsorption phase, the ionic strength, due to both buffer and salt concentrations, must be low enough to allow the protein to be adsorbed. Thus the selectivity of the chosen matrix depends on both the pH of the buffer and on the concentrations of buffer and other component of the mobile phase.

Although it is not critical, it is a good practice to choose a buffer system with Na^+ counter-ions for cation exchangers and with Cl^- for anion exchangers. Some useful buffer systems are listed in Table 13.4

Table 13.4: Buffers for ion exchange chromatography.

Anion exchangers			Cation exchangers		
Substance	pH interval	Counter-ion	Substance	pH interval	Counter-ion
L-Histidine	5.5–6.5	Cl^-	Citric acid	2.6–3.6	Na^+
Bis-Tris	6.0–7.0	Cl^-	Formic acid	3.3–4.3	Na^+
Triethanolamine	7.3–8.3	Cl^-	Acetic acid	4.3–5.3	Na^+
Tris	7.6–8.6	Cl^-	MES	5.6–6-6	Na^+
Ethanolamine	9.0–10.0	Cl^-	Phosphate	6.7–7.7	Na^+
Piperidine	10.6–11.6	Cl^-	BICINE	7.8–8.8	Na^+

Matrices with diethylaminoethyl (DEAE) and quaternary ammonium (Q) are commonly used for anion exchangers, whereas carboxymethyl (CM), sulphopropyl (SP) and methyl sulphonate (S) are found in cation exchangers. The selectivity of the matrix depends on the pH of the buffer and the ionic strength but also on the porosity and shape of the matrix. Q, S and SP are considered as strong ion exchangers. The charges of the weak exchangers DEAE and CM are pH dependent. The carboxy group of CM becomes protonated and uncharged at pH below 5.

A typical IEX separation is shown in Figure 13.4. There are four steps: equilibration of the column with the loading buffer, loading of protein sample, elution of unbound proteins and, finally elution of adsorbed proteins by a salt gradient or pH gradient.

The length of the gradient affects the peak separation; with a steeper gradient protein elutes closer to each other. A reasonable length to start with is 10–20 column volumes. If eluted peaks are baseline separated, it is possible to use a steeper gradient.

To ascertain that all proteins are eluted from the column, it is a good practice to wash the column with 2 column volumes of a high salt buffer before re-equilibration.

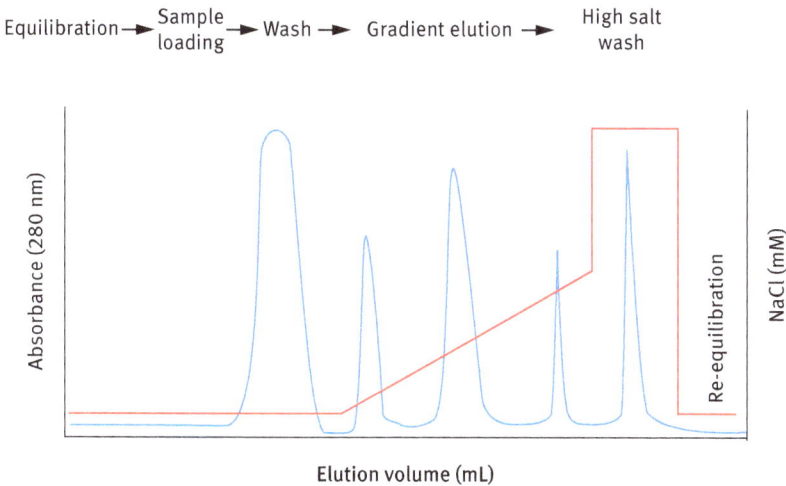

Figure 13.4: Ion exchange chromatography. Before the sample is injected, the column is equilibrated with 2–5 column volumes of the loading buffer. After injection, the flow of the loading buffer is continued until no more proteins elute from the column to wash out all unbound proteins. Next a continuous salt gradient (red line) is applied to the column to elute any adsorbed proteins, followed by a re-equilibration step. The protein concentration is followed by measuring the absorbance at 280 nm (blue line).

With knowledge of the concentration of salt to elute the target protein, a stepwise gradient can be used instead of a continuous gradient. This may speed up the chromatography procedure (Figure 13.5).

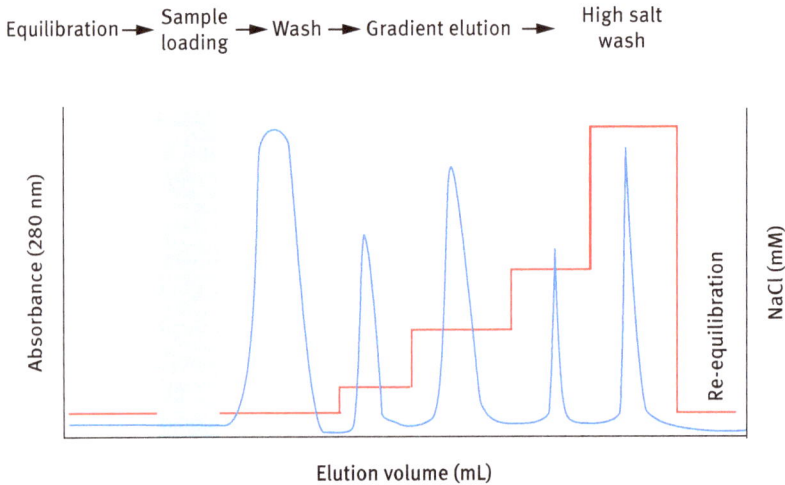

Figure 13.5: Stepwise elution. The protein concentration is followed by measuring the absorbance at 280 nm (blue line) and salt concentration by conductivity (red line).

13.4.2 Hydrophobic interaction chromatography (HIC)

The workflow when performing HIC is the same as for IEX. The only difference being the matrix, which is functionalized with hydrophobic ligands, such phenyl, butyl or octyl groups, instead of charged ones and the initial equilibration conditions, which must include a high concentration (several molar) of ammonium sulfate or sodium chloride (Figure 13.6).

The addition of molar concentrations of $(NH_4)_2SO_4$ or NaCl to the protein solution causes exposure of hydrophobic patches, as mentioned before. These hydrophobic patches interact with the hydrophobic groups on the matrix and the protein is adsorbed. Protein with no hydrophobic patches will not be bound to the matrix and elutes therefore with the flow-through volume. Lowering the salt concentration reduces the size of the hydrophobic patches and leads to dissociation of the bound protein. Proteins with small hydrophobic areas will elute before proteins with large patches.

Similar to IEX, a stepwise gradient can also be used for HIC, going from a high concentration of salt to a low concentration.

13.4.3 Affinity chromatography (AC)

The basic principle of affinity chromatography (AC) is a reversible interaction between a protein or a group of proteins and a specific ligand attached to the stationary phase, as illustrated in Figure 13.7. AC can be used whenever a suitable ligand that

Equilibration → Sample loading → Wash → Gradient elution

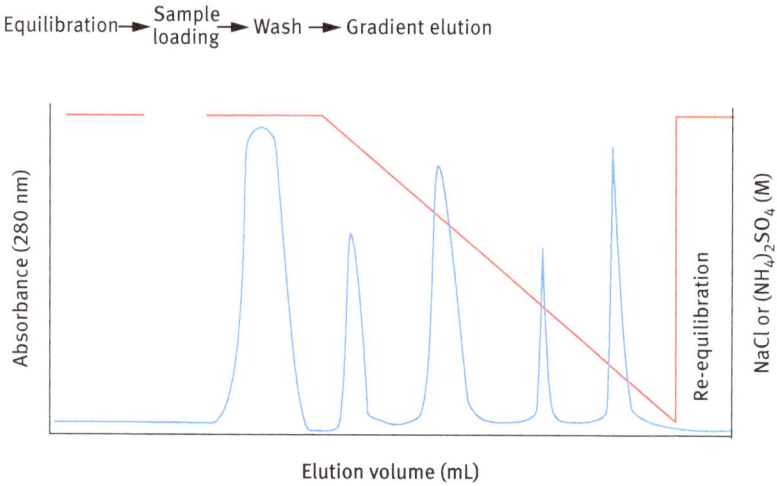

Figure 13.6: Hydrophobic interaction chromatography. Before the sample is injected, the column is equilibrated with ca. 2–5 column volumes of the loading buffer, containing molar concentrations of $(NH_4)_2SO_4$ or NaCl. After injection, the flow of the loading buffer is continued until no more proteins elute from the column to wash out all unbound proteins. Next a continuous salt gradient (red line) is applied to the column to elute any adsorbed proteins, followed by a re-equilibration step. The protein concentration is followed by measuring the absorbance at 280 nm (blue line).

can be attached to the matrix is available. The selectivity of AC is very good and it has usually a high capacity.

Recombinant proteins are often expressed as fusions with a purification tag at the N- or C-termini (Table 13.5). The purification tag can be a short amino acid sequence or a small protein. Independent of the tag, the expression principle is the same: a nucleotide sequence coding for the tag is inserted either before or after the protein coding sequence. The tag is used to adsorb the fusion protein on the stationary phase that has been functionalized with a ligand that interacts with the tag. A proteolytic cleavage site is usually inserted between the tag and protein to allow for the removal of the tag after purification.

Immobilized metal AC (IMAC) is commonly used for the purification of recombinant expressed fusion proteins. IMAC depends on the affinity of the amino acid histidine for divalent metal ions, such as Ni^{2+}, Co^{2+}, Cu^{2+} and Zn^{2+}. The fusion tag is a polyhistidine with 6–12 residues, at either end of the protein. The stationary phase contains a matrix that is functionalized by iminodiacetic acid, nitrilotriacetic acid or tris(carboxymethyl)ethylenediamine (Figure 13.8). These matrices coordinate the divalent metal ion with three (tridentate), four (tetradentate) and five (pentadentate) bonds leaving three, two and one valences, respectively, free for imidazole ring binding. Nitrilotriacetic acid matrices have turned out to be most popular due to capacity and low leaching of the coordinated metal ion.

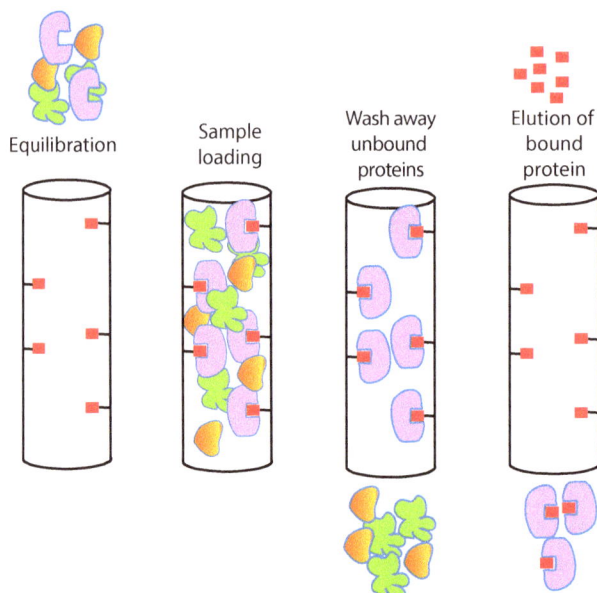

Figure 13.7: Affinity chromatography. The solid matrix contains a ligand (red square) with affinity for the target protein. After equilibration, the sample is loaded on the column and all unbound proteins are washed away with the loading buffer. When all unbound proteins are eluted, the bound protein is eluted with a buffer containing the ligand in excess.

Table 13.5: Common purification fusion tags.

Fusion tag	Number of residues	Matrix	Elution conditions
Poly-Histidine tag	6–12	metal ions (Ni^{2+}, Co^{2+}, Cu^{2+}, Zn^{2+})	20–250 mM imidazole or low pH
Streptavidin-binding tag	8	streptavidin	2 mM biotin
FLAG tag	8	anti-FLAG antibody	pH 3.0 or 2–5 mM EDTA
Calmodulin-binding domain	26	calmodulin	EGTA
Maltose-binding protein	396	amylose	10 mM maltose
Glutathione-S-transferase	211	glutathione	5–10 mM reduced glutathione

Since the metal ion is coordinated by carboxylic groups, low pHs (less than ~5) must be avoided during the purification due to protonation of the carboxyl groups and dissociation of the metal ion. The structure of the His-tag is not critical, as long as it is exposed; therefore, it is possible to purify a His-tag fusion protein from inclusion bodies after dissolving the aggregate by urea or guanidinium hydrochloride.

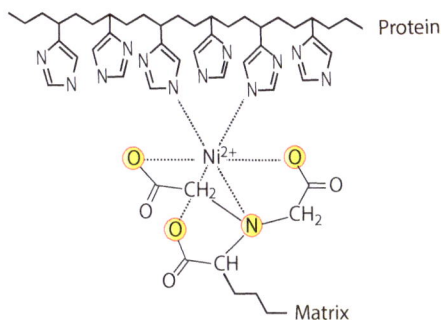

Figure 13.8: Coordination of Ni^{2+} to nitrilotriacetic acid.

A low concentration of imidazole (10–50 mM) is usually included in the loading buffer to reduce unspecific binding. After washing away all unbound protein, bound protein is eluted with a gradient of imidazole (up to 250 mM) or by low pH. A typical purification of a His-tagged fusion protein by IMAC is illustrated in Figure 13.9.

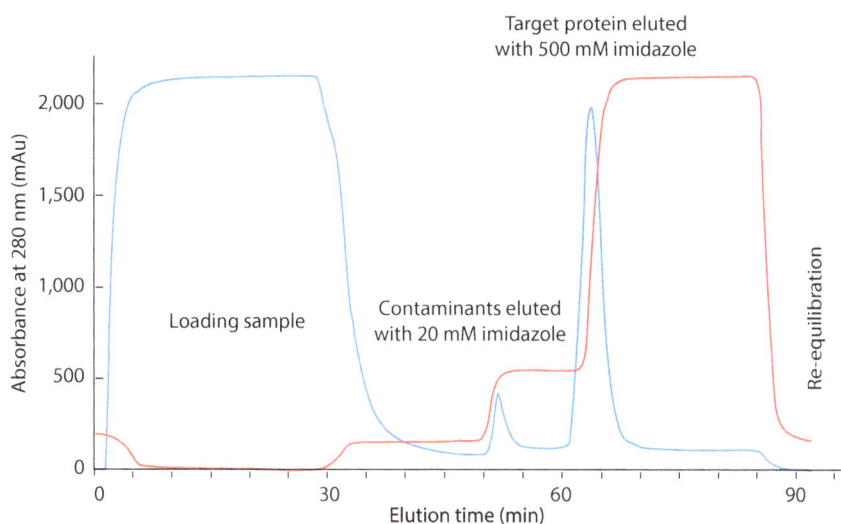

Figure 13.9: IMAC purification of the C-terminal calcium-binding domain of human α-actinin. Cells were lysed by sonication in 25 mM sodium phosphate buffer, pH 7.6, 150 mM NaCl and 10 mM imidazole and the clarified samples was loaded on a Ni^{2+}-column (His-Trap FF). Contaminating proteins were eluted by including 20 mM imidazole in the elution buffer. The target protein was eluted by increasing the imidazole concentration to 500 mM. Protein concentration was followed by absorbance at 280 nm (blue line) and the gradient by conductivity (red line).

13.4.4 Reverse-phase chromatography (RFC)

Reverse-phase chromatography separates proteins on basis of hydrophobicity. In contrast to normal phase chromatography, the stationary phase used in RFC is apolar and the mobile phase usually contains an organic solvent (such as acetonitrile or methanol). Most proteins are denatured by organic solvents and loose both structure and activity. RCF is therefore not useful when a folded and active protein is required. However, if the protein does not denature or can be refolded, RCF is a highly selective technique.

13.4.5 Gel filtration (GF)

In gel filtration (molecular sieve chromatography or size exclusion chromatography), molecules partition between the mobile phase and the stationary phase depending on its size, or rather its Stoke radius. The porous matrix allows some molecules to enter the pores but excludes larger molecules. The molecules diffuse in and out of the pores, and at the same time, the flow transports the molecules along the length of the column (Figure 13.10). When the molecules are inside the pores, they do not move with the buffer flow in contrast to molecules outside the pores. This causes the initial zone of molecules loaded on the column to spread out. Some molecules move before others and some are delayed due to diffusion into the pores. The size of the zone is also increased by diffusion along the column.

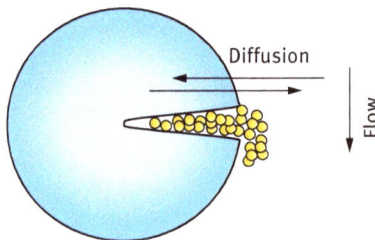

Figure 13.10: The porous matrix. Small molecules can diffuse in and out of the pores. The flow transport molecules along the length of the column.

As large molecules cannot enter the pores, these molecules will be transported faster through the column than smaller molecules that can diffuse into the matrix. The result is that large molecules elute from the column before small molecules. The column with its matrix is in reality a molecular sieve, as illustrated in Figure 13.11.

The selectivity and resolution in GF depend primarily on the pore size of the matrix, the buffer flow rate and the length of the column. Selectivity is determined by the size of the pores. For instance, the separation range of globular proteins by Sephacryl S-100 is 1–100 kDa compared to 20–8,000 kDa for Sephacryl S-400. Independent of the choice of matrix, all matrices separate molecules as well as proteins with

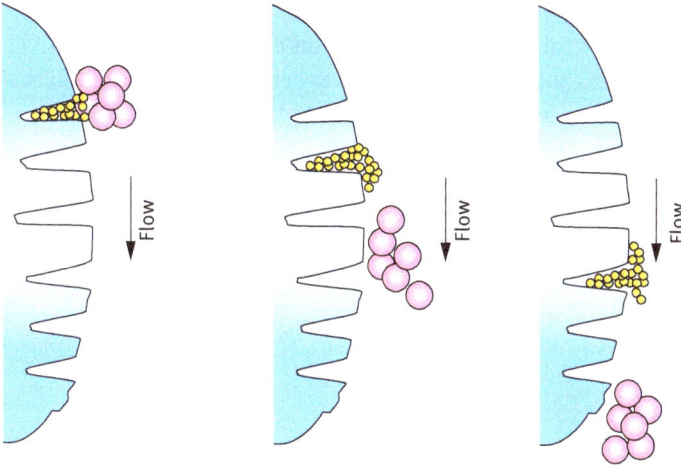

Figure 13.11: A molecular sieve. As only small molecules can enter the pores of the matrix, small molecules are delayed during their transport through the column in contrast to large molecules that are transported with the flow. Large molecules elute therefore from the gel filtration column before small molecules.

large differences in size. As mentioned before, gel filtration can be used to change buffer of a protein solution quickly. However, it is not possible to separate proteins that are close in size to each other; a 30 kDa protein cannot be separated from a 40 kDa protein with gel filtration.

The flow rate does not affect the selectivity, only the resolution. A very high flow rate will not allow molecules to diffuse into the pores and equilibrate between mobile and stationary phases, leading to poor result. A low flow rate allows diffusion in and out of the pores but also leads to unnecessary band broadening due to diffusion along the column. An optimal flow rate maximizes horizontal diffusion into the pores and minimizes vertical diffusion along the column and give best possible resolution. If a certain flow rate gives baseline separation, the flow rate can be increased to speed up the separation process.

The length of the column is important. A long GF column always gives better separation than a short on. If the elution pattern displays overlapping peaks without baseline separation, a twice as long column will improve the resolution but only by a factor of 1.4 and at a cost of band broadening.

The sample volume is not important for IEX or any other type of adsorption chromatography as long as the capacity of the matrix to bind protein is not exceeded. This is contrary to GF, where the sample volume has a large influence on the resolution. In general, the sample volume should not exceed 3–5% of the column volume. Larger sample volumes lead to overlapping peaks and insufficient separation.

Gel filtration can not only be used for purification purposes but also for analytical purposes. From the elution volumes of known globular proteins, it is possible to calculate a partition coefficient, K_{AV}. A graph of K_{AV} versus the logarithm of the molecular mass gives a straight line. By calibrating a gel filtration column by the elution volumes of a set of know proteins, the graph can be used to determine the native molecular mass of an unknown protein as illustrated in Figure 13.12.

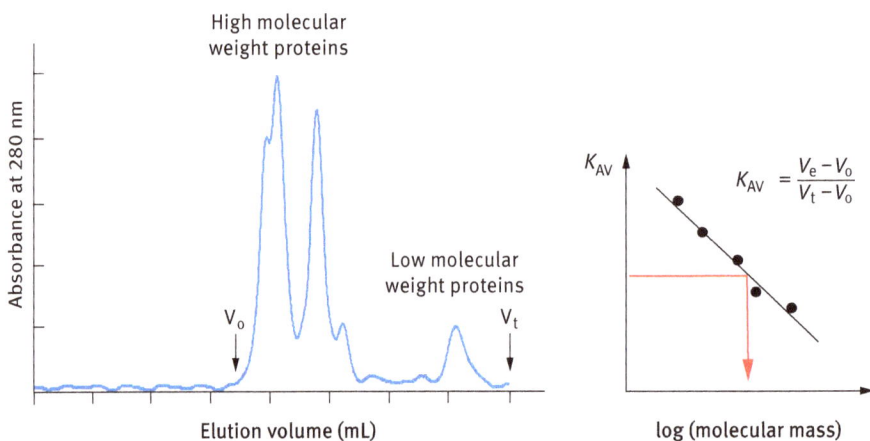

Figure 13.12: Gel filtration. (Left) An elution profile of a protein sample containing both high and low molecular mass proteins. V_o is the void volume, the elution volume of a protein that is too large to diffuse into the pores (the volume outside the matrix). V_t is the total volume of the column. (Right) K_{AV} of globular proteins with known molecular mass plotted versus the log molecular mass. The determined K_{AV} from the elution volume (V_e) of an unknown protein can be used to obtain the molecular mass, as indicated by the red line.

13.5 Protein purification in a nut shell

There are many factors that determine what methods to use in order to purify a protein of interest. It is essential to collect as much information on the protein before trying to purify it. As mentioned before, from the protein sequence it is possible to extract information like molecular mass, isoelectric point, an estimate of the molar attenuation coefficient (formerly molar absorptivity or molar extinction coefficient) that can be used to estimate the amount of the protein by measuring its absorbance and solubility in aqueous solutions. Knowledge of the cysteine content may indicate that there is a need to add a reducing agent (such as dithiothreitol) to the protein solution to keep the cysteine residues in a reduced state.

Figure 13.13 illustrates some logical combinations of purification steps. Irrespective of the protein to be purified, all purification protocols begin with a homogenizing step, to lyse the cells and a clarification step. Next step in the protocol will depend on the protein as well as on the solvent used.

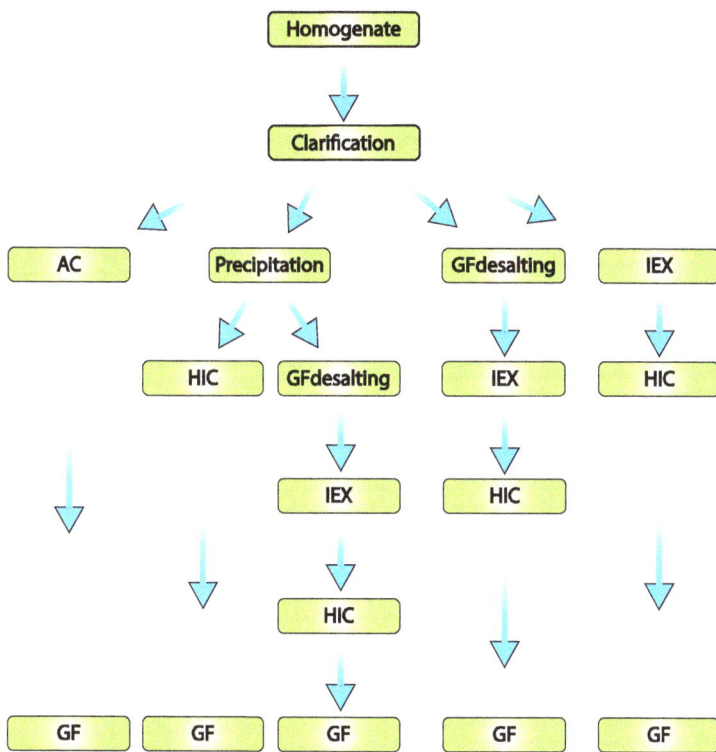

Figure 13.13: Logical combinations of purification steps. Dialysis can equally well be used for desalting as gel filtration, albeit it takes more time. AC: affinity chromatography; GF: gel filtration; IEX: ion exchange chromatography; HIC: hydrophobic interaction chromatography.

13.6 Where is the protein?

During the whole purification process, it is necessary to follow the target protein and determine the purity after each purification step. Measuring the absorbance at 280 nm is a simple way to estimate the total protein concentration, assuming that the molar attenuation coefficient (ε) is the same for all proteins (which is not true). By setting $\varepsilon_{280\ nm}$ to 1 mL/mg, the absorbance at 280 nm gives the protein concentration directly in mg/mL (from $A_{280} = c \cdot \varepsilon_{280\ nm}$). Later in the purification process, when the target protein is the major protein, the correct molar attenuation coefficient can be used to better estimate the concentration.

However, the absorbance at 280 nm does not distinguish between the target protein and other proteins. Therefore, a method is required that measures only the target protein. If the target protein is an enzyme, its enzymatic activity can be used to follow the target protein in each purification step. As an extra bonus, as long as the protein is active it is also folded correctly and not denatured or degraded.

For proteins without enzymatic activity, it is more difficult to follow the target protein in the purification process. A useful alternative, which also indicates the purity of the protein, is gel electrophoresis, in particular under denaturing and reducing conditions.

13.6.1 Sodium dodecyl sulfate polyacrylamide gel electrophoresis (SDS-PAGE)

Gel electrophoresis separates proteins in an electric field according to size. The gel is usually formed by polymerizing acrylamide or by agarose. For electrophoresis of proteins, polyacrylamide gels are usually used. Agarose gel electrophoresis is mostly used for electrophoresis of DNA and RNA.

The most common gel electrophoresis systems are variants of the discontinuous sodium dodecyl sulfate polyacrylamide gel electrophoresis (SDS-PAGE) introduced by Ulrich K. Laemmli. The anionic detergent sodium dodecyl sulfate (SDS) has a negative charge (the sulfate group) at one end and a hydrophobic tail at the other end of the molecule. As SDS binds to the peptide bond (ca. 1.4 g SDS per 1 g protein, corresponding to around 1 SDS molecules per two amino acid residues), the charge-to-mass ration will be nearly the same for any protein independent of size. By boiling the protein sample with SDS, all proteins will denature. If a reducing agent (such as dithiothreitol or β-mercaptoethanol) is included, any disulfide bonds will be reduced and oligomeric proteins will be dissociated. The treatment unfolds each protein into a linear chain with a negative charge proportional to its length.

Polymerization of acrylamide gives a gel with a porous mesh-like structure. The size of the pores depends on the concentrations of acrylamide and bisacrylamide used for pouring the gel; higher percentage of acrylamide or bisacrylamide gives a tighter gel, with narrower pores.

The discontinuous SDS-PAGE system consists of an upper stacking gel and a lower resolving gel. The stacking gel has a lower percentage of acrylamide and a different pH than the resolving gel. In the stacking gel, all proteins move fast and stack into a narrow band before entering the resolving gel. In the resolving gel, proteins of different molecular masses (or sizes) move at different speeds. Small proteins have a higher mobility than large proteins.

Since the percentage acrylamide determines the resolving power, the gel can be optimized for the size range of proteins present in the sample; see Table 13.6. By using a gradient gel, the size range becomes much larger and both small and large proteins can be resolved in the same gel.

The denatured protein sample usually also contains a dye to make it easier to load the sample on top of the gel and to stop the electrophoresis before the dye band reaches the bottom of the gel.

To visualize the resolved proteins, the gel must be stained. This is usually done by submerging the gel in a solution of Commassie Brilliant Blue in methanol and acetic

Table 13.6: Resolving range and percentage acrylamide used for SDS-PAGE.

Percentage acrylamide # (%)	Molecular mass range (kDa)
7	50–500
10	20–300
12	10–200
15	5–100

The acrylamide solution contains 3.3% bisacrylamide.

acid. The sulfonic acid group of Coomassie binds to amino groups in the proteins through ionic interactions and van der Waals attractions. After destaining, to remove excess stain, blue bands appear on an otherwise clear gel (Figure 13.14).

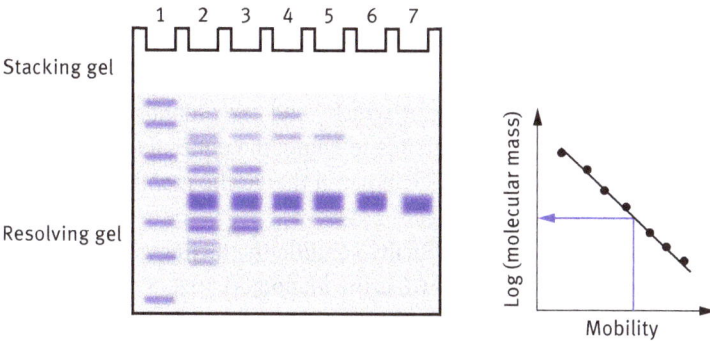

Figure 13.14: SDS-polyacrylamide gel electrophoresis. (Left) A Coomassie Brilliant Blue stained gel after SDS polyacrylamide gel electrophoresis. Collected fractions from each purification step were loaded on the gel. Lane 1: molecular size markers; lane 2: clarified homogenate; lane 3: dissolved pellet from ammonium sulfate precipitation; lane 4: sample after cation exchange chromatography; lane 5: sample after hydrophobic interaction chromatography; lane 6: sample after gel filtration; lane 7: concentrated target protein from gel filtration. (Right) Plotting the logarithm of results in a set of reference proteins with known molecular mass (or size) versus the mobility of each reference protein results in a straight line results. By determining the mobility on the same type of gels of an unknown protein, the molecular mass can be obtained directly from the graph (blue line).

In Figure 13.14, samples from each purification step were loaded on a polyacrylamide gel. When stained, a number of blue bands appeared in each lane. The purity of the target protein becomes better and contaminating proteins become fewer with each purification step. Thus, SDS-PAGE is a simple and quick method to assess purity.

With Commassie Blue staining the detection limit is around 100 ng or better. As Commassie binds to amino groups, the sensitivity of the staining varies with the amino group content of the proteins; some proteins stain very well, whereas others stain weakly.

Silver staining is an alternative method when Commassie staining is not sensitive enough. Silver staining can detect protein bands down to a few nanograms. The basis is rather simple. Proteins bind silver ions and when the bound ions are reduced to silver metal a visible image develops and shows the protein band in the gel. As this staining technique is very sensitive, extra care must be taken to avoid artifacts from finger prints and other contaminants.

Neither Commassie Blue nor silver staining is specific; they both stain any proteins though with different sensitivity. Western blot can be used for a specific detection of the target protein after gel electrophoresis. In this case, the electrophoretic resolved bands are first transferred to a membrane, usually by electroblotting. The membrane is then incubated in a solution containing an antibody directed toward the target protein. The primary antibody is then visualized by a labeled secondary antibody that recognizes and binds the primary antibody. The secondary antibody can be labeled with a fluorescence dye or an enzyme that produces a color signal.

Commassie Blue or silver staining gives information on the purity. The aim of the purification is to obtain a protein sample that gives a single band on a stained gel. Immunostaining makes it possible to determine which band corresponds to the target protein. With that knowledge it would be possible to obtain the molecular mass of the target protein by determine the mobility and compared that to the mobility of known proteins, as illustrated in the right part of Figure 13.14.

Since proteins are denatured under reducing conditions, causing any oligomers to dissociate, the sizes determined by SDS-PAGE is the monomeric size. Comparing the size determined by gel filtration with that determined by SDS-PAGE, the oligomeric state, if any, can be determined.

Further reading

Backman, L. and Persson, K. (2018). The no-nonsense SDS-page. *Methods Mol Biol* 1721:89–94.

Burgess, R.R. and Deutscher, M.P. (2009). Guide to protein purification. 2nd ed. *Methods Enzymol* 463: xxv–xxvi.

Burgess, R.R. (2018). A brief practical review of size exclusion chromatography: Rules of thumb, limitations, and troubleshooting. *Protein Expr Purif* 150:81–85.

Du, M., Hou, Z., Liu, L., Xuan, Y., Chen, X., Fan, L., Li, Z. and Xu, B. (2022). Progress, applications, challenges and prospects of protein purification technology. *Front Bioeng Biotechnol* 10:1028691.

Laemmli, U.K. (1970). Cleavage of structural proteins during the assembly of the head of bacteriophage T4. *Nature* 227:680–685.

Liu, S., Li, Z., Yu, B., Wang, S., Shen, Y. and Cong, H. (2020). Recent advances on protein separation and purification methods. *Adv Colloid Interface Sci* 284:102254.

Remans, K., Lebendiker, M., Abreu, C., Maffei, M., Sellathurai, S., May, M.M., Vanek, O. and de Marco, A. (2022). Protein purification strategies must consider downstream applications and individual biological characteristics. *Microb Cell Fact* 21:52.

Roe, S. (2001). Protein purification techniques. Oxford University Press. Oxford.

Structural Genomic Consortium et al. (2008). Protein production and purification. *Nat. Methods* 5:135–146.

Strategies for Protein Purification, GE Healthcare (Can be downloaded from www.cytivalifesciences.com)

A guide to polyacrylamide gel electrophoresis and detection, Bulletin 6040, Bio-Rad laboratories, Inc. (Can be downloaded from www.bio-rad.com)

14 Silly season

Protein purification is in principle very simple: isolate the target protein in an active state without any contaminating proteins or other molecules. Some proteins can easily be isolated in large quantities and with high purity, whereas other proteins require elaborated procedures to even obtain the faintest amounts.

Protein purification usually involves several steps that exploit differences in protein size, charge and solubility, binding affinity and biological activity. These steps may include techniques such as centrifugation, chromatography, filtration and precipitation. The choice of purification methods depends on the protein's properties, the scale of purification and downstream applications.

Recombinant protein expression is a very convenient way to produce a certain protein in a suitable host organism, such as bacteria, yeast, insect cells or mammalian cells. This requires that the gene (or gene fragment) of the protein is inserted into an expression plasmid. The plasmid, with the inserted gene, is then introduced into a host organism, and induced to produce the protein.

Often an affinity tag is added to the N- or C-terminal of the protein to facilitate the isolation and purification of the recombinant protein. The tags can be a short unique amino acid sequence or a protein. The presence of a terminal tag allows for selective binding of the tagged protein to a specific affinity matrix, enabling efficient purification. By adding a protease cleavage site between the tag and the protein, the tag can be removed.

Some of the most widely used affinity tags are polyhistidine (His-tag), maltose-binding protein (MBP) and glutathione-S-transferase (GST). His-tagged proteins bind to matrices with immobilized nickel or cobalt ions. GST has affinity for immobilized glutathione, whereas MBP binds to amylose matrices. A His-tagged protein can be released from the matrix by increasing concentrations of imidazole that also has affinity for nickel or cobalt ions.

14.1 Risk assessment

Before beginning any practical work in the laboratory, possible risk associated with the planned work must be assessed. Risk assessment is used to identify and analyze potential hazards and their potential consequences, to ensure the safe handling, use and disposal of the used chemicals and apparatus.

The risk assessment should result in appropriate risk management strategies, which may include personal protective equipment (such as gloves and safety goggles), safe handling procedures, apparatus control and proper storage and disposal methods.

https://doi.org/10.1515/9783111350684-014

The risk assessment should also ensure compliance with the safety regulations and standards. It is important to realize that the risk assessment must be reviewed and updated regularly to account for any changes, new information and evolving risks.

14.2 Green fluorescent protein (GFP)

Green fluorescent protein is a ca. 27 kDa protein containing 238 amino acids (Figure 14.1). GFP forms a β-barrel with 11 antiparallel β-strands. It was first isolated from the jellyfish *Aequorea victoria* but has since been found in several other organisms.

Due to its spectacular properties, GFP turned out to be an excellent reporter molecule for monitoring protein localization in vivo and in situ as well as in real time. When exposed to light in the ultraviolet or blue range, GFP fluoresces bright green. In contrast to other bioluminescent reporters, GFP does not require additional proteins, substrates or cofactors to emit light.

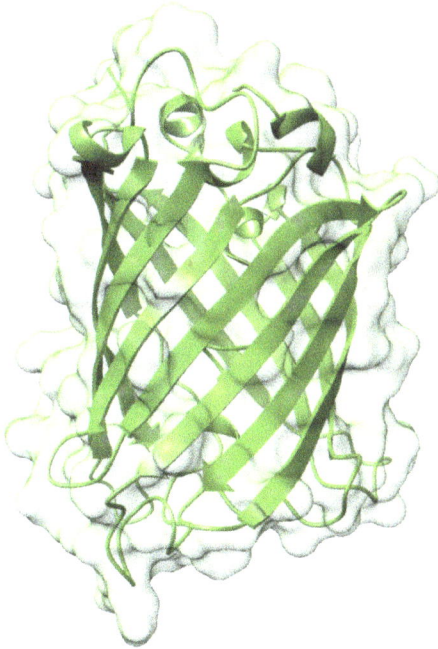

Figure 14.1: Crystal structure of Green Fluorescent Protein (GFP). The β-barrel contains 11 β-strands (pdb: 2B3P).

Later, several variants of GFP have been created that cover the whole rainbow of colors. The discovery of GFP was awarded the Nobel Prize in Chemistry 2008.

In this experiment, GFP will be isolated from a bacterial cell culture. For this *Escherichia coli* first was transformed with a plasmid containing the GFP gene and was then induced to synthesize large quantities of the protein. Finally, the cell culture was collected by centrifugation.

Since GFP is produced inside the bacterial cells, the cells must first be lysed or broken up, to access the intracellular proteins. This is accomplished by sonication, which will rupture the bacterial cell wall and release the cell soup. Alternatively, lysis can be induced by addition of lysozyme followed by freeze–thaw cycles. After clarification by centrifugation, the generated lysate contains a mixture of GFP and endogenous bacterial proteins.

In contrast to most other proteins in the bacterial cell soup, GFP is rather hydrophobic, which can be utilized to purify GFP from the other less hydrophobic (more hydrophilic) bacterial proteins by hydrophobic interaction chromatography (HIC).

By modifying the GFP by adding a His-tag at the N-terminal, immobilized metal affinity chromatography (IMAC) can be used to isolate GFP. A single passage through an IMAC column removes most of the impurities and is often enough to obtain a reasonable pure protein.

The purity of the final purified GFP can be controlled by sodium dodecyl sulfate polyacrylamide gel electrophoresis (SDS-PAGE).

14.2.1 Equipment

- Ultrasonicator
- High speed centrifuge
- A chromatography system (capable to form gradients)
- Nickel-chelated column
- Handheld ultraviolet (UV) light lamp
- Electrophoresis equipment, to run SDS-PAGE

14.2.2 Material

- Bacterial culture with expressed His-tagged GFP
- 25 mM Sodium phosphate buffer, pH 7.6, 150 mM NaCl
- 25 mM Sodium phosphate buffer, pH 7.6, 150 mM NaCl, 500 mM imidazole
- 1 M imidazole, pH 7.6
- Precast SDS-PAGE gel
- Electrophoresis buffers and sample buffer

14.2.3 Procedure

1. Prepare the chromatography system by mounting the nickel-column and equilibrate it with 25 mM sodium phosphate buffer, pH 7.6, 150 mM NaCl, 10 mM imidazole.
   Set the detection unit to measure the absorbance at 280 nm.
2. Thaw the bacterial culture.
3. Sonicate the thawed culture.
   Alternatively, use lysozyme and DNAse I in combination with freeze-and-thaw cycles to break the cells open.
4. Centrifuge the lysate to remove whole cells, cell debris, etc.
   Keep the supernate and add imidazole to 10 mM.
5. Load the supernate on the nickel-column. Make sure to collect the eluate!
6. When all unbound protein is eluted, as indicated by the absorbance, initiate a continuous imidazole gradient, from 10 mM to 500 mM.
   A discontinuous gradient can also be used; by increasing the imidazole concentration by 10 or 20 mM in each step.
7. Use the UV lamp to check the collected fractions for GFP.
8. Check the purity of the fractions containing GFP by SDS-PAGE.

14.2.4 Report

The final report should include a short introduction, describing the background, a section with the methods used, a section with results, containing obtained results and data analysis (data, graphs, etc.) for the chromatography and SDS-PAGE and a discussion of the results.

14.3 Alcohol dehydrogenase

Alcohol dehydrogenase (ADH) catalyzes the reversible oxidation of primary or secondary alcohols to their corresponding aldehydes or ketones. The reaction of short primary alcohols is more efficient than for long primary alcohols or secondary alcohols. The acceptor of the electrons (in the form of hydride ions) is *nicotinamide adenine dinucleotide* (NAD$^+$), which is reduced to NADH in the reaction (Figure 14.2).

$$\text{Alcohol} + \text{NAD}^+ \rightleftharpoons \text{Aldehyde} + \text{NADH} + \text{H}^+ \tag{14.1}$$

A zinc ion in the active site is important for the reaction. The zinc ion positions the hydroxyl group of the alcohol in the right conformation to allow the reaction to occur.

NAD$^+$/NADH has very useful optical properties, which can be used to follow the reaction. In contrast to NAD$^+$, NADH absorbs light at 340 nm (Figure 14.2). As oxidation

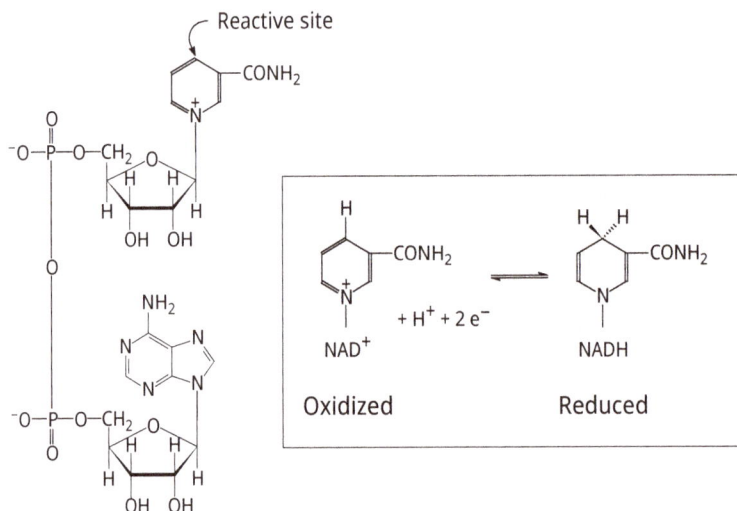

Figure 14.2: Nicotine adenine dinucleotide. Chemical structure of NAD$^+$. (Insert) Reduction of NAD$^+$ to NADH.

of 1 mol of alcohol occurs concomitant with the reduction of 1 mol NAD$^+$, the reaction can be determined by following the change in absorbance at 340 nm. Thus, $\Delta A_{340}/\Delta t$ is proportional to how many moles alcohol that is oxidized per second (minute).

The reaction rate, v_0, is usually expressed as change of concentration per second, $\Delta c/\Delta t$, with the unit M/s. The relation between absorbance and concentration is given by Bouguer–Beer–Lambert's law:

$$A = \varepsilon \cdot c \cdot l \tag{14.2}$$

where A is the absorbance, ε is the molar attenuation coefficient (formerly molar absorptivity or molar extinction coefficient) and l is the light path through the cuvette, usually 1 cm. At 340 nm, ε_{NADH} is 6220 M^{-1}·cm^{-1}.

Thus, the reaction rate can be determined from the change in absorbance from

$$\frac{\Delta A}{\Delta t} = \frac{\Delta c}{\Delta t} \cdot \varepsilon \cdot l \Rightarrow \frac{\Delta c}{\Delta t} = \frac{\Delta A}{\Delta t} \cdot \frac{1}{\varepsilon \cdot l} \tag{14.3}$$

The initial reaction rate's v_0 dependency on the substrate and enzyme concentrations is given by the Michaelis–Menten equation:

$$v_0 = \frac{V_{\max} \cdot [S]}{K_M \div [S]} = \frac{k_{cat} \cdot [E]_T \cdot [S]}{K_M + [S]} \tag{14.4}$$

where k_{cat} is the first-order rate constant for the rate-determining reaction step, which gives the maximal number of substrate molecules the enzyme can transform per second. V_{max} is the reaction rate at infinite substrate concentration, K_M is the Michaelis–Menten constant and $[E]_T$ and $[S]$ are the total enzyme and substrate concentrations, respectively.

The Michaelis–Menten equation describes how the activity of an enzyme depends on the substrate concentration, as Figure 14.3 illustrates.

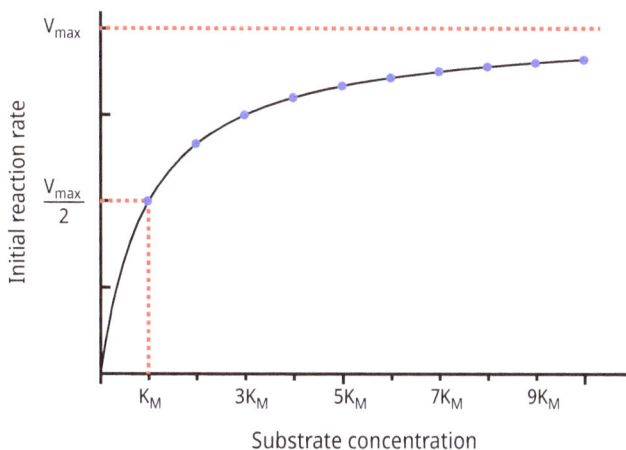

Figure 14.3: Graph of the initial reaction rate versus the substrate concentration of a Michaelis–Menten enzyme. K_M corresponds to the substrate concentration at which the initial reaction rate is equal to $V_{max}/2$.

It is important to realize that the determined reaction rate must be the initial rate, as strictly the Michaelis–Menten equation is valid only when no product is present, that is, before the reaction has started. As soon as the reaction starts, substrate is consumed and product is formed, which reduces the reaction rate, as Figure 14.4 shows. Therefore, it is necessary to determine the absorbance change during the first few seconds. For this reason, $\Delta A_{340}/\Delta t$ must be determined from the slope of the tangent at time zero. The obtained $\Delta A_{340}/\Delta t$ can then be used to calculate $\Delta c/\Delta t$, and thus the initial reaction rate, v_0.

There are two reasons why a large excess of substrate over enzyme is used when measuring the activity of an enzyme. First, with a large excess of substrate, all enzyme molecules have a substrate molecule in the active site, which allows the enzyme to work at its maximal rate. Second, which is a more practical point, the initial absorbance trace appears linear for a longer time, making it easier to draw the tangent and to determine the initial rate of absorbance change.

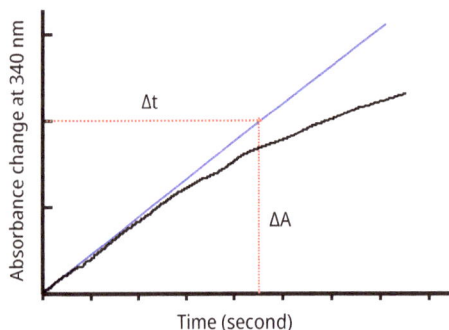

Figure 14.4: Typical data obtained from measurement of the activity of alcohol dehydrogenase. The trajectory of the change of absorbance with time (black line). Tangent to the trajectory at time point zero (blue line), which gives $\Delta A_{340}/\Delta t$ and thus also v_0.

14.1.1 Equipment

- Spectrophotometer, for absorbance measurement at 340 nm.
- 1 ml, 200 µl and 20 µl micropipettes.
- Cuvettes (quarts or UV-transparent disposable plastic cuvettes), with 1 cm light path.
- Parafilm, to use when mixing the solution in the cuvette.

14.1.2 Material

- Yeast alcohol dehydrogenase (can be obtained from Sigma-Aldrich or Roche), diluted to 0.2 mg/mL in 50 mM sodium phosphate buffer, pH 7.2. Should be kept on ice. Molecular weight yeast alcohol dehydrogenase is 147.4 kDa (g/mol).
- 5 mM NAD^+, dissolved in water and neutralized by sodium hydroxide. NAD^+ solutions are unstable and should be kept at −80 °C until used.
- 2 M ethanol and appropriate dilutions.
- 50 mM Tris-HCl, pH 9. Other buffers (such as sodium phosphate, MES, MOPS and many more) can be used to measure the activity at different pHs.

14.1.3 Procedure

To measure the activity of alcohol dehydrogenase (ADH), mix in the cuvette (in this order)
- 2 mL buffer
- 0.5 mL ethanol
- 100 µl NAD^+ solution
- 20 µl ADH

As quickly as possible, cover the cuvette with parafilm and mix by inverting 3–4 times, add the cuvette to the spectrophotometer and measure the increase in absorbance at 340 nm for 1–2 min.

14.1.4 Experimental design

With this background information it is time to design the experiments!

Obviously, one experiment must be designed to determine K_M, V_{max} and k_{cat} of ADH.

There are several other properties and parameters that can be studied and determined:

- Can ADH catalyze the oxidation of other alcohols? Is methanol or butanol a better substrate than ethanol?
- How is pH or temperature affecting the activity of ADH?
- Are there substances that may inhibit ADH? If so, is the inhibition competitive or noncompetitive?
- What happens with the activity if the zinc ions are removed?

14.1.5 Report

The final report should include a short introduction, describing the background, a section with the methods used, a section with results, containing obtained results and data analysis (data, graphs, fits, etc.) for all experiments and a discussion of the results and how they relate to what is known about the enzyme.

Further reading

Bendinskas, K., DiJiacomo, C., Krill, A. and Vitz, E. (2005). Kinetics of alcohol dehydrogenase-catalyzed oxidation of ethanol followed by visible spectroscopy. *J Chem Edu* 82:1068–1070.
Chalfie, M. (1995). Green fluorescent protein. *Photochem Photobiol* 62:651–656.
Silverstein, T.P. (2016). The alcohol dehydrogenase kinetics laboratory: Enhanced data analysis and student-designed mini-projects. *J Chem Edu* 93:963–970.
Shaner, N.C., Campbell, R.E., Steinbach, P.A., Giepmans, B.N., Palmer, A.E. and Tsien, R.Y. (2004). Improved monomeric red, orange and yellow fluorescent proteins derived from discosoma sp. Red fluorescent protein. *Nat Biotechnol* 22:1567–1572.
Tsien, R.Y. (1998). The green fluorescent protein. *Annu Rev Biochem* 67:509–544.

Index

https://doi.org/10.1515/9783111350684-015

www.ingramcontent.com/pod-product-compliance
Lightning Source LLC
Chambersburg PA
CBHW061357210326
41598CB00035B/6010